Schooling for Sustainable Development in Europe

Schooling for Sustainable Development

Volume 6

Series Editors
John Chi-Kin Lee
Michael Williams
Philip Stimpson

This book series addresses issues associated with sustainability with a strong focus on the need for educational policy and action. Current attention and initiatives assume that Education for Sustainable Development (ESD) can be introduced successfully and gradually into schools worldwide. This series explores the issues that arise from the substantial and sustainable changes to be implemented in schools and education systems.

The series aims to counter the prevailing Western character of current research and enable cross-cultural comparisons of educational policy, practice, and project development. As a whole, it provides authoritative and comprehensive global coverage, with each volume providing regional/continental coverage. The volumes present data and insights that contribute to research, policy and practice in ESD-related curriculum development, school organization and schoolcommunity partnerships. They are based on ESD-related project experiences, empirical studies that focus on ESD implementation and teachers' perceptions as well as childhood studies that examine children's geographies, cultural characteristics and behaviours.

More information about this series at http://www.springer.com/series/8635

Rolf Jucker • Reiner Mathar
Editors

Schooling for Sustainable Development in Europe

Concepts, Policies and Educational
Experiences at the End of the UN Decade
of Education for Sustainable Development

 Springer

Editors
Rolf Jucker
SILVIVA
Foundation for Experiential
  Environmental Education
Zurich, Switzerland

Reiner Mathar
Co-ordinating expert for Education
  for Sustainable Development
Ministry of Education and Training
Hessen, Germany

ISBN 978-3-319-09548-6   ISBN 978-3-319-09549-3 (eBook)
DOI 10.1007/978-3-319-09549-3
Springer Cham Heidelberg New York Dordrecht London

Library of Congress Control Number: 2014952315

Preface by the Series Editors

Education for Sustainable Development (ESD) has rapidly become part of educational discourses worldwide. Within its global attractiveness lie both its strength and its weakness. Its strength lies in its capacity to alert educators, broadly defined, to a shared concern for the future of both the planet and local communities. Its weakness lies in its lack of shared meaning and, stemming from this, the enormous difficulties encountered in trying to bring ESD into the mainstream activities of educational institutions.

In designating the period 2005–2014 as the International Decade of Education for Sustainable Development, the United Nations sought to bring to the fore the need for politicians, policy makers and practitioners to seek ways by which ESD could become part of the fabric of formal and informal education. At the heart of the numerous initiatives that have been stimulated by this designation is the assumption that ESD should be introduced and can be introduced successfully into schools worldwide. It is assumed that children, older students and adults can be educated formally to act now in the interests of a sustainable future and to act internationally.

What is evident is that different nations have adopted different approaches to ESD, sometimes interchanging the term with environmental education, another term subject to a wide range of interpretations. These differences are evident in educational practice in regions, districts and individual schools as well as in academic studies and commentaries. Obviously, this is not to say that there is no common ground in policies and practice, it is simply to keep to the forefront the recognition that, even when nations make pronouncements about aspects of ESD, these should not be treated as authoritative statements about what is happening at the school and classroom levels. Broad statements have a value in highlighting issues and trends, but they need to be treated with caution. The same caution needs to be applied to pronouncements emanating from academic sources. Academics have their own agendas, and care must be taken when reading what appear to be authoritative statements about developments in ESD occurring within their own communities and nations.

Our series addresses the array of issues arising from attempts made to convert assumptions about, and definitions of, ESD into substantial and sustainable changes principally in schools. Underpinning the series is a concern for identifying those cultural forces that impact national, regional and local adaptations to approaches to ESD that have international currency. In this, the editors of the books in the series, each based on experience in a single continent or an extensive region, seek to counter the strong Western (Australian, North American, European) character of much research and writing in the broad field of ESD. Research and scholarly studies are commonly underpinned by values and assumptions derived from Western culture, broadly defined. The design of the series as a set of broadly continent-scale books seeks to bring together experts from various countries in each continent. The books bring out contrasting experiences and insights with a range of explanations of policies and practice.

Within the broad cultural contexts of the continents and regions included in the series, authors provide evidence of policies, formal curriculum developments and innovations, and informal school-related activities. Some authors have paid close attention to policy making at various levels, others have addressed whole school organisational issues and yet others have provided detailed case studies of localities and individual schools.

Children and young people live in distinct worlds of their own. They have very distinctive cognitive and affective characteristics that vary from one culture to another, at whatever scale that culture is defined. They are also often targets for environmental campaigns that wish to promote particular behavioural changes. ESD is often construed as an attempt to change habits, to encourage children and young people to 'think globally and act locally'. This series demonstrates how this and other slogans are translated in education systems and schools worldwide.

For this volume, *Schooling for Sustainable Development in Europe*: *Concepts, Policies and Educational Experiences at the End of the UN Decade of Education for Sustainable Development*, the editors Rolf Jucker and Reiner Mathar have brought together an array of chapters highlighting the recent developments and issues related to ESD in Europe. The chapters have been written against a background of, amongst other things, economic austerity experienced by most countries, regeneration of post-industrial communities, the intercontinental and extra-continental migration of people, pressures for both centralisation and decentralisation of political decision making and changes in international institutions. These are set in a broad range of natural environmental contexts. Educational policy makers have been engaged in wide-ranging reforms designed to modernise institutional organisational arrangements, curricula and examinations. Schools, colleges, teacher training institutions and universities have all been subject to these reforms.

With regard to ESD, the nations of Europe have developed many distinctive and innovative features. What is evident in this volume is that much effort has been spent in clarifying the relationships between environmental education, itself a marginal aspect of school curricula in many countries, and education for sustainable development. Many of the contributors to this book have been actively engaged for decades in the innovative programmes initiated under the aegis of the Environment and School Initiatives (ENSI). How ENSI members and other ESD practitioners have responded to the multiple challenges emanating from the UN Decade of Education for Sustainable Development is, as the subtitle indicates, the recurring theme through the chapters.

The book is divided into two sections. Part I focuses on general and cross-national issues, and it provides the context for a series of national case studies, comprising Part II. These case studies highlight aspects of educational policy making and educational practice at various levels from national governments to individual classrooms. They illustrate the variety of experience gained in recent decades mainly in primary and secondary schools. This experience has been set alongside not only policy making but also theory building. Whole school programmes; school-community collaboration; local, regional, national and inter-national networks; and intra-curricular and extra-curricular activities are all highlighted in the chapters.

Hong Kong Institute of Education   John Chi-Kin Lee

Faculty of Education and Health Studies   Michael Williams
Swansea University

University of Hong Kong   Philip Stimpson

May 2014

Acknowledgements

Without the help and support of the ENSI (Environment and School Initiatives) network, this book would never have seen the light of day. The editors express their deep gratitude for this support. In addition, a deeply felt thank you is due to éducation21, the Swiss Foundation for Education for Sustainable Development (ESD): without the belief into the necessity of this book and considerable support in terms of time, finance, administration and encouragement from the foundation, and especially its director, Jürg Schertenleib, this book would not exist. In particular, we would like to thank Florence Nuoffer, Christina Jacober and Nadia Lausselet for their invaluable help. In many ways, the visionary support by éducation21 is testimony to its mission to spread the lessons learnt from the past decades in ESD as widely as possible, so that good practice is mainstreamed and continues to improve.

Special thanks to Professor Michael Williams, formerly of the Faculty of Education and Health Studies, Swansea University, in the UK, who, besides encouraging our work, worked hard in reviewing all of the chapters of this book. Also thanks to Professor John Chi-Kin Lee of the Hong Kong Institute of Education and Dr. Philip Stimpson, formerly of the University of Hong Kong, who supported this book in their capacity as series editors.

There would be many other people to thank, but rather than risking forgetting someone, we would like to extend our thanks and gratitude to all those who did help. For us, it was a wonderful journey into the diversity of ESD in Europe.

Contents

Part II   Case Studies from Individual Countries or Regions

Contributors

Mauri K. Åhlberg is Emeritus Professor of Biology and Sustainability Education at the University of Helsinki. He is Visiting Professor at the University of Exeter and senior advisor of ENSI International. He conducts research on sustainability education, focusing on good environment, sustainable development and good life. He has written extensively, most recently on biodiversity education. He is one of the founders of NatureGate, an inquiry-based learning environment for species and biodiversity education promoting sustainability (http://www.naturegate.net).

Mervi Aineslahti is a classroom teacher at Sorrila school, Valkeakoski, Finland. She has been coordinating several ESD projects in her school. As part of the ENSI network, her doctoral thesis focused on school development through ESD.

Annukka Alppi is a classroom teacher and the principal of Mahnala Environmental School in Hameenkyro, Finland. She has been active both nationally and internationally in EE and ESD. She is a PhD student in Professor Mauri Åhlberg's Research and Development Group on Sustainability Education (University of Helsinki). In 2002, she joined the international network of ENSI.

Arnau Amat is an Assistant Professor of Science Education at the Universitat de Vic (UVic). He is a biologist and a PhD student in Science Education at the Universitat Autònoma de Barcelona (UAB) and a member of the Gresc@ Research Group (Education for Sustainability, School and Community) and the LIEC Research Group (Language and the Teaching of Science), both from the UAB. He has worked as an environmental educator for six years in local and national ESD programmes as well as private environmental education companies. His main research fields are science teacher education, classroom discourse and school/community development in ESD. He has recently contributed to the ESD network CoDeS funded by the European Union with an exemplary case study based on his experience in school-community collaboration for ESD at a local level.

Mari Ugland Andresen was awarded an MSc in Environmental Sustainability from the University of Edinburgh and currently is regional coordinator for climate change mitigation at the Norwegian Public Roads Administration. In her previous role as a university lecturer at the Norwegian Centre for Science Education (2008–2013), she was involved in national and international work on ESD, including policy and strategy work, research and development projects and the development of learning activities for ESD. At that time, she was a member of the expert group on sustainable development Baltic 21 and the Norwegian representative in ENSI. In 2011–2012, she was Vice President of ENSI.

Søren Breiting is an educational researcher and Associate Professor Emeritus at the Department of Education, Aarhus University, Denmark. For many years, his main interest has been in democratic environmental education and, more recently, in how the idea of ESD could be developed in theory and practice both in the Nordic countries and globally. The conceptual development of students' 'action competence' has been a focus of his research, linked to understanding the mechanisms that generate 'feelings of ownership' for innovations, ideas, problems and challenges. This is highly relevant for the practical development of ESD in schools and elsewhere.

Alba Castelltort is a graduate in Environmental Sciences and currently works as a researcher in the Department of Educational Sciences at the Universitat Autònoma de Barcelona. She has worked as an environmental educator in various organisations and institutions that have organised activities for schools and teachers. From 2004 to 2011, she has worked for the Agenda 21 School Programme of Barcelona. She is a member of the Societat Catalana d'Educació Ambiental (Catalan Society of Environmental Education).

Ingrid Claus is a policy advisor in the Environmental Education Unit of the Flemish government. She has facilitated the implementation of the UN Decade of ESD in Flanders with a particular focus on sustainable technological development and the role of education.

Katalin Czippán is an environmental education expert. During her career, she worked, inter alia, at the Ombudsman for Future Generations in Hungary and served as Director for the Environmental Education and Communication Programme Office in Hungary. She is an invited expert of the UNECE Expert Group on Competences in ESD. She is currently the European Vice Chair of the Commission on Education and Communication of IUCN.

André de Hamer is co-founder and coordinator of Duurzame PABO, the Dutch ESD network for teacher training institutes and primary schools.

Martin de Wolf works at Fontys University of Applied Sciences as a geography teacher trainer and coordinator of ESD. He has been involved in many ESD projects in the Netherlands and has written geography schoolbooks and a book for higher education about ESD.

James Dillon is the Communications Manager for the UNESCO Centre based within the School of Education at the University of Ulster. He was previously at the Sustainable Development Commission in Northern Ireland where he worked as the Communications and Engagement Manager. He works alongside research teams on programmes focussing on education for human rights, pluralism and democracy, both in the context of Northern Ireland and at a global level.

Patrick Dillon is Professor of Applied Education in the Faculty of Philosophy at the University of Eastern Finland and Emeritus Professor at the University of Exeter. He has a first degree in Biology and doctorates in Economic History and Education. He has worked in higher education since 1981 and in his early career taught and researched in the fields of ecology and landscape studies. His current research is in cultural ecology and is concerned with interactions between people and their environments which give rise to generative, transactional, relational and co-constitutional ways of knowing and being.

Maryvonne Dussaux was awarded a PhD in Science Education. She is a senior lecturer at Paris-Est Créteil University and at the Department of Sciences Techniques Education Formation (STEF) of the École Nationale Supérieure in Cachan, France. Her research focuses on mediation between school and territory with regard to education for sustainable development.

Mariona Espinet is Professor of Science Education and coordinator of the Gresc@ Research Group (Education for Sustainability, School and Community) and the LIEC Research Group (Language and the Teaching of Science) at the Universitat Autònoma de Barcelona (UAB) in Catalonia, Spain. Her research and innovation interests focus on ESD for science teachers, classroom discourse and critical literacy in science education and ESD as well as school-community development in ESD. She is a member of ENSI (Environmental and School Initiatives) and has been the national coordinator of several European EE/ESD network projects funded by the European Union, such as SEED, CSCT, SUPPORT and at present CoDeS. She has been directly involved for the last eight years in local ESD in school agroecology, developing a model of school-community collaboration for ESD.

Peter Higgins is Professor of Outdoor and Environmental Education at the University of Edinburgh and teaches academic and practical elements of these fields. He is a member of a number of national and international panels and advisory groups on outdoor and sustainability education and advisor to Scottish ministers. He is a Scottish representative on the UNESCO programme 'Reorienting Teacher Education to Address Sustainable Development' and is Director of the UN-recognised Regional Centre of Expertise (RCE) in ESD for Scotland. His current research focuses on the theory, philosophy and practice of outdoor environmental and sustainability education.

Nina Høgmo holds a master's degree in Pedagogy from Lillehammer University College, Norway. She has diverse experience in teaching at the undergraduate level and currently holds a position as a regional network advisor for education counsellors at Buskerud County Municipality. In her previous role as a university lecturer at the Norwegian Centre for Science Education (2008–2013), she was involved in national and international work for ESD, including policy and strategy work, research and development projects and the development of learning activities for ESD. During those years, she was a member of the expert group on sustainable development Baltic 21.

Dániel Horváth is a PhD student on ESD and an expert at the Library and Information Centre at the Hungarian Academy of Sciences. His doctoral thesis concerns the development of the Hungarian Eco-school Network. He has worked at the Hungarian Institute for Educational Research and Development, contributing to a number of development and research projects on eco-schools and textbooks concerning ESD.

Lea Houtsonen is a counsellor of education at the Finnish National Board of Education. She has been a researcher and lecturer in environmental education at the University of Helsinki. She is a member of the Finnish Matriculation Examination Board and the leader of its Geography Section. She is the former Chair of the Commission on Geographical Education of the International Geographical Union. Lea Houtsonen has been an ENSI country coordinator for Finland.

Rolf Jucker is currently Director of SILVIVA, the Swiss Foundation for Experiential Environmental Education (www.silviva.ch). Until the end of 2013, he ran the Department of Conceptual ESD Development at the Swiss Foundation for ESD, éducation21 (www.education21.ch). From 2008 to 2012, he was Director of the Swiss Foundation for Environmental Education (www.umweltbildung.ch), an independent body tasked with the mainstreaming of environmental education and ESD in the Swiss school system. Having gained an MSc in Education for Sustainability, he has worked extensively as an international ESD advisor and has widely published on the subject (see rolfjucker.net), amongst others *Our Common Illiteracy. Education as if the Earth and People Mattered* (2002).

Mercè Junyent is a trained biologist with a PhD in Pedagogy. She is Associate Professor of Science Education and a member of the Complex Research Group based at the Universitat Autònoma de Barcelona. Her main research interests are environmental education in teacher training, curriculum greening in higher education and professional competences in ESD. She has been involved in the Spanish/Latin American PhD programme, which brought together researchers who had an interest in sustainability research, and has mentored many master's and doctoral research studies in ESD and environmental education. She is a member of the UNESCO/UNITWIN International Network Reorienting Teacher Education to Address Sustainability and co-coordinator of the Catalan Education for Sustainability Network (Edusost). Recently, she has been part of the project 'The curriculum greening process in schools: reflection, communication and action', in the context of the Catalonian Network of Schools for Sustainability (XESC).

Chrysanthi Kadji-Beltran is an Assistant Professor for ESD, Department of Primary Education, at the Frederick University in Cyprus. She teaches EE/ESD at both graduate and postgraduate levels. She has participated in several research projects and coordinates research concerning teacher education in ESD through induction programmes and mentoring systems at her institution. Her research interests also concern biodiversity education. Her work has been presented at several international conferences and in scientific journals.

Niels Larsen has a PhD in Pedagogy and a BA in Cultural Sociology. He has researched teachers' competences in environmental education in schools in Kenya and Thailand. He is currently a consultant at the Kijani Institute, Nairobi, Kenya, and Vice Chairman in the Danish UN-approved Regional Centre of Expertise on ESD.

Carl Lindberg is special advisor to the Swedish National Commission for UNESCO in ESD. He was for seven years a member of UNESCO's High-Level Panel on the UN Decade 2005–2014 for ESD. During the period 1994–2004, he was Deputy State Secretary of the Ministry of Education. In 2012, he became an honorary doctor at Uppsala University.

Jürgen Loones is a policy advisor in the Environmental Education Unit of the Flemish government. He coordinates and facilitates the implementation of the UN Decade of ESD in Flanders. He has been involved in the development of an implementation plan for ESD which was the outcome of a participatory process on an 'ESD consultation platform'. He is also a member of the UNECE Steering Committee of the UN Decade of ESD.

Katrine Dahl Madsen is a postdoctoral student at the Department of Education, Aarhus University. She has recently finished her PhD thesis: *A Study of Learning Communities on ESD in Denmark and Ireland* (2012). Currently, she is involved in the research centre Schools for Health and Sustainability, Aarhus University, Denmark (www.shs.au.dk).

Stephen Martin is Visiting Professor in Learning for Sustainability at the University of the West of England and is an Honorary Professor at the University of Worcester. From 2009 to 2013, he was the Chair of the Higher Education Academy's Sustainable Development Advisory Group and a former member of the UK's UNESCO Education for Sustainability Forum. He is the co-founder and President of Student Force for Sustainability (now called Change Agents UK).

Reiner Mathar is currently heading the coordination of ESD in the state of Hessen (Germany), is German representative of the KMK (Standing Conference of Ministers of Education in Germany) at the Steering Committee of UNECE (United Nations Economic Commission for Europe) and member of the National Round Table for the UN Decade on ESD in Germany. Since 1997, he has been involved in all nationwide programmes on ESD in Germany, in particular in the process of developing the concept of 'Gestaltungskompetenz', which is the leading concept of ESD in Germany. In the field of ESD with a focus on global development, he is one of the editors of the national curriculum framework. In 2010, he was co-opted as an expert on ESD onto the international esd-expert.net (India, Germany, Mexico and South Africa).

Michela Mayer is a Physics graduate with a PhD in Science Education. For more than 25 years, she has been involved in the ENSI network as a national representative for Italy, acting as President from 2000 to 2002. She has contributed as a researcher and/or evaluator to many European projects in ESD and in science education (such as SEED, SUPPORT, CoDeS, Form.it, kidsINNscience, S-Team, Traces). She was a member of the UNECE expert group for the development of Quality Indicators for the UNECE ESD Strategy. As researcher at the Italian National Institute for the Evaluation of the Educational System (INVALSI), she has worked as an international expert for the OECD PISA Science Group for the construction of the 2006 and the 2015 frameworks. At present, she is a member of the UNESCO Scientific Committee for the DESD in Italy and of the ENSI Steering Committee. Her research focus is on sustainability science, sustainable education and sustainable evaluation.

Florence Nuoffer is currently a member of the ESD development team at the Swiss Foundation for ESD, éducation21. With a background in biology, she was awarded a PhD in Development Studies and focused her research on the interface of natural and social sciences. Her main interests lie in natural resources co-management and indigenous ecological knowledge, using the critical institutional economics approach, an integrated and heterodox perspective to analyse sustainability. She has worked with young people and adults in various educational contexts in Switzerland (environment, global ecology and sustainability, gender equality).

Anna Maaria Nuutinen is a teacher and a coordinator of UNU IAS RCE in Espoo, Finland. She is a member of the Research and Development Group for Sustainability of the Department of Teacher Education (University of Helsinki). She is interested in cumulative collaborative knowledge building and inquiry-based learning that is directed to transform education and to improve school-community collaboration for SD. She is contributing to the new Finnish national curriculum 2016, led by the Finnish National Board of Education.

Günther Pfaffenwimmer is Head of the Sub-department for Environmental Education at the Austrian Ministry for Education and Women's Affairs. He has teaching degrees in Biology and Environmental Sciences and a PhD in Limnology from the University of Vienna. He was a classroom teacher from 1975 to 1985, and since 1986 he has been employed at the Ministry of Education. He is the Austrian representative in ENSI International and coordinator of the Austrian ENSI Team and was President of ENSI from 2004 to 2008.

Franz Rauch holds a doctorate and a habilitation in Education. He is currently Associate Professor and Head of the Institute of Instructional and School Development at the Alpen-Adria-University in Klagenfurt, Austria. His areas of research and development, publication and teaching are environmental education, ESD, networking, school development and school leadership, science education, continuing education for teachers and action research.

Mónika Réti is a researcher at the Hungarian Institute for Educational Research and Development, involved in development projects in science education, ESD and learning environments. As a secondary school teacher, she taught biology and chemistry for 12 years. She serves as a Secretary of the Hungarian Research Teachers' Association and is involved in a variety of international collaborations on ESD and science education.

Michel Ricard has a doctorate in Science and is a Professor of Ecology and Biology. He is the President of the French Digital University on Sustainable Development (UVED) and the Head of the Francophone UNESCO Chair on ESD. From 2002 onwards, he focused on ESD as the Chair of two bodies attached to the French Prime Minister's office, the French Council on Sustainable Development and the French Committee of the UN Decade for ESD. He is also a member of the UNECE Steering Committee of the ESD UN Decade.

Simon Rolls works at the Department of Education, Aarhus University, Denmark, where he has been centrally involved in the process of establishing, consolidating and coordinating a UN-approved Regional Centre of Expertise on ESD in Denmark.

Torben Ingerslev Roug is the Communication Officer at the Faculty of Science, University of Copenhagen. He is involved in developing and coordinating school visits at the university in relation to climate change education and ESD.

Arto Salonen is an Adjunct Professor at the University of Helsinki. He wrote his doctoral dissertation on *Sustainable Development and its Promotion in a Welfare Society in a Global Age*. He has written several publications concerning sustainable development, global ethics and social change. His current research is on values, attitudes and behavioural change as well as on the flourishing of life.

Astrid Sandås was trained in physics and pedagogy at the University of Oslo. She has held a range of consultative and administrative positions for the Norwegian Ministry of Education and the Norwegian Directorate of Education and Training. She has been responsible for strategies and actions to promote ESD in Norway, and she has been an active partner in the development of the Norwegian Network for Environmental Education since the 1990s. Over a period of 20 years, she has participated in various Nordic and international programmes for ESD as well as written several textbooks on teaching and learning.

William Scott is Emeritus Professor of Education at the University of Bath, where he was founding editor of *Environmental Education Research*, and Director of the Centre for Research in Education and the Environment. He researches the role of education (viewed broadly) in sustainable development and blogs at http://blogs.bath.ac.uk/edswahs.

Glenn Strachan has taught at all levels of formal education in the UK. His posts have included Head of Geography in a secondary school, Advisory Teacher for Political and International Understanding, Deputy Head of Community Education and Training in an FE College and Co-director of the Education for Sustainability MSc course at London South Bank University. As an independent consultant, Glenn has worked with the Welsh Government, developing guidance documents in ESDGC (ESD and Global Citizenship) for schools and for the further education sector.

Katrien Van Poeck as a postdoctoral researcher at the Laboratory of Education and Society (University of Leuven) seeks to contribute to the development of a public pedagogy in the face of sustainability issues. Her research centres around questions of democracy, politicisation, citizenship, community building and participation with regard to sustainability issues. Her research interests are situated at the intersection of educational and sustainability research as well as between politics and education and between theory and practice. She is also a policy advisor in the Environmental Education Unit of the Flemish government.

Paul Vare is a senior lecturer at the University of Gloucestershire, UK, and Executive Director of the South West Learning for Sustainability Coalition, a network of over 130 organisations. His recent doctoral thesis explored the inherent contradictions of promoting sustainability within the English school system. For over a decade, he has represented the European ECO Forum, an NGO coalition, on various expert groups of the United Nations Economic Commission for Europe (UNECE) drafting the UNECE Strategy for ESD, a set of ESD indicators and recommendations for ESD educator competences.

Attila Varga is a senior researcher at the Hungarian Institute for Educational Research and Development, working in the area of development of education, especially ESD. At the start of his career, he worked as a lecturer in psychology at Berzsenyi Dániel Teacher Training College, and currently he lectures on ESD and environmental awareness at the Technical University of Budapest and at Eötvös Loránd University. He is a member of the Hungarian National Commission for UNESCO.

Arjen E.J. Wals is a Professor of Social Learning and Sustainable Development at Wageningen University in the Netherlands and a UNESCO Chair in the same field. He is an Adjunct Faculty member of the Department of Natural Resources of Cornell University and the Adlerbert Research Professor at the University of Gothenburg. Currently, he is also the Director of Wageningen University's Centre for Sustainable Development and Food Security. A central question in his work is how to create conditions that support new forms of learning, research and outreach that take full advantage of the diversity, creativity and resourcefulness that is all around us, but so far remains largely untapped in our search for a world that is more sustainable than the one currently in prospect.

Aravella Zachariou is coordinator of the Environmental Education Unit at Cyprus Pedagogical Institute. She is also a Visiting Professor at Frederick University, Nicosia, Cyprus, teaching on the master's programme for Education for the Environment and Sustainable Development. She has participated in many international networks for environmental education and ESD as well as in EE/ESD research programmes. Her research interests concern the integration of ESD in formal and non-formal education and teacher education on ESD. Her work has been published in international scientific journals and presented at international conferences.

Abbreviations

CERI	Centre for Educational Research and Innovation
CoDeS	Collaboration of Schools and Communities for Sustainable Development (EU Comenius project)
CSCT	Curriculum, Sustainable development, Competences, Teacher training (EU Comenius project)
DESD	United Nations Decade on ESD 2005–2014
EE	Environmental Education
ENSI	Environment and School Initiatives
ESD	Education for Sustainable Development
EU	European Union
GC	Global Citizenship Education
IAS	Institute for the Advanced Study of Sustainability
ITC	Information and Communication Technology
n.d.	no date (in references)
NPM	New Public Management
OECD	Organisation for Economic Co-operation and Development
PISA	Programme for International Student Assessment
RCE	Regional Centre of Excellence
SEED	School Development through Environmental Education (EU Comenius project)
SD	Sustainable Development
SUPPORT	Partnership and Participation for a Sustainable Tomorrow (EU project)
UN	United Nations
UNECE	United Nations Economic Commission for Europe
UNEP	United Nations Environment Programme
UNESCO	United Nations Organization for Education, Science and Culture
UNU	United Nations University

Part I
General and Cross-National Issues

Chapter 1
Introduction: From a Single Project to a Systemic Approach to Sustainability— An Overview of Developments in Europe

Rolf Jucker and Reiner Mathar

1.1 Introduction

It is almost inevitable to start with some general reflections on the nature of paradigm change and the role education can play in this, if we attempt to summarise and introduce the rich harvest which the following 19 chapters provide on the theory, politics, conceptual development and country and region specific implementation in Europe of what is called education for sustainable development (ESD)—even though authors do not necessarily share the same understanding of the concept.

ESD is defined by UNESCO, the United Nations (UN) body responsible for education, as an approach to learning and teaching that "allows every human being to acquire the knowledge, skills, attitudes and values necessary to shape a sustainable future" (UNESCO 2014a). The need and understanding for such an approach to education has grown out of an increasing, worldwide concern for such issues as climate change, environmental degradation, loss of biodiversity, hunger and poverty. Since the 1970s, but even more so after the so-called Rio Earth Summit in 1992, politicians, the general public and educators have realised that sustainable development (SD)—i.e. development which allows future generations to lead a meaningful life supported by a functioning biosphere—is key to the future of humankind. In addition, it became increasingly clear that "education is essential to sustainable development. Citizens of the world need to learn their way to sustainability.

R. Jucker (✉)
SILVIVA, Foundation for Experiential Environmental Education, Zurich, Switzerland
e-mail: rolf.jucker@bluewin.ch; http://www.silviva.ch; http://rolfjucker.net

R. Mathar
Co-ordinating expert for Education for Sustainable Development,
Ministry of Education and Training, Hessen, Germany
e-mail: reiner.mathar@t-online.de

© Springer International Publishing Switzerland 2015   3
R. Jucker, R. Mathar (eds.), *Schooling for Sustainable Development in Europe*,
Schooling for Sustainable Development 6, DOI 10.1007/978-3-319-09549-3_1

Our current knowledge base does not contain the solutions to contemporary global environmental, societal and economic problems. Today's education is crucial to the ability of present and future leaders and citizens to create solutions and find new paths to a better future" (UNESCO 2014b).

On the instigation of Japan, at the Rio+10 conference in Johannesburg in 2002, the United Nations committed to what is called the UN Decade of Education for Sustainable Development (2005–2014) (DESD). The overall goal of the DESD "is to integrate the principles, values and practices of sustainable development into all aspects of education and learning" worldwide (UNESCO 2014c). ESD and the DESD are seen as controversial by some people because of their emphasis on critical thinking, participation, democratic citizenship and equality. ESD practitioners, on the other hand, insist on the importance of ESD in the face of overwhelming scientific evidence that our societies and economies are currently unsustainable.

It therefore comes as no surprise that the authors of this volume start almost unanimously from the assumption that ESD is something bigger than just a little add-on to normal school education. If ESD practitioners, so they seem to argue, take the challenges of sustainability or sustainable development (SD) seriously— i.e. the serious imbalances humankind has brought to its life-insurance system planet earth, to justice between its people and between present and future generations—the only sensible conclusion seems a paradigm change in European societies, economies and educational systems. In schools, this does not mean adding additional SD content to existing lessons, but developing the contributions of all subjects and stages of education to SD. In other words, the authors advocate education *for* SD instead of education *about* SD.

There is also the underlying assumption that ESD is equal to a comprehensive understanding of good quality education *per se*, and that such an education can, nowadays, only be transdisciplinary both in its pedagogical approaches and its content.

1.2   Part I

Several chapters, particularly in part I, engage with this bigger perspective and ask some searching questions at the end of the UN Decade of ESD (DESD). If future oriented education needs to focus on the necessary competencies for learners to enable them to face the sustainability challenges ahead, asks Mathar in Chap. 2, does this not mean that the concept of sustainable development (SD) should underpin all of school education, but in a holistic way, so that all aspects of a school are guided by SD principles?

Mayer and Breiting take the quality discussion a step further in Chap. 3. They argue that what they call the empowerment perspective of ESD has a genuinely socio-political dimension: only if ESD manages to imbue the learners with real ownership of whatever change processes might be needed locally to increase sustainability, will ESD finally mature from an issue-focused campaigning tool to a real participatory learning journey. There is a need for a clear and informed understanding of quality criteria so that progress on the journey can be critically assessed. It is important, so the authors stress, not to see such quality criteria as a given structure or check list to follow and tick slavishly. Quality criteria for ESD should be designed to encourage learners to ask reflective questions about SD and ESD and the implementation of (E)SD in concrete practice.

In Chap. 4 William Scott provides one of the conceptual centre-pieces of this book. His critical look at the promises and results of ESD, and at some of the conceptual tensions that arise between the need for real-world change towards sustainability and the necessary openness of educational processes is insightful. Any advocates of self-proclaimed ESD programmes which cannot convincingly answer questions about the contributions these make to sustainable development in the real world need to engage in critical reflection. Equally, if ESD is viewed by teachers or teacher trainers as something to be taught and disseminated, rather than as an open learning process, serious doubts should be raised. Also—since many authors state that social, political and economic change should not be delegated to schools, but be the obligation of the relevant actors in society—schools should prioritise facilitating the learning of students, rather than institutional change, even though the latter can clearly help and support the former. Scott cautions readers against too lofty or grand ideas of ESD as a force for bringing about socio-economical transformation. He argues that, in the best cases, ESD can create the conditions for transformation, but only on a small-scale and on the ground.

From the eagle-eyed perspective of a European member of the UNESCO high-level panel on ESD, a number of disturbing questions are highlighted by Lindberg in Chap. 5: decisive leadership and the political will to tackle SD issues depend on a well-educated public, but the decisions are needed now, not in the future when effective ESD programmes might have made their impact on the wider population. What needs to be done? Why, in many countries, do environmental or development ministries continue to be the main drivers and financiers behind ESD, and not education ministries? How do we get out of the double-bind underlying ESD, namely that people with the best education world-wide, i.e. in Western countries, have by far the biggest individual ecological footprints? How can education really create deep understanding of production and consumption patterns and their destructive impact on the planet? How do we merge top-down govern-mental and bottom-up grass-roots processes so that they reinforce each other, rather than block each other? These are all pertinent questions addressed in this book and finding answers will be an ongoing task in the post-DESD period.

Wals raises this analysis to a different level in Chap. 6. His is not primarily an inside look at the ESD community, but an outside perspective on what is happening to learning in general in our societies and how this might interact with develop-ments in ESD. Based on a number of trends in business, society and education, he shows that relevant real-world learning increasingly happens in boundary-crossing contexts. Here it is difficult to delineate formal from non-formal and informal learning, school learning from learning in other social contexts. He argues that ESD should become hybrid social learning, where new partnerships and co-operations facilitate rich and exciting learnings which otherwise could not happen. Wals cautions that these new forms of learning are demanding and difficult to organise.

Wals makes an important point in relation to the transformative and transdisciplinary nature of ESD which—as we have stated—is shared by most authors in this volume. If we are to take these two elements seriously, then ESD has to grow up and move out of the confined spaces of traditional schooling which is generally based on the same foundations as when it was invented as a handmaid for the Industrial Revolution. To move away from this requires new, often temporary learning environments with a whole host of different stakeholders and actors. Wals suggests that if ESD really wants to lend a helping hand to tackling sustainability issues in an integrative, critical and systemic way than it needs to grow into cooperative social learning which deserves the name: unless ESD takes in, cooperates with and reflects current learnings in many relevant fields, such as social media, technological development, power structures, economic systems, and much more, it will just remain an insignificant little bubble, mainly concerned with itself, rather than the world out there.

Wals also raises, but does not really address, two main issues of SD which are rarely ever touched upon in ESD, namely power and inequity. Without a deep understanding of how our societies work and function (and that is primarily a discourse about power), any transformation strategies ESD might come up with are severely limited in scope and impact.

The perspective taken by Dillon in Chap. 7 is important for overcoming the still ongoing trench war—manifest in some of the country chapters in Part II—between environmental education, ESD and other 'some-issue'-educations. By using a cultural ecology frame it becomes manifestly clear that it is nonsense to separate humans from the environment, poverty, heath issues or social justice. All of these SD dimensions are clearly co-created by humans, and this is true on a social, psychological, institutional, economic or environmental level. By focusing on connections and differences, Dillon manages to pinpoint where meaningful change and learning might occur, reinforcing Wals' message. The creative space happens in boundary encounters where differences between stakeholders, disciplines, and ways of knowing merge, at a given moment and in a particular place, into something new—which equals learning. This creates relational, rather than fixed knowledge or practice. Interestingly, Dillon also concludes by suggesting that only such relational, interdisciplinary explorations in learning can adequately address the most pressing SD questions, such as the structure of power, the distribution and allocation of resources, fairness, justice and moral responsibility, and not least reconciling individual with communal needs.

1.3   Part II

Part II starts with Mathar's analysis of developments in Germany, by many seen as one of the European countries where the DESD has made most progress. Yet Mathar draws a cautiously optimistic conclusion: despite all the tangible successes, such as a cross-party politically mandated National DESD committee and action plan, structural integration of ESD in German states and a sound programme for training multipliers, many tools and an elaborate national ESD website, not to forget more than 1,700 certified individual ESD projects, there is still a lack of understanding and acceptance of ESD amongst politicians and the general public, there is insufficient structural and institutional anchoring of ESD in schools and only the beginnings of comprehensively ESD focused regional educational landscapes.

Jucker and Nuoffer, in Chap. 9, take the paradigm shift, implied by all UN documents on ESD, seriously and ask what this means for education. They raise similar questions as Lindberg in Chap. 5. They suggest that a fundamental rethink of education is needed to foster a sustainable world, given that current educational systems—particularly if successful in conventional terms—strengthen the *un*sustainability of the current state of the world. By analysing Switzerland's performance during the DESD against criteria suggested by the DESD Monitoring and Evaluation group, they conclude that despite progress on the surface, much work still needs to be done. They argue, like Wals, that the way forward lies in a broader understanding of learning, widening the types and areas of learning as well as the actors, i.e. the learners involved. These are not any more just the pupils or students, but all the participants in such a sustainability learning process, including teachers, facilitators, decision makers and leaders at all levels of the system.

Two further conclusions follow from this: the post-DESD aim should not be to mainstream ESD into the existing unsustainable education system, but to co-create a new educational system—along the lines of Wals' and Dillon's hybrid boundary-crossing social learning communities of practice—which equals ESD. Secondly, they caution, akin to Scott, against grand ideas of 'changing the world'. Meaningful and effective sustainable transformations will only happen locally and on a small scale.

Rauch and Pfaffenwimmer focus in Chap. 10 on another crucial driver for ESD, namely networks. Austria's lessons learnt from previous decades strongly indicate that networks at all educational levels can strengthen and nurture the mutual exchange of experiences (rather than one-way transfer) amongst ESD practitioners. In addition, they can build the trust necessary for success. Yet experience also shows that the main challenges for successful networks lie in finding the delicate balance between structure and process, stability and flow.

A long tradition of research and practical implementation of ESD in Catalonia leads Espinet, Junyent, Amat and Castelltort, in Chap. 11, to reinforce the Austrian message that networks are key to successful ESD implementation. Yet they add a couple of distinct new elements: a number of Catalan research groups and networks focus on collaborative research models, which start from the assumption that school community networks need to be underpinned by research that takes seriously the contribution made by a diversity of agents such as pupils, school teachers, researchers and others. Apart from emphasising the importance of the collective construction of knowledge and experiences in regional, national and international school networks, the authors also focus on the vertical dimension of implementation in schools. They show that a successful participatory approach to an ESD school cannot so much be deduced from SD topics in the curriculum, but rather from the institutionalised structures of participation. Catalonia, as other countries such as Hungary, is also a good example showing that the entire discussion around ESD cannot be separated from the wider social and economic context, especially since the recent financial crisis has drastically impacted on what is and is not possible.

The exploration of the way environmental education (EE) slowly develops into ESD is an interesting journey into systemic analysis. Réti, Horváth, Czippán and Varga show in Chap. 12 that only a multitude of combined elements brings about systemic change—including slow step-by-step changes, horizontal knowledge exchange, reward and support systems for stakeholders, empowerment of practitioners as well as institutionalisation of structures. Apart from visualising the various understandings of ESD in relation to EE, they make one more crucial point: even if there is a rather comprehensive national strategy on ESD, this does not amount to much on the ground, if support and rewards for ESD are not built into existing, regular support structures.

Chapter 13 on Finland takes a different route. By focusing on the concrete examples of two schools, Åhlberg, Aineslahti, Alppi, Houtsonen, Nuutinen and Salonen home in on what they think is the essence of ESD: a good life based on respect for a systemic, scientifically grounded, holistic world-view. What they call an ecosocial approach to education enables an understanding of the world that clarifies the interdependence of human life, including the economy, on healthy ecosystem services, sustainable use of resources and biodiversity. The perspective from the Finnish National Board of Education adds an interesting element: SD was emphasised in the most recent revision of the national core curriculum, yet if pupils are interviewed on related learnings, it transpires that they have a fairly good grasp of a very narrow set of technical ecological knowledge (such as on recycling or energy-saving), yet no real grasp of systemic understandings or biodiversity.

Andresen, Høgmo and Sandås refine and complement a number of findings from previous chapters through an analysis of ESD in Norway (Chap. 14). Norway has been comparatively fortunate to have a strong national strategy on ESD, some exemplary national programmes on EE and ESD such as the projects SUPPORT and Extreme Weather as well as the Environmental Toolbox. Yet in summarising the lessons learnt during the DESD, the authors identify a number of obstacles to good ESD implementation which also exist in other countries: the pressure to focus on literacy, numeracy and ICT diverts necessary energies from transdisciplinary ESD projects; teachers and schools are often overstretched to adapt the national core curricula into a truly local ESD curriculum; the kind of boundary-crossing collaborations mentioned above are very demanding for all involved, with the result that they do not happen as often as they should; there is a strong correlation between pupil's learning outcomes and the level of collaboration with outside stakeholders, reinforcing Wals' and Dillon's messages for cross-stakeholder engagements: it actually increases the quality of learnings; many ESD projects have not been able yet to break out of specific boxes, such as a natural science orientation; and lastly there is, in Norway at least, no vertical integration or feedback loop from local experiences into national educational policy making.

What transpires from Norway is also true for other countries, such as Switzerland, Germany, or the UK: even if there are national policy documents or support structures for ESD, this does not, on the whole, translate directly into meaningful ESD in individual schools, particularly if the national guidance documents are not localised and not backed up by an enforced system of reporting and evaluation.

This view is also enforced in Chap. 15 on Denmark. Rolls, Dahl Madsen, Roug and Larsen present three very diverse and interesting individual examples of successful ESD implementation in their country. Yet, paradoxically, these promising examples also show how unfamiliar the concept of ESD is amongst Danish teachers and how uneasily it sits with current teacher's practices and organisational structures which are firmly rooted in strict traditional disciplinary approaches. The richest mutual learnings in all three case studies can be attributed, again, at cross-boundary stakeholder engagement: collaboration does challenge all involved and allows participants to move forward into new, more nuanced understandings. In Denmark, again, there is a tension between the relative autonomy of teachers (which allows ESD, but might leave it in the margins with a few enthusiasts) and the tendency of the national educational system to focus on literacy, numeracy and ICT. Whether reform or paradigm shift will be on the agenda, the authors suggest that collaborative, cross-curricular and cross-stakeholder solutions will be an important element either way.

Van Poeck, Loones and Claus present a unique perspective in Chap. 16, with their reading of the situation of ESD in Flanders. They provide an important piece of self-reflective soul-searching that would benefit ESD practitioners in most other countries as well. It is an analysis focused on Flanders, but very good arguments could be made to place the chapter also in Part I amongst the more generally conceptual analyses. The main insight they provide is that what we often perceive as an autonomous discourse framed by ESD theorists and practitioners is, in fact, framed by broader developments in society and the environment, such as the impact of the ESD discourse on EE, the framing of social and political problems as learning problems and what they refer to as ecological modernisation. Their discourse analysis makes clear that by defining ESD in certain ways, by enshrining it in policy documents and demanding certain practices we not only enable or encourage certain practices but always also disable and discourage other practices, which, in fact, might be far more sensible or locally adequate in view of sustainable solutions. In other words, they highlight the tension between managerial policy processes and open-ended learning experiments, embedded in a specific local practice based on voluntarism and commitment. Implicitly, they pose the question whether a managerial problem-solving approach, based on strategies, national policy documents and implementation guidelines, is actually compatible with an educational, i.e. learning approach. This echoes Scott's as well as Mayer and Breiting's concern that a focus on institutional change or issue campaigning might undermine education's prime concern, namely to foster learning and growth in understanding.

In Chap. 17, Ricard and Dussaux trace developments in France where, perhaps not surprisingly given France's centralist history, there is a successful example of a top-down approach to ESD implementation. There have been numerous efforts, government decrees and programmes which have led to good progress also on the ground. Yet it seems that even here, there are similar problems as elsewhere. There seems to be no clear roadmap with regard to teacher education and training, and there seem to be many open questions with regard to the vertical, in-depth dimension of ESD implementation. To phrase the question with previous observations in mind: is France's success a managerial one only, or have the policies and programmes led, as Scott demands, to a real world contribution to SD? It also seems an open question whether the trench war mentioned above between ESD and EE has been fought or abandoned.

Cyprus is an interesting case of a small country on the periphery of Europe. Zachariou and Kadji-Beltran argue in Chap. 18 that this country has come relatively late to EE and ESD, but has managed to make fast progress with a number of important national policy documents, such as a national action plan on ESD and the integration of ESD into the national core curriculum. Yet similar tensions operate in Cyprus as elsewhere: the centralised system, with an emphasis on a managerial, instrumental approach, to be implemented by schools, stands in stark contrast to the situation on the ground where a whole school approach demands a dynamic transformational process in individual schools, focusing on key aspects of SD, such as participation, cooperation, quality of life, equity and justice. An additional problem—to be overcome in the post-DESD years—is the very vague understanding by key actors in schools (headteachers and teachers) of what ESD means. In Cyprus the question posed by Sterling (2001) comes to the fore again: what kind of schools do we want, reproducing the current order of society and economy (i.e. at the very most tinkering at the edges with first order change), or schools which co-create and construct a sustainable community (i.e. third order change or transformation)?

The very specifically grounded but converging observations by Martin, Dillon, Higgins, Strachan and Vare in Chap. 19 on the situation of ESD in the United Kingdom adds more weight and conviction to many of the findings presented by previous chapters. Highlighting the three most important ones is sufficient here: firstly, ensure that regional reality is examined before judgement is passed on any country. Implementation on the ground manifests itself in a complex mix of policy, culture, socio-economic situation and resilience of individual key actors. Secondly, all of the UK's devolved administrations in Wales, Northern Ireland, Scotland and England, have to varying degrees impressive national policy documents, even ESD checks in school inspections, with some good evidence that ESD implementation in schools increases educational quality. Yet, as elsewhere we are very far from a changed educational landscape, as originally implied by the UN's DESD implementation documents. Thirdly, the inverse of what we stated above, that much of the true picture of ESD implementation depends on local initiatives on the ground, is also true: the national policy context has a clear impact on what happens on the ground. While previously England and especially Wales have been at the forefront of ESD implementation worldwide, changes in government have altered the picture, so much so, that it now seems that Scotland has leapfrogged these countries.

Finally, in Chap. 20, de Wolf and de Hamer describe developments in the Netherlands, highlighting a specific feature of the Dutch educational system, namely that it is very easy to run private or independent schools. This gives a large degree of freedom to headteachers and teachers with regard to ESD. Yet at the same time the fact that national standards do not systematically integrate ESD, this freedom is clearly limited and shifts ESD again into the corner of 'nice to have', so that only enthusiastic teachers tend to engage in it. The authors also note, based on the expressed preferences of Dutch schools, that the willingness to engage in transformative approaches to learning is very limited. As in other countries, there is a strong tendency to stick with the traditional approaches, focusing on reproduction of knowledge and application of rules. Despite the fact that there have been a number of successful ESD programmes which the authors detail in the chapter, the impact of these programmes has been limited since they have been focused only on one educational level (here primary education). Additionally, such central programmes have a tendency not to be demand-oriented, since they are not established with intimate integration of the end-users, i.e. schools, right from the start of the design phase. A final point the authors make is again one which is reflected in many other chapters, i.e. that a distinct focus on teacher training is necessary if a shift in pedagogical approaches is to be achieved any time soon.

1.4   Conclusions

If we summarise the findings of the volume and also add the editors' perspectives we would like to highlight the following points:

- There is a very good understanding that meaningful ESD needs to go beyond both managerial approaches, ticking policy boxes or tokenistic, issue-campaign based waste management actions. In the best of cases, ESD is boundary-crossing, multiple-stakeholder, communal social learning in a specific place and focused on a real-world problem which needs solving. It is a learning process which involves all participants and is open-ended, i.e. it enables a deepened understanding through the educational activity. It needs to be a multidisciplinary exploration of social, psychological, economic, political, technological and environmental dimensions. Yet, given that this would really place any ESD endeavour in the midst of where our state-of-the-art understanding of these issues are, we see a distinct lack of the ESD discourse to engage in such truly boundary-crossing explorations of knowledge. Most ESD practitioners still seem to be content in their educational corners (see more extensively Jucker 2014). For example, there is a notable lack in the ESD literature of explorations of the implications of the new power structures established by new control technologies embedded in social media and what this means with regard to our generally shared hype of using ICT tools and social media in ESD (see Baumann and Lyon 2013).

- Several authors in this book highlight this, amongst them Scott, Wals and Espinet *et al.*: In ESD, despite all the efforts during the DESD, advocates are keeping their fingers crossed and hope that their programmes, activities, learning endeavours, projects and lessons will yield their desired results. However, there is little evidence-based research to convincingly show that ESD works or that the reasons for any success are understood.
- There is considerable agreement amongst the chapter authors that sustainable change and transformation has to be local and small-scale. Maybe—rather than, once again, dreaming up big schemes like the DESD or self-contradictory claims of main-streaming ESD (How do you want to mainstream something into a system whose ideology, construction principles, guiding values and understanding of education are diametrically opposed to sustainability?)—small steps are more likely to be successful: face-to-face, hands-on learning as a committed local community exploiting to the fullest respective spheres of influence.
- There is a really big tension between ESD1 and ESD2, as Scott and Vare (2007) call it, or, to phrase it even more paradoxically: how do we reconcile the fact that any ESD which does not contribute, palpably in the real-world, to more sustainability cannot consider itself ESD, *and* any ESD which forgets that it is primarily a learning process and not social transformation is also hardly ESD? The question is on the table whether theory, policy and practice of ESD have found meaningful and sufficiently rich and complex answers to this paradox yet.
- The chapter on Flanders, particularly, highlighted the question of whether ESD practitioners are sufficiently aware of the paradox of modernity that every so-called step towards progress has its drawback. So, any gain in visibility, importance and compulsory embedding of ESD in national policy documents or curricula is also always legitimising certain actors to claim to be the true voices of these discourses, thereby de-legitimising, even silencing other discourses. This can be seen in many countries where the ascendancy of the ESD discourse has silenced or sidelined EE approaches with arguably at least as much credibility and validity.
- Given the emphasis on social learning, collaboration, communal learning and involvement of (external) stakeholders shared by most chapter authors it is almost self-explanatory that networks—not just of schools but also of practitioners, community actors and professions—are seen as crucial elements in any ESD implementation. They not just enable the integration of a diversity of perspectives, but also of different systemic levels of educational systems whose members often do not talk to each other. It is wise, though, to head the experiences of Austria in this regard: networks are hard work and demand skilled balancing between flow and openness versus stability and structure.
- There is value in exploring approaches like cultural ecology in order to move beyond the trench wars mentioned above. In too many countries there are still territorial and hegemonial fights going on over who has the right to the 'correct' history and to define ESD. This is often fuelled by funding streams coming

from specific government departments, e.g. health, environment or development ministries. So, health education, global education, political education or environmental education challenge each other not to be the 'right' ESD. But advocates of these different types of 'hyphen'-educations often forget that they should be seen and accepted as different doorways to ESD, realizing that all of them have a specific history and background (DCSF 2008: 41–45). Such positionings and posturings can only be overcome if the transdisciplinarity of ESD and its embeddedness in the complexity of real-world issues is taken seriously. Such issues can never be reduced to any one of these limited perspectives. The open and solution-oriented frames of cultural ecology or hybrid social learning could help transgress these counterproductive and resource-gobbling in-fights.

- There has been no real progress in the sense of the necessary paradigm change. There is a need to step outside the world-view and mental models (including the mental model of education) of the Industrial Revolution and advanced capitalism, if we are to move towards a sustainable society. Bateson has neatly summarised this: "The raw materials of the world are finite. If I am right, the whole of our thinking about what we are and what other people are has got to be restructured." (Bateson 2000: 468) But the fact that the paradigm change has not occurred is, as Welzer states, no surprise: the discourse of ESD has, until now, been an integral part of the narrative of modernity, couched in terms of 'progress' ('lifelong learning'), 'competition' ('student achievement') and 'growth' ('personal development') (Welzer 2013: 65–66). Only if we manage to create a new political, social and economic paradigm in the real world, will the education system, and therein ESD, follow as veritable learning processes.

- There is no real high-level policy commitment to ESD. This again is not surprising, since the decision makers of all parties still function in the unsustainable 'industrial growth model'. So the economic growth model still determines overall governmental, social and educational policies. Only a political movement to change these priorities in society could achieve such a change.

- We need to remember how resilient mental models are and how susceptible professionals generally are to 'old ways of thinking'. So there is a need to think outside the box and this often means working with outsiders, as elaborated above when referring to boundary-crossing approaches. We need solutions (new or old) that work and they are often found outside the education system. Engaging experts, wisdom and experiences from outside can help this process. Whether this is called transformative education (WBGU 2011: 351–357) or design thinking (Design Thinking 2013) does not matter, as long there is respect for the principle that a wealth of perspectives, experiences and personal involvement and relevance generate better results.

- Most importantly there is a need to agree on, and routinely reconfirm the common aim: transition towards sustainability, i.e. one-planet living. There is a need to always reconnect to the overall picture, the holistic overall aim, in order to ensure continued travel in the desired direction. Day-to-day activities and especially highly motivated work can easily disconnect from—or turn into something contradictory to—the overall aim.

References

Bateson, G. (2000). *Steps to an ecology of mind: Collected essays in anthropology, psychiatry, evolution, and epistemology*. Chicago: University of Chicago Press (Original 1972).

Baumann, Z., & Lyon, D. (2013). *Liquid surveillance*. Cambridge: Polity Press.

DCSF (Department for Children, Schools and Families). (2008). *Planning a sustainable school: Driving school improvement through sustainable development*. Nottingham: Department for Children, Schools and Families. https://www.education.gov.uk/publications/eOrderingDownload/planning_a_sustainable_school.pdf. Accessed 14 Apr 2014.

Design Thinking. (2013). *Design thinking*. http://en.wikipedia.org/wiki/Design_thinking. Accessed 10 June 2013.

Jucker, R. (2014). *Do we know what we are doing? Reflections on learning, knowledge, economics, community and sustainability*. Newcastle upon Tyne: Cambridge Scholars Publishing.

Scott, W., & Vare, P. (2007). *Learning for a change: Exploring the relationship between education and sustainable development*. Paper presented to the UNECE Expert group on indicators for education for sustainable development, fifth meeting, 20–22 June 2007, Vienna. http://www.unece.org/fileadmin/DAM/env/esd/inf.meeting.docs/EGonInd/5meet/Learning_Change_Vare_Scott.pdf. Accessed 31 May 2013.

Sterling, S. (2001). *Sustainable education: Re-visioning learning and change*. Darrington: Green Books [Schumacher Society Briefing No. 6].

UNESCO. (2014a). *Education for sustainable development (ESD)*. http://www.unesco.org/new/en/education/themes/leading-the-international-agenda/education-for-sustainable-development/. Accessed 4 June 2014.

UNESCO. (2014b). *Education for sustainable development*. http://www.unesco.org/new/en/education/themes/leading-the-international-agenda/education-for-sustainable-development/education-for-sustainable-development/. Accessed 4 June 2014.

UNESCO. (2014c). *Mission UN decade of education for sustainable development*. http://www.unesco.org/new/en/education/themes/leading-the-international-agenda/education-for-sustainable-development/mission/. Accessed 4 June 2014.

WBGU (German Advisory Council on Global Change). (2011). *World in transition. A social contract for sustainability*. Berlin: WBGU. http://www.wbgu.de/fileadmin/templates/dateien/veroeffentlichungen/hauptgutachten/jg2011/wbgu_jg2011_en.pdf. Accessed 13 June 2013.

Welzer, H. (2013). *Selbst denken. Eine Anleitung zum Widerstand* [Think for yourself. A guide to resistance]. Frankfurt/M: S. Fischer.

Chapter 2
A Whole School Approach to Sustainable Development: Elements of Education for Sustainable Development and Students' Competencies for Sustainable Development

Reiner Mathar

2.1 The History of Environmental and Nature Education— A Success Story

In the eighteenth and nineteenth century thinkers, writers and educationists like Rousseau, Goethe, Froebel, Dewey, Montessori and Steiner published important studies and publications on the integral links between education and the environment.

In the eighteenth century Jean Jacques Rousseau called for a "Return to nature": He argued that education should include a focus on the environment and that one of the main things a teacher should do is facilitate opportunities for students to learn (Sarabhai 2007: 1).

Friedrich Froebel, believed in developing the inborn moral, social and intellectual capacities of the child through nature studies, gardening and play. He emphasised sympathy and oneness with nature (Sarabhai 2007: 1).

The progressive education movement of the 1920s carried forward the idea of experimental education which was defined as a process through which a learner constructs knowledge, skills, and values from direct experience (Sarabhai 2007: 2). A similar approach is reflected in Mahatma Gandhi's conception of basic education.

This element of environmental education (EE) is very important as a basis for understanding sustainability and should be mainly focussed on in elementary and primary education, to help students develop a basic understanding of nature and its impacts their own lives as well as their life's impact on nature. This remains an important part of EE and ESD even today.

The term "Environmental Education" was used for the first time during a conference of the International Union for Conservation of Nature (IUCN), which was held in Paris in 1948 (Sarabhai 2007: 2). In 1965 a meeting of the IUCN called

R. Mathar (✉)
Co-ordinating expert for Education for Sustainable Development,
Ministry of Education and Training, Hessen, Germany
e-mail: reiner.mathar@t-online.de

© Springer International Publishing Switzerland 2015
R. Jucker, R. Mathar (eds.), *Schooling for Sustainable Development in Europe*,
Schooling for Sustainable Development 6, DOI 10.1007/978-3-319-09549-3_2

for: "environmental education in schools, in higher education, and in training for the land-linked professions" (Sarabhai 2007: 3). In 1968 the UNESCO Conference in Paris went further and asked for translating the concept into practice:

> EE is aimed at producing a citizenry that is knowledgeable concerning the biophysical environment and its associated problems, aware of how to solve these problems, and motivated to work toward their solution. (Stapp 1968, quoted in Sarabhai 2007: 4)

In 1970, a frequently mentioned definition of Environmental Education went as follows:

> Environmental Education is the process of recognising values and clarifying concepts in order to develop skills and attitudes necessary to understand and appreciate the inter-relatedness among man, his culture, and his biophysical surroundings. Environmental education also entails practice in decision-making and self-formulation of a code of behaviour about issues concerning environmental quality. (IUCN 1991: 17)

Other milestones included:

- June 1972: UN Conference on the Human Environment, Stockholm
- 1972 Founding of UNEP (United Nations Environment Programme) Life is one and the World is One
- 1975: IEEP: International Environmental Education Programme
- 1975: Belgrade Charter: A Global Framework for EE
- 1976: UNEP launches of regular newsletter on EE
- According to the Belgrade charter the goal of environmental education is:

> To develop a world population that is aware of, and concerned about, the environment and its associated problems, and which has the knowledge, skills, attitudes, motivation and commitment to work individually and collectively toward solutions of current problems and the prevention of new ones. (Sarabhai 2007: 22)

2.1.1   Tbilisi 1977—A Defining Milestone for EE and ESD

The first international conference on EE brought together 265 delegates and 65 observers from 66 UNESCO-member States. The Conference Agenda (UNESCO 1978) included:

- Major environmental problems
- Role of education in facing challenges
- Current efforts at the national and international levels of EE
- Strategies for developing EE on national and international level
- Regional and international cooperation EE
- Needs and modalities

Recommendations from Tbilisi:

- EE is an integral part of the education process. It should be centred on practical problems and be of an interdisciplinary character.

- EE is a lifelong process and should not remain confined within the formal system.
- In formal education, at all levels, account should be taken of all ingredients of the education process (programmes and curricula, books and textbooks, teaching aids and resources, methods) and interdisciplinarity gradually achieved.
- EE should not be just one more subject to add to existing programmes but should be incorporated into all programmes intended for all learners, whatever their age.
- The central idea is to attain a practical education oriented towards a solution of the environmental problems, or at least to make pupils better equipped for their solution by teaching them to participate in decision-making.
- EE is a task, which requires the application of new concepts, new methods and new techniques as part of an overall effort stressing the social role of educational institutions. To this end, legislative measures may be taken providing the State with a legal framework in which to draw up an EE system for the entire community.
- EE should not give rise to competition with other subjects; it should represent a means of introducing a certain unity into the educational process.
- The training of qualified personnel was considered to be a priority activity.
- The basic training of all environmental specialists will need to include the study of the principals of EE, sociology and ecology which are necessary to enable the learners to foresee the consequences of their actions.
- The importance of EE is stressed, including the necessity in-service training of all educational staff (UNESCO 1978: 27–29).

This conference and the recommendations mark an important milestone on the way to implementing EE into educational legislation and the core curricula. Looking at the European development, starting in 1980 most of the European countries made basic formulations and started to develop concepts of EE.

The German case is a typical example:

- The connection to the environment has become existential for every individual as well as mankind.
- Therefore environment has to become a main topic for all schools.
- Schools have to help children to develop an understanding of nature and the environment and to be able to protect the environment by actively involving them (the children).
- EE has to become a basic concept for science and social science education at all levels of schooling.
- EE must integrate outdoor experiences and partners from outside the schools (KMK 1980: 2, translated by Reiner Mathar).

As a result of these post-Tbilisi insights, environmental issues, and solutions for environmental questions and problems were integrated into school curricula and school practice. The main measures where:

- more outdoor experiences
- environment as a topic in science education (focus on biology and chemistry) and in civics or politics education.

To support work at the school level a number of environment and nature centres were established and they helped to open up the school development process through EE. This combination of school development elements with partners from outside school was the success factor for implementing EE in the late 1980s and the 1990s.

On this basis, the work of the Brundlandt commission and the Rio-Process marked an important change in the field of EE. Beside questions of nature protection and the use of nature and environment in general, the social, political and economical impacts on nature and our planet in general became the main parts of the Agenda 21. By combining environmental issues with the development of humankind, elements of intra- and intergenerational justice as well as chances for all human beings to have decent and fair living conditions worldwide, the general questions of EE were to be introduced and understood in a broader way.

By defining SD as the leading concept of future development on planet Earth, the majority of governments worldwide changed the underlying concept of international politics. This change necessitated a change of the basic ideas of EE and education in general as well. From this point onwards the information about and discussion of solutions of environmental problems was not enough to prepare students to develop a sustainable lifestyle.

To prepare children and adults for a sustainable future, competencies became a core concern, especially competencies for:

- the conservation of natural resources for human consumption
- socially and environmentally acceptable modes of economic activities, work and life.
- overcoming poverty worldwide;
- the participation of all people in education, democracy and good governance and the abilities to determine one's own life.

In 1997 the OECD members launched the Programme for International Student Assessment (PISA) aimed at monitoring the extent to which students near the end of compulsory schooling have acquired the knowledge and skills essential for full participation in society.

> The assessment of student performance in selected school subjects took place with the understanding, though, that student's success in life depends on a much wider range of competencies. (OECD 2005: 3)

This important background to all PISA Studies was not really realised by the public and political discussion following the first published results. The discussion focused only on the results of three subjects: literacy, mathematics and science. There was no further discussion on the main question of competencies, which were very difficult to assess. Therefore, in 2002, the OECD launched a process on the "Definition and Selection of Competencies" (DeSeCo).

The general idea was to launch a broader concept and understanding of basic competencies for a sustainable life in the future, by defining the broad categories:

- Use tools interactively (e.g. language, technology)
- Interacting in heterogeneous groups
- Act autonomously (OECD 2005: 5)

In the understanding of OECD experts, these categories are interrelated and together form a basis for identifying and mapping key competencies.

Parallel to this process at the OECD expert level, different processes and programmes worked in the field of education to define sets of competencies for sustainability and especially education for sustainability. All concepts were based on the discussion of 'dynamic qualities', key competencies—developed in the 1980s and 1990s.

ESD which emerged post-Rio is based on different 'traditions' such as:

- Environmental Education
- Global Education/Education for global responsibility
- Civics/political education
- Education against violence and racism
- Health education

In the following years different sets of competencies (OECD 2005; KMK 2007; Transfer 21 2007; UNECE 2012) were developed and discussed.

A description of some of the main examples can be found below:

DeSeCo (OECD 2005)	Gestaltungskompetenz (having the competency to shape the future, 2008) (Transfer 21 2008)	Kompetenzen für den Lernbereich Globale Entwicklung / Competencies for the learning area: Global Development (KMK 2007)
Interactive use of media and methods (tools)		
Ability to use language, symbols and text interactively	Gather knowledge with an openness to the world and integrating new perspectives	Gather information on questions of globalisation and development and process them thematically
Ability to use knowledge and information interactively	Think and act in a forward-looking manner	Recognise social-cultural and natural diversity
Ability to use technologies interactively	Acquire knowledge and act in an interdisciplinary manner	Look at a variety of methods to evaluate development aid measures, taking diverse interests and frameworks into account, and make individual evaluations
Interacting in social heterogeneous groups		
Ability to maintain good and durable relationships with others	Ability to identify and reflect on risks, threats and uncertainties	Subject globalisation and development processes to qualified analysis, applying the guiding principle of SD
Ability to cooperate	Ability to plan and act together with others	Recognise the different social structural levels from the individual to the global and identify their respective functions for development processes
Ability to overcome and resolve problems	Ability to reflect on action strategies and goal conflicts	
	Ability to be part of decision making processes	
	Ability to motivate oneself and others to get active	

(continued)

DeSeCo (OECD 2005)	*Gestaltungskompetenz* (having the competency to shape the future, 2008) (Transfer 21 2008)	*Kompetenzen für den Lernbereich Globale Entwicklung*
		Competencies for the learning area: Global Development (KMK 2007)
Acting autonomously		
Ability to act within the wider context	Ability to reflect upon one's own principles and those of others	Become conscious of, appreciate and reflect on one's own and other values and their meaning for life choices
Ability to form and implement a life plan and personal projects	Ability to reflect on questions about equity and use it for decision making	Form opinions after critical reflection on globalisation and development issues, informed by international consensus on SD and human rights
Awareness of rights, interests, boundaries and requirements.	Ability to plan and act autonomously	Recognise areas of personal responsibility for humankind and the environment and take up the challenge
	Ability to show empathy and solidarity with disadvantaged	Overcome social-cultural and special interest barriers to communication, cooperation and problem resolution
		Pupils can act in times of global change, especially in personal and professional life, through openness and the willingness to innovate as well as through a reasonable reduction of complexity, and are able to withstand the uncertainty of open situations
		Pupils as a result of their autonomous decisions, are able and willing to follow SD objectives in their private lives, at school, and at work, and can work towards their implementation on the social and political level

These different sets of competencies define the competencies a student should have developed at the end of school education. Therefore, the most important next step is to operationalize and adapt them to school education in practice and subject-based education at all levels of education. The ESD based competencies are on the level of cross-curricular and cross subject competencies. In order to avoid add-ons, there is a need to interlink them with subject-based competencies so as to formulate concrete curriculum points and contributions of (all) subjects to ESD. The Standing Conference of Ministers of Education in Germany (KMK) established a team of experts to formulate those subject contributions. The first set of subjects (primary education, politics, ethics and religion, geography, economics, biology) was published in 2007 (KMK 2007). It is planned to publish the contributions for all other subjects in 2015.

Apart from competencies, the concept of multiple perspectives is another backbone of ESD. If students should really understand general questions of SD they must have the ability to assess such questions from different perspectives:

- From their own perspective
- From the perspective of different generations
- From the perspective of people from other regions
- From the perspective of different cultural backgrounds
- From the perspective of different disciplines and domains
- From the perspective of different historical periods.

School education and all types of formal and informal education should offer opportunities and examples of this multi-perspective learning arrangement.

Furthermore, the different components and elements of SD must be integrated into ESD and be part of the process. Figure 2.1 illustrates the main components and elements of SD:

The foundations of ESD as described above—competencies, multiple perspectives, components and elements of SD—need, as a next step, ideas and principles of how to choose appropriate thematic areas and topics.

In my view, thematic areas and topics in ESD:

- should offer the possibility of different perspectives, especially global perspective
- should be based on the concrete living situation of participants
- should be based on the experiences of participants
- must offer real participation
- must offer the possibility for student action
- must integrate partners from outside school.

Beside these general guidelines, thematic areas should integrate important and accepted areas of SD. Additionally, reflection on non-sustainable developments should also be included. Relevant thematic areas include:

- Diversity of values, cultures and living conditions
- Globalisation of religious and ethical guiding principles
- History of globalisation: From colonialism to the 'global village'
- Consumer goods from around the world: Production, trade and consumption
- Food and agriculture
- Illness and health
- Education
- Globalised leisure-time activities
- Protection and use of natural resources and energy production
- Opportunities and dangers of technological progress
- Global environmental changes
- Mobility, urban development and traffic
- Globalisation of economy and labor
- Democratic structures and developments
- Poverty and social security
- Peace and conflicts

Components and Structural Levels of Development

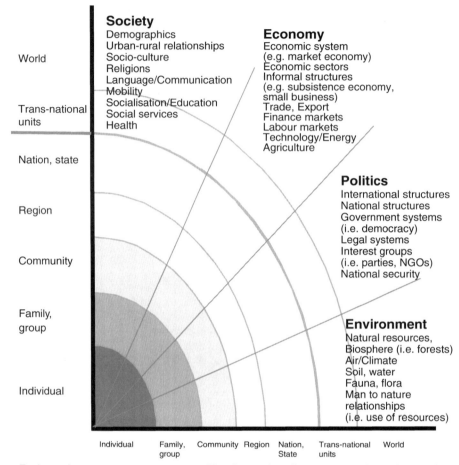

Each quadrant represents one structural level spanning all components. It may be practical to define components, levels and elements of development differently, depending on the analytical purpose and current developments.

Fig. 2.1 The principal elements of sustainable development (Source: KMK 2007)

- Immigration and integration
- Political power, democracy and human rights (Good governance)
- Development cooperation and institutions
- Global governance—world order (KMK 2007: 63)

SD as discussed and formulated 20 years ago in Rio is not possible without the participation of all people of all ages. Only participation can help them create and accept the different—hopefully best—ways to a more sustainable society. Education, understood not as instruction, but as a process of involvement in the process of future orientation, future planning and creation of a sustainable future, is one of the main strands of development in the field of SD. The world summit in Johannesburg in 2002 focussed mainly on this question and led to the UN-Decade on ESD 2005–2014. This means that ESD should influence education in general on all stages of education, at kindergarten, schools, adult education and all types of non-formal education. As SD is not a closed concept with given solutions, but an on-going process of looking for and discussing new and best solutions accepted by the people, ESD should introduce the concept of lifelong learning to everyone at school. During the time children stay at school, they must have the opportunities to develop and realise their specific concept of lifelong learning. Education understood on the basis of this concept must change teaching, learning and students' participation as well as how the school is organised institutionally. It also changes the nature of cooperation with the local community and partners in the society. Instruction by adults—mostly teacher—must be replaced by co-construction between students, teachers, parents, partners and experts from outside schools. Another aspect of the development from EE, GE and other such areas to ESD is to link those elements to a structured concept of ESD. Beside EE, health education and the question of equity are the main strands to combine. Taking this into account the concept of a whole school approach follows the worldwide development initiated by the world healthy school association (BAG 2010). The impact of the concept of a whole school approach will be shown by describing the elements in detail later on.

Within this general change of education towards ESD, the school must be seen as a role model for SD. The students at all stages spend an increasing part of their life at school. This means that part of the general education of real life experiences must be offered and realised during school time and at school. This includes questions of food and consumer education, social learning, energy use and personal resource management. On the other hand, school is an optimal place to reach a whole generation in a protected, safe space where young people can test and develop their main life skills and their own lifestyle. The understanding that sustainability must be the general guideline of this discussion can lead this development at school:

> The key message that comes from the story of Eco-Schools' success has to be that for change to happen, power must be disseminated to the point of implementation. Schools are dominated by students. They are the ones who act as the eyes and ears of behavioural change. Develop the schools' processes and systems to support student-led change. Eco-Schools highlight that ESD is not just about curriculum content, but a whole of school body, whole of school mind set and whole-school action process. The case study also acknowledges that change is slow, incremental and is only sustainable if genuine models of participatory learning and decision-making form the basis of the process.
>
> The greatest gift a school head teacher can give to his/her students, therefore, is the gift of freedom for self-directed and purposeful learning, supported by structures and processes that empower and engage with real-life ecological issues.
>
> The lessons of Eco-Schools also highlight that those who create the ecological footprint need to have opportunities to reflect and understand what it means to be part of the environment, the effects one has in all the different interconnected cycles and biomes of life and to be involved in and in control of remedial action or proactive measures.
>
> Ultimately, Eco-Schools are a process that becomes a way of life, a cultural paradigm for school administrators to master through delegation and a belief in their teachers' and students' capacity to change the school from the ground up. (UNESCO 2012: 71)

This discussion must take into account the already existing studies on school development in the field of ESD and ideas of a whole school approach, like the study of Ferreira et al. (2006) and the S3 concept of the British school inspection. Especially the S3 self evaluation instrument on ESD practice at school offers a great possibility to interlink different aspects of SD and ESD, by naming different doorways to ESD at school:

- focus on food and drink
- focus on energy
- focus on water
- focus on travel an traffic
- focus on purchasing and waste
- focus on school buildings
- focus on school grounds
- focus on inclusion and participation
- focus on local well-being
- focus on the global dimension. (DCSF 2008: 3)

On the way of moving from different areas of ESD to a whole concept, it must be acknowledged that there are different 'cultures' of cross-curricular education in the fields of environmental education, education for global development, health education, citizenship education. They all contribute to ESD and are linked in several areas of cooperation and common fields of action. A recent UNESCO-report on ESD (2012) gives a clear picture of this linking.

The concept of a whole school approach as understood in our context is not a given structure to be followed or a concept for a sustainable school in itself, but it offers a platform to evaluate the practice of the individual school and to formulate challenges, possible co-operations, combining different fields of activities and help to develop new fields of practice and activities towards a more sustainable practice at school. This concept follows the experience of the former OECD based network ENSI, which defines ESD as mainly an open concept and based on dynamic qualities and competencies. As ENSI formulates:

> A School engaged in ESD is engaged in learning for the future, by inviting students and teachers to enter a culture of complexity, by using critical thinking to explore and challenge, in clarifying values, reflecting on the learning value of taking action and of participation, revising all subjects and the pedagogy in the light of ESD. (Breiting et al. 2005: 10)

Following these ideas of conceptualising ESD a first notion to give a picture of all elements of whole school approach to sustainability emerges:

This model should help to raise a discussion at every school on their practices in the field of ESD and all their ideas to improve ESD at their school on the way to a more sustainable school in the community.

Elements of a Whole School Approach (Fig. 2.2):

In addition to the development and expansion of thematic references in subjects and areas of learning, involving the educational institution as a whole is becoming

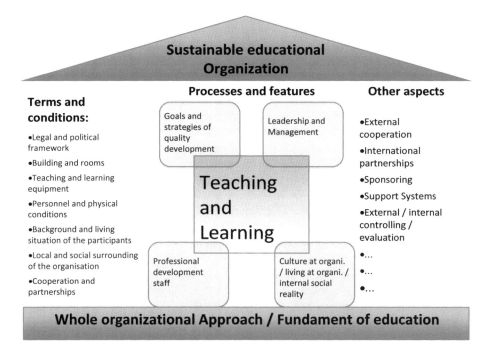

Fig. 2.2 Source: Reiner Mathar, unpublished

increasingly important. The various elements of the educational institutions need to come together to combine the contributions of subject areas and to make cross-curricular projects.

School management and organisation of life at school must be guided by the principle of SD in terms of the Rio process and schools are seen as role models for sustainable design of all areas of life. Pupils can learn sustainable living in school and are supported in the development of their own sustainable lifestyle.

In addition to lessons, the core task of the school in the educational process, this moves all other areas of the school to the foreground:

- The educational material cycles and resource management
- Cooperation and partnerships
- School life and identity of school
- The roles and cooperation between the stakeholders in the school
- The school management as a linking element.

The concept "ESD as a task for the whole school" is not about a self-contained structural requirement, but about an open platform. Sustainable educational institutions are consolidating and developing existing fields and the development and establishment of new areas of education in order to support school development.

The core business of the school is teaching and learning in lessons, projects and extra-curricular projects. In this area there are a variety of approaches for the subject areas to implement ESD.

School planning should be concentrated on the contributions of as many subjects for the competencies development of students in the field of SD and outline the individual elements in a school curriculum or a school programme, and increase the interconnections and collaborations between subjects.

If students are to be actively supported in the development of competencies for shaping their own lives in terms of SD, the design of the internal social structure and cooperation in the school is of particular importance. The involvement of pupils in the planning and design of social interaction and also the lessons has a high priority. Regulations and agreements for the design of social life at school must include elements of sustainable lifestyles and should be developed in terms of participatory and creative approaches to SD in a democratic process of negotiation.

In school life skills programmes must, as a first priority, provide and implement the social aspect of SD. Again, it is always about the connection of existing approaches (learning democracy, mediation, violence prevention, schools without racism) with their own priorities of the school in the field of ESD.

2.1.2   Frame and Conditions of the School

Legal and Policy Framework

In addition to direct curricular requirements, educational standards as well as other documents and legal requirements promote the dimensions and the acceptance of ESD. They also provide, in dialogue with the local school environment, transparency and mutual understanding of the basic principles of SD. For example:

- Basic documents of the United Nations and its institutions,
- The constitutions of countries, school laws with their respective approaches to parent education and educational tasks.
- More internationally discussed documents (e.g. Earth Charter)

Buildings and Facilities

The state of the building and the rooms of a school, as well as their existing and desired features in the context of SD, can be an important field for action and reflection for the students. Through the development of concepts of sustainable design of material cycles and the utilisation of resources and dialogue about it with school authorities and the wider community of the school, the students can develop skills for their own sustainable lifestyle and apply them in the demarcated area of action at school. They can also reflect on the conditions and reactions to their activities.

The school also offers the advantage of a confined space and manageable change. So the design of sustainable energy use or design of spaces can be tested, but students are also confronted with the conflicts and limitations of the implementation and designs. These range from sustainable and efficient use of limited resources to fair and climate-neutral sourcing to sustainable alignment of all biogeochemical cycles of the school. Working groups, elective offers for students and student companies can be formative elements in the everyday life of the schools here.

Human Resources

For the design of the social and cultural dimension of SD, the staffing of the schools, i.e. the staff-student ratio and recruitment of teachers, also may be of particular importance and should it be used accordingly. The design of providing care, the integration of school social work is a learning field of action for SD and frequently offers numerous opportunities for design by the students and the involvement of other partners. This is particularly true for the practical design of those areas at the schools which are used all day.

The integration of migrant women organisations provides the opportunity to experience cultural diversity and cultural differences and understanding. Proven organisational forms are parents' cafes, cooking classes, gaming and cultural festivals.

In the area of human resources the reflection on the necessary competencies of teachers and their development through training and education is particularly important. Incentives must actively be used in the field of ESD and can be an element of active personnel planning. In Germany, the school board plays a significant role in the design of ESD, hence the qualification and training of school principals is crucial. They should have a basic understanding of the principles of SD so that they are able in their management role to design the school with a focus on SD and accompany the ensuing processes.

Background and Living Conditions of Students

The development of skills for SD and an understanding of how global developments take place on the platform of social and cultural lives of students is central. This includes the necessity to obtain information about the cultural background of students and to consider how to include this a school.

The school should pay attention to learning processes, and align those to these backgrounds in the context of ESD. Through a conscious reflection on different cultural backgrounds, different perspectives can be made clear and used as an opportunity for dialogue and change. Using pupils with an immigrant background, aspects of change in perspective can be analysed in a global context and the influence of different ways of life can be related to SD.

Local and Social Integration of the School into the Community

An essential element of the design of SD are, in addition to the global perspective, the regional roots. The school social environment—beyond the narrow sphere of each pupil's own experience—opens up new and extended fields and opportunities. This is all about the reflection on local/regional cultural characteristics, economic development and social and environmental challenges.

Schools should and could realise that the knowledge and competencies orientation also applies to regional development and opening up new fields of competencies and action for the students.

This includes school activities as well as the cooperation with business and civil society and is part of the creation of partnerships and collaborations.

Examples beyond the established collaborations include:

- The active creation of twinning
- Citizens' solar panels on school roofs
- Neighborhood cafes and branch libraries
- Translation and secretarial services for the district
- Public bike shops.

Partnerships and Collaborations

The systematic inclusion of collaborations and partnerships in the school concept of ESD opens up more possibilities for the school. Many schools have partnerships in their region and partnerships with schools in Europe and in the countries of the South. However, these are often built on the initiative of an enthusiastic individual and not integrated into an overall view of the school. The study area global development here provides opportunities to strengthen these partnerships with a content and thematic focus. When schools work in partnerships on concrete examples, both sides manage the integration of the topics in question into the core business of the school—teaching and learning—and can also help to achieve planning reliability and a long-term edge. If students are working in partnership on specific topics, it facilitates the change of perspective and a personal exchange from student to student. To realise that looking at problems and challenges and their management from different perspectives is not only possible but also necessary, and helps students to develop the appropriate skills. Southern Partners open up possibilities, contribute their perspective and thus change the often inadequate images of their life situation. Existing facilities and internet platforms offer a variety of support options here.

Communication via internet and mobile network-based instruments are already part of students daily lives, not only in countries like Germany but also increasingly in the countries of the South, but it is still underutilised in schools. Systematic use of such electronic tools in the educational process can also give the pupils guidance for their own use of social media here.

There are different types of co-operation between schools from so-called developed countries/schools of the North and from developing countries/Schools of the South (including schools from Africa, Asia, Latin America and South-East Europe). An important distinction is the distinction between sponsorships and partnerships. While a sponsorship focuses on one-sided help from the North to the partner school, a partnership aims to promote dialogue between more or less equals. Here, mutual learning from each other is in the foreground and financial support is only one of many aspects.

In the context of partnerships and collaborations the non-governmental organisations (NGOs) of development cooperation have an important role to play. They can competently advise schools and also contribute concrete projects and proposals. Using Global Development Education or ESD as a task for the whole school can lead to a systematic cooperation, developed and established as an integral part of the school. The development of regional networks can act as a mediator and pilot.

2.2   Other Aspects of School Development With the Main Focus on Sustainable Development

The systematic integration of school and student competitions in the organisation of ESD in a school can help encourage pupils to engage in individual or group work. By participating in competitions, students can learn effectively, but above all, gain appreciation for their work that goes far beyond the work at school. The information about scholarship programmes for young people to gain authentic experience of global development in the countries of the South should be systematically supported in school.

In the area of school inspection and external quality evaluation of schools of Global Development Education and its implementation in the school should be included in the focus of the evaluation and assessment. This could feed all of the elements described above with the quality considering the external evaluation.

2.3   Conclusion

At the end of the UN Decade on ESD 2005–2014 the concept of competence-orientation has become one of the main backbones for ESD. Linked to all main developments in the fields of education, which also focus on the development of the skills and competencies of the learner coherent with the idea of lifelong learning, ESD contributes directly to concepts of futures education. Fed by different traditions, which are mainly content driven, ESD manages to be education *for* sustainable development more than education *about* sustainable development. Following this concept the future question—some steps on the way have been made already—

is to integrate the concept of sustainable development as a guiding principle into all stages of education and to become the core part of all educational institutions in the sense of a whole institutional approach to sustainable development.

References

BAG (Bundesamt für Gesundheit). (2010). *Linking health, equity and sustainability in schools*, 10–11 July 2010, international conference, Geneva. http://www.ecoles-en-sante.ch/data/data_612.pdf. Accessed 14 Apr 2014.

Breiting, S., Mayer, M., & Mogensen, F. (2005). *Quality criteria for ESD schools, guidelines to enhance quality of education for sustainable development*. Vienna: BMUKK.

DCSF (Department for Children, Schools and Families). (2008). *Planning a sustainable school driving school improvement through sustainable development*. Nottingham: Department for Children, Schools and Families. https://www.education.gov.uk/publications/eOrderingDownload/planning_a_sustainable_school.pdf. Accessed 14 Apr 2014.

Ferreira, J., Ryan, L., & Tilbury, D. (2006). *Whole-school approaches to sustainability: A review of models for professional development in pre-service teacher education*. Canberra: Australian Government Department of the Environment and Heritage and the Australian Research Institute in Education for Sustainability (ARIES). http://aries.mq.edu.au/projects/preservice/files/TeacherEduDec06.pdf. Accessed 17 Apr 2014.

IUCN (International Union for Conservation of Nature). (1991). *Caring for the Earth. A strategy for sustainable living*. Gland.

KMK (Kultusministerkonferenz). (1980). *Empfehlungen zur Umweltbildung*. Bonn: KMK.

KMK. (2007). *Global development education: A cross-curricular framework in the context of education for sustainable development*. Bonn: KMK.

OECD (Organisation for Economic Co-operation and Development (OECD). (2005). *The definition and selection of key competencies*. Paris: OECD. http://www.oecd.org/pisa/35070367.pdf. Accessed 29 Mar 2013.

Sarabhai, K. V. (2007). *Tiblisi to Ahmedabad, the journey of environmental education*. Ahmedabad: Centre for Environmental Education (CEE).

Transfer 21. (2007). *Guide education for sustainable development at secondary level—justifications, competencies, learning opportunities*. Berlin: Freie Universität.

Transfer 21. (2008). *Gesaltungscompetenz: Lernen für die Zukunft—Definition von Gestaltungskompetenz und ihrer Teilkompetenzen*. http://www.transfer-21.de/index.php?p=222. Accessed 11 May 2014.

UNECE (United Nations Economic Commission for Europe). (2012). *Learning for the future, competencies in education for sustainable development*. Geneva: UNECE.

UNESCO. (1978). *Final report: Intergovernmental conference on environmental education*. Organized by UNESCO in Co-operation with UNEP (Tiblisi, USSR, 14–26 October 1977). Paris: UNESCO.

UNESCO. (2012). *Shaping the education of tomorrow—Full length report on the UN Decade on Education for Sustainable Development* (DESD Monitoring and Evaluation). Paris: UNESCO. http://unesdoc.unesco.org/images/0021/002164/216472e.pdf. Accessed 17 Apr 2014.

Chapter 3
Quality Criteria for ESD Schools: Engaging Whole Schools in Education for Sustainable Development

Søren Breiting and Michela Mayer

3.1 Introduction

In 2005 a booklet with the title *Quality Criteria for ESD Schools* was published with the subtitle *Guidelines to enhance the quality of Education for Sustainable Development* (Breiting et al. 2005). The booklet is now available in around 18 languages, among them 16 European languages.[1] Different versions are widely circulated around the world in print and online (www.ensi.org, accessed 8 May 2014). They seem to have fulfilled a need during the UN Decade of Education for Sustainable Development (ESD) 2005–2014 (DESD).

In this chapter we go beyond the publication and highlight the main driving forces behind the strategy of focusing on 'quality criteria' for ESD development, on the one hand, and experience of using the booklet, on the other hand. We started work on the booklet at a time when trends from 'new public management' (NPM) were already hampering grassroot development in many schools in Europe. It analyses why it is essential to emphasise the democratic nature of education, especially when this is related to sustainable development (SD), and to avoid the instrumental view of schools, teachers and learners, often held by authorities outside educational institutions. To impose on schools the solving of community problems that the community

---

[1] English, Portuguese, Spanish-Castellan, Spanish-Catalane, French, German, Italian, Dutch, Danish, Norwegian, Swedish, Finnish, Romanian, Hungarian, Greek, Russian.

S. Breiting (✉)
Department of Education, Aarhus University, Campus Emdrup,
Tuborgvej 164, DK-2400 Copenhagen NV, Denmark
e-mail: sorenbreiting@gmail.com

M. Mayer
UNESCO DESD Italian Scientific Committee, UNESCO Italian Commission, Rome, Italy
e-mail: michela.mayer@gmail.com

© Springer International Publishing Switzerland 2015
R. Jucker, R. Mathar (eds.), *Schooling for Sustainable Development in Europe*,
Schooling for Sustainable Development 6, DOI 10.1007/978-3-319-09549-3_3

and politicians are not able to solve counteracts the empowerment effect on learners that is so much needed in ESD, as will be shown below.

An important background for the publication *Quality Criteria for ESD Schools* was a survey by Mogensen and Mayer (2005) about explicit and implicit quality criteria in operation related to the development of 'Eco-Schools' in European schools, see Eco-Schools (2014) as an example. The findings revealed weak points in the general interpretation of Eco-Schools when seen in the light of a socio-critical paradigm and of "sustainable education", with reference to Robottom and Hart (1993) and Sterling (2001). In that way it emphasised important issues to avoid when attempting a transfer into a tool for schools' development towards sustainability. At the same time it was found to be imperative to distinguish between environmental management initiatives at a school, often initiated by adults, on the one hand, and, on the other hand, the mechanisms of open learning processes related to the complexity and controversial nature of development issues spanning for example environmental concerns and social issues related to present and future generations.

After the publication of *Quality Criteria for ESD Schools* a need arose to collect experience derived from the use of the publication, and to take a critical look at the limitations of the publication and the activities flowing from it. This chapter provides an overview of the principal results of such a critical evaluation of the ideas behind the publication and their interpretation.

3.2 The Eco-schools Survey's Focus and Main Scenarios

As a background to the booklet a survey was carried out in 2003–2005 by the ENSI (Environment and School Initiatives) network and the European Comenius 3 Project SEED (School Development through Environmental Education). The focus was on collecting and discussing different visions of 'a sustainable future' embedded in the 'whole school approaches' existing at that time and that had been roughly identified with the 'Eco-Schools programmes'. The survey (Mogensen and Mayer 2005) was conducted mainly among European countries (11 countries: Austria, Belgium, Denmark, Finland, Germany, Greece, Hungary, Italy, Norway, Spain and Sweden) with the participation of Australia and Korea, and gathered information concerning 28 different 'Eco-Schools' programmes,[2] involving over 3,500 schools.

National experts were asked to present not only the 'explicit' but also the 'implicit' criteria guiding the programmes and/or used for their evaluation. In the guidelines prepared for the survey, implicit criteria were considered to be more representative of the general vision guiding the programmes than the explicit ones

---

[2] Not only the well-known international 'Eco-School programme' from the Foundation for Environmental Education (FEE) but many others with different names and focus, such as Green schools, Model schools, Future concerned schools).

and were defined as "aims and general values proposed; importance given to a set of explicit criteria compared with others" (Mogensen and Mayer 2005: 52). As an example: Programmes that ask to fulfil ten or more criteria for the care of the school's physical environment (from waste collection to energy saving) but only ask for just one or two criteria related to changes in the teaching and learning processes, or in the relationships with the local community, give evidence of being centred mainly on environmental management and behavioural change (change of individuals) rather than on a change of school pedagogy or of the school's local community relationships.

The analysis of the national reports and of the case studies prepared by the national experts was guided by a 'critical framework', inspired by the basic ideas of the ENSI network (CERI-OECD 1991, 1995). In the ENSI vision EE and ESD should act as "an example for the quality of education in general" (CERI-OECD 1995: 100). They are embedded in a culture of complexity where the notion of uncertainty asks for critical reflection and for a democratic exchange of views, conscious of the fact that each vision of world problems is oriented by values. Students should be helped to develop their action competence and become autonomous: i.e. critical and constructive thinkers, able to intervene and to be proactive (and not only to react) in emerging and unforeseen situations (Breiting and Mogensen 1999). Education in this vision has an 'emancipatory' and 'transformative' role, which implies that its understanding and practice needs to change in order to become 'sustainable', as proposed by Sterling:

> The term 'sustainable education' implies whole paradigm change, one which asserts both humanistic and ecological values. By contrast, any 'education *for something*', however worthy, such as for 'the environment', or 'citizenship' tends to become both accommodated and marginalized by the mainstream. So while 'education for sustainable development' has in recent years won a small niche, the overall educational paradigm otherwise remains unchanged. (Sterling 2001: 14; emphasis in the original)

The survey analysis was then oriented to find indications and traces of 'emerging changes', and to construct different 'scenarios' on those changes as tools for reflection: "Scenarios are stories about the way the world might turn out tomorrow, that help us to recognise changing aspects of our present environment" (Schwartz 1991: 5) Three alternative scenarios were sketched, none of them being the best or the more probable one, but each one gathering some trends found during the analysis:

- "An Eco-School as an ecological enterprise" scenario, in which SD is mainly a matter of technological advancement and effective behaviours: The school works as a functional enterprise where the focus is on excellence and where the main changes brought about by ESD are guided by up-dated information about the world's situation and the ways to react (Mogensen and Mayer 2005: 92–93);
- "An Eco-School as a family, fond of nature" scenario, where SD is mainly a matter of reconstructing meaningful relationships between people and with nature, and where the school comes across as a "core social centre" (ibid.: 91) for community initiatives. Empathy with nature, feeling of belonging, creativity, are the key words of such a scenario (ibid.: 93);

• "An Eco-School as an 'Educational research' community" scenario, where SD is conceived as a social and cultural challenge, and the school aims to act as a "learning organisation" fostering democratic dialogue, accepting internal conflicts as a 'stimulus' to the community in order to progress together. "The assumption is that a critical attitude will prepare for continuous changes in a 'not as yet definable' sustainable future" (ibid.: 93–94).

Using the three scenarios as an ex post frame for analysis, there was some evidence that the first one was at that time, in 2005, the most commonly used, especially in secondary schools and in national awards schemes, while the second one was present mainly in primary school programmes. The third one was considered the deepest change (a "third order learning" in the words of Bateson 2000: 301–306), but it was just emerging and to be found only in a few cases with a strong school development approach.

3.3   The Challenge of Evaluation in ESD: The Quality Criteria Proposal

The second focus of the survey was the definition and the use of an evaluation strategy consistent with the "sustainable education" approach and with the "socio-critical paradigm" guiding ESD (Robottom and Hart 1993; Liriakou and Flogaitis 2000).

ESD processes are indeed complex and dynamic in nature and a meaningful evaluation must embrace this complexity and accept the controversial nature of dealing with development issues. Evaluation should not be used for 'quality control', as we often see it under the initiatives of NPM, but for 'quality enhancement'.

Such an understanding of quality and evaluation ran counter to the language and culture of (educational) quality evaluation, used by politicians and administrators in Europe, which had dominated the 20 years up to the survey. 'Evaluation' was used as a tool for strong educational control and not for school development towards ESD. For this reason it was and still is important to rethink the concept of educational quality in order to build a new meaning, useful for all members of a school community, and consistent with the importance of accepting uncertainty and complexity as part of ESD.

In a culture of complexity, in effect, evaluation cannot reduce the quality of educational processes to a 'set of standardised procedures', 'outcomes' or 'performances' so much in focus in NPM. Rather, it should take into account the educational values, the cultural characteristics of the local community as well as the emotions and perceptions. Robert Pirsig—author of the famous motorcycle novel *Zen and The Art of Motorcycle Maintenance*—makes a distinction between "static quality", the one which pushes a system to achieve defined benchmarks and standards, and "dynamic quality", the quality that a system needs when something new happens, when it is necessary to proceed in uncertainty where standards do not exist. To acknowledge the value of the distinction between static and dynamic qualities has been a consistent part of the ENSI network research (CERI-OECD 1991; 1995); both are relevant and necessary: "without dynamic quality an organism cannot develop, without static quality it cannot last" (Pirsig 1992: 375).

The collected case studies by Mogensen and Mayer (2005) made clear that a standardised evaluation was limiting many Eco-School programmes to "a mere physical improvement in the school environment (. . .), lacking the perception of its educational effects" (Sun Kiung Lee, Korean National Report, in Mogensen and Mayer 2005: 86). On the other hand, when the programme evaluation was based on reflection, self-evaluation and peer evaluation, the evaluation itself could become an essential part of school quality development. In a socio-critical paradigm, quality criteria should not be confused with 'performance indicators', again a concept from NPM. The notion of quality should build on values and principles that inspire engagement with sustainability issues, and provide indications or general descriptions that help to turn values into educational actions, behaviours and choices. Moreover, quality criteria should be seen as an instrument for change and not as an instrument for assessment, focusing the attention not only on foreseen results but also on emerging, unexpected outcomes. The criteria thus bring theory and visions closer to practice, and can be used as links for moving from ideal values to the reality one wishes to change. They constitute a way

> to travel within the valued ESD vision through reflection. ENSI has enforced for many years a vision of ESD school change as action research. (. . .) Promoting ESD quality reflection in schools is then a central part of ESD quality development. (Espinet 2012: 99)

Taking account of these considerations about assessment and evaluation of school changes, a list of ideas for quality criteria was provided in the booklet. They were presented not as prescriptions, but rather as an inspiration for creating a list fitting a school's own needs. The criteria were seen as a "starting point for schools that wish to make use of the focus on ESD as a vehicle for the school's own development" (Breiting et al. 2005: 9). The booklet tried to summarise and in a way to specify an educational philosophy for school development with respect not only to ESD but to deep educational change: The criteria are organised in three main groups, each group divided in smaller areas. Each area is introduced by a concrete example taken from school practice—many of them collected during the survey—to help imagine what the criteria could mean in the everyday educational process. After the example a short explanation—'a rationale'—presents the debate and the trends concerning the area in question, and justifies the choice of a specific number of criteria. These can be used by teachers, students or school communities as a tool, not so much for assessment, but mainly for fostering reflections on, and the self-evaluation of, the school journey towards sustainability. The criteria listed should be used as suggestions, to be discussed and modified in the understanding that every school should define its own quality criteria. For this reason each quality criteria area ends with 'open dots' asking for new or revised ones.

The three groups of criteria are

- The quality of teaching and learning processes (nine areas)
- The quality of the school policy and organisation (four areas)
- The quality of the school's external relations (two areas).

It is evident that the teaching and learning processes are at the centre of the changes envisioned: the rationale for this choice is that no SD will be possible without profound learning and changes not only in institutional and individual behaviour, but also in the way people look at things, in competences and attitudes relating to the natural and social world. Some of the suggested quality criteria concern core aspects of working with SD issues, such as "Students get involved in comparing short term and long term effects of decisions and alternatives" (from "Quality criteria in the area of perspectives for the future", Breiting et al. 2005: 19). Others are of a more general nature but still seen as particularly relevant for ESD: "The teachers facilitate students' participation and provide contexts for the development of students' own learning, ideas and perspectives" (from "Quality criteria in the area of teaching-learning approaches/processes", ibid.: 15).

In the following box some extracts from one of the areas are presented.

Quality Criteria in the Area of Critical Thinking and the Language of Possibility

Example: Students in seventh grade were working on a project dealing with problems and questions related to the use of pesticides (. . .) half of the class was asked to take one position (. . .) while the other part of the class took the opposite position (. . .) It was a task for the students to present the position they were advocating in a panel (. . .) 'informers' from the local community (. . .) were invited to be present at the panel debate (. . .) students were asked to take a stand on the issues (. . .) but also to identify alternatives and possible actions (. . .).

Rationale: Students are exposed to an overwhelming amount of information every day (. . .) In order to become active and responsible citizens students need to be able to think for themselves (. . .) By combining critical thinking with the language of possibilities it is emphasised that to be a critical human being does not mean to be negative or sceptical (. . .) but to couple the critical process of reflection and inquiry with an empathetic and optimistic vision (. . .) searching for solutions (. . .).

Quality Criteria:

- Students work with power relations and conflicting interests
- Students are encouraged to look at things from different perspectives and to develop empathy (. . .) (Breiting et al. 2005: 24–25).

3.4 The Use of the Quality Criteria Booklet

The ENSI network has monitored the use of the booklet within different countries, by editing different translations and by supporting presentations and teacher training activities in national and international contexts (such as Comenius courses and meetings). It has also organised occasions for reflections: the ENSI Louvain Conference in 2009 had a plenary and two workshops dedicated to the use of the Quality Criteria (QC) booklet as 'a travelling guide on the school journey toward sustainability'.

In preparation for the conference, an internet based survey on the QC booklet was prepared and distributed in 12 different European countries and translated into the different languages.

> The purpose of the survey, that includes qualitative as well as quantitative items, was to explore the potentials and the constraints that relevant stakeholders in the countries may have experienced in working with the QC booklet. Thus the respondent was asked, among others, to give their opinion on the relevance of the booklet and the ideas presented in it, the possible use they have made of it and suggestions for improvements. (Mogensen 2012: 69)

The debate carried on at the Louvain Conference (Réti and Tschapka 2012) and resulted in a number of conclusions and suggestions:

- The booklet was highly appreciated—as shown by the number of translations—but not frequently used by individual schools and teachers. The material had been predominantly used by researchers, teacher training institutions, teachers associations and environmental associations, and mainly as an inspiration for building more contextualised lists of criteria, or indicators (sometime to be used for external certification), and not so much as a tool for internal school quality enhancement;
- Some of the criteria—more demanding both from an epistemological point of view (e.g. the ones in the area of a 'culture of complexity') and from school policy and organisation point of view (including the schools' external relations)—seemed still to be far removed from the current, at that time, 'whole school approach to ESD' in Europe;
- More examples connected with what is considered 'SD' in the curricula are needed for a better appeal to teachers or schools and could be added in future editions.

The findings from the survey were rather unexpected: on the one hand, the booklet was strongly appreciated, mainly by researchers and by educational planners. It was, for example, chosen as one of the case studies in the UNESCO publication *ESD. An expert review of processes and learning* (Tilbury 2011). It has also been quoted and used by the UNECE international programme on Indicators for ESD, and by many national programmes for the evaluation of ESD schools and ESD programmes (e.g. in Germany, Italy and Finland). Slovenia has even used the booklet as is for its national quality assessment and enhancement scheme (UNECE 2011: 12). On the other hand, only few teachers, schools and school networks have tried systematically to use the booklet in practice at the time of the survey.

Although the main idea was to propose a tool to encourage "the integration of ESD in the normal life of the school and consider engagement in ESD not as an extra burden for teachers and headmasters but as an opportunity for improving the existing teaching and learning and to provide innovation useful for the whole school" (Breiting et al. 2005: 11), the Quality Criteria booklet was not so easy to use at school level as originally intended. One of the possible reasons for this unexpected outcome is that in order to be a frame of reference and a binding element of a programme or a school, all the participants need to construct and accept jointly the Quality Criteria proposal. This means that, to be effective, the proposal requires an already innovative school, having an open mind with a view to change and well prepared teachers.

In fact, the 'resilience' of the school system should be taken into account when such invitations to deep changes are presented. Many teachers and many parents do not want to change their understanding of education as 'transmission' of information, procedures, techniques (as reported also by the TALIS survey OECD 2009). As Paul Vare noted during the debate at the Louvain conference, it seems far easier—as is true for the general social and economic system—to 'adjust' or to 'add' something here and there within the curricula, than to change the entire system. Vare and Scott (2007) proposed to not lose the vision of 'Sustainable Education' but to combine the two visions of ESD which they have termed ESD 1 and ESD 2:

> ESD 1 is characterised by the promotion of changes in what we do, often facilitating behaviours and ways of thinking where the need for this is clearly identified and agreed (e.g. conserving energy within the school). This is learning **for** sustainable development. (. . .) In contrast to this emphasis on content and 'good' sustainable practice behaviour, ESD 2 focuses on building capacity to think critically about [and beyond] what experts say and to test sustainable development ideas. (. . .) This is learning **as** sustainable development. (Mayer and Vare 2012: 57)

The ENSI Quality Criteria proposal is clearly inspired by ESD2 vision. It could also be said that the QC booklet was intended to make a clear distinction between a focus on environmental management at school and a focus on students' learning processes concerning their engagement and understanding of development issues related to sustainability. The criteria therefore are a response to the need for a profound transformation of the schools' learning and teaching practices relating to sustainability. This does not mean that simple examples, more clearly related to ESD1, could not be added—paying attention not to reduce ESD to a pure 'addition' of new content—in order to ease the access of 'normal' schools and teachers to ESD2. Agents outside the school are often occupied with visible physical changes of schools, such as waste management, infrastructure and buildings, but such changes might have no educational value in themselves. They might naturally contribute to a more sustainable local community but contribute very little to the quality of ESD at the school.

The two main strategies for educational changes in schools towards sustainability have also been acknowledged by Wals in the 2012 UNESCO report on the results of the DESD:

> The challenge of ESD and related educations is to work with two main strategies: 1) the add-on and integration strategy and 2) the whole-system redesign strategy. Whereas the former seeks to widen the space within existing national curricula for ESD, the latter challenges the entire system more fundamentally (. . .). (UNESCO 2012: 41)

While the data collected by UNESCO from DESD initiatives, e.g. see UNESCO (2009) suggest that this 'add-on strategy' is generally prevalent, nevertheless the 'whole institutional approach' has succeeded in becoming the prevalent school development trend in many European countries. The most recent UNECE report on the implementation of the ESD strategy states that: "Sixty-three percent of all countries say to have adopted a "whole-institution approach" which tends to refer to the simultaneous infusion of sustainability in a school's curriculum, reduction of its institutional ecological footprint, strengthening students 'participation, and improving school-community relationships" (UNECE 2011: 10). The data generated in this second phase of the DESD gives evidence that, in spite the difficulties of implementation, ESD 'people' are strongly in favour of learning processes and multi-stakeholder interactions that foster deep changes and involve the creation of alternative ways of learning and teaching, of new schools and different professional development, as well as of new forms of monitoring and evaluation.

In addition to this first challenge linked to implementation difficulties, the authors of the booklet faced a second challenge related to evaluation approaches which proves to be important still: UNESCO's ESD monitoring and evaluation process emphasises critical reflection and reciprocal learning. Quality criteria and indicators have been proposed by countries, NGOs and university based researchers as possible tools both for "highly process-oriented schemes (using indicators of participation, self-evaluation, own initiative, creativity, etc.) or more outcome-based schemes (using checklists to determine whether the school has taken specific measures such as becoming CO_2-neutral and integrating sustainability topics in the curriculum)" (UNESCO 2012: 45).

Amongst the final challenges reported in the DESD report (2012), there is the "need for continuous monitoring, evaluation, research and flexibility" (UNESCO 2012: 81) and the "need for currency and up-to-datedness with emerging paradigms and concepts in the sustainability discourse"(ibid.: 82). The UNECE report equally calls for "continued attention to developing appropriate monitoring, evaluation and indicator schemes and support of related ESD research" (UNECE 2011: 22).

Quality and evaluation are still central needs within ESD: the quality of processes has been considered in the DESD monitoring as important as the quality of products (or even more so, assuming a learning point of view), and a reflective approach to evaluation and monitoring is the only one consistent with the 'learning changes' the Decade aims at:

> Efforts (. . .) only have value when used to improve the quality of processes and products. On the other hand, an imbedded reflexive approach helps to build in ways of continuously reviewing past actions and learning in order to enable better, more meaningful and transformative processes to achieve the same goal. (UNESCO 2012: 81–82)

3.5   The Empowerment Perspective of ESD

With hindsight, ESD has been marred by a number of challenges. The basic question about what ESD is and what function it should have in different forms of learning and education has been raised since its inception (see for example McKeown and Hopkins 2007; Stevenson 2006; Stables and Scott 2002). The debate on the position of ESD regarding environmental education (EE) is ongoing: should ESD be seen as a progression from EE, as a competitor, or even as a step backwards from the goals and achievements of EE? The authors of *Quality Criteria for ESD Schools* saw ESD as a progression from EE with the ambition to enhance its democratic potential, building on experience from research and practice in EE, as well as from other relevant areas, thus underlining EE's orientation towards future challenges.

So why is the democratic potential of ESD so important for its conceptualisation as well as for its outcomes? It is not possible to teach SD starting from an understanding that SD is a given thing. It is often quite easy to analyse a development trend and to identify it as non-sustainable, i.e. that it cannot continue forever. But in reality it is not possible to define SD once and forever in a specific way. This has to do with the uncertainty of future 'development' or change.

But even more fundamentally it has to do with the whole conception of SD. In the classic formulation, well-known from the Brundtland report, SD "seeks to meet the needs and aspirations of the present without compromising the ability to meet those of the future" (World Commission on Environment and Development 1987: 39, §49). At first sight this formulation can be read as a harmonious consensus situation, but some afterthought will reveal that many opposing interests are involved. This becomes obvious when addressing questions like:

- What does it mean to satisfy the needs of a generation?
- What are the minimum requirements for a person to have a fulfilling life and access to resources?
- Which levels of risks are acceptable and can be imposed on the next generations?
- How can people, nations and the international community agree on a direction of development that avoids activities which already now can be labelled as non-sustainable?

Introducing such questions in education underlines the need for learners to develop their abilities to think critically, constructively and to be able to deal and be interested in dealing with complex and controversial issues related to development, on a local, national or international level. Without such competencies learners will not be prepared to participate in shaping the future for themselves and for others, including giving their voice to what they think is right and wrong and to argue for their own interest and positions. Therefore the approach to quality criteria in the booklet was to introduce these concerns to the practice of real school life in an inductive way by giving inspiration to the schools' own development in such a direction. Combining the overall ethos of a school with its intentions of adopting SD as a serious pedagogical challenge was regarded by the authors as the most promising strategy for ESD improvement.

A number of more factual aspects from natural science, from social science, from history and culture have to be a part of the learning process. But the point is that learners will not be adequately empowered to engage in SD without questioning aspects of dilemmas and issues, as indicated by the questions above (see also Schnack 1998; Breiting et al. 2009). In other words, the content of ESD is not only focussing on 'matters of fact' but also on 'matters of opinion', i.e. the goal of ESD has to be a 'Bildung' (education)-perspective as argued in Mogensen and Schnack (2010).

Over the years there have been many contributions to the content matter and approach of ESD (e.g. Rauch and Steiner 2006). It is clear that the content of ESD has to be shaped to fit the local and cultural situations of the learners and their living conditions. On the other hand campaigning for SD issues in schools is not seen by the present authors as a solution.

Figure 3.1 visualises the outcome from campaigns and compares their effect with empowerment strategies. We often see campaign approaches when agencies outside the educational system turn to schools to implement changes in society that are controversial and difficult to achieve among the general population. We find many good examples of such well-meant efforts from ministries of the environment in many countries, e.g. concerning waste and recycling.

The slim curve in Fig. 3.1 indicates the effect over time of a campaigning strategy to achieve some kind of change, often related to behaviour, for example the buying behaviour of consumers, or aspects of a healthier lifestyle, or a more environment friendly behaviour. It can be a part of teaching or in the general community. If the campaign is successful it will generate some effect, but this will disappear after a rather short time. Accordingly, new campaign initiatives are needed to keep the momentum. A campaign is always building the change on external motivation, even if it often intends to internalize the external motivation into the consumer, who might want to be like the role model in the campaign. The bold curve expresses an empowerment strategy, for example the action competence approach in education. Here there is not a simple message to live up to but a much more sophisticated process to engage in and learn from. The effect will often

Campaigns for specific behavioral change or developing action competence?

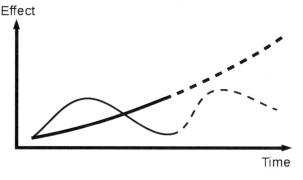

Fig. 3.1 A tentative visualisation of the impact or effect of two different approaches to inducing change: The *slim curve*: A campaigning approach. The *bold curve*: An empowerment approach like the action competence approach

come slowly and take time. But when successful the effect will tend to be a self-enhancing mechanism having more effect on the learner as the learner reaps the benefits from the enhanced empowerment.

From commercial campaigns it is well-known that a successful campaign typically only has a temporary effect (see for example Seth et al 2013). And many campaigns have actually very little effect. Just think of the many 'stop smoking' campaigns by governments and agencies, financed by enormous amounts of money. The temporary nature of a campaign is the lifeblood of commercial agencies and PR-people. The effect has to be rejuvenated again and again to keep the momentum, and typically the effect will diminish rather shortly after the efforts of the campaign.

Empowerment strategies are different as they let participants decide and identify their concerns and the solutions they expect. Such approaches are, therefore, building on intrinsic motivation, and not motivation induced from the outside. This was exactly the approach chosen in using the notion of 'quality criteria' as the main focus, inspiring schools to develop their own ideas of what could be regarded as quality related to ESD development and encouraging them to shape it into the daily school life. The approach should build on genuine participation among all stakeholders at an individual school and not be seen as an instrument for agents outside the school to manipulate the individual school in a specific narrow direction. By fertilising cooperation within the school and between the school and the local community the aim was to stimulate long-term changes beneficial to all actors in the school.

In accordance with the characteristics of ESD, the much needed pedagogical empowerment approach aims—similar to the action competence approach—to create frames for learners to identify their own concerns, develop their insights and engagement in complex and controversial issues in order to support their intended actions and finally to harvest their experience and learning coming from their actions (Jensen and Schnack 1997; Breiting 2008; Mogensen and Schnack 2010). The difference between a campaigning strategy and an empowerment strategy is profound, be it at the level of organisational development or of education. In education, it is meaningful to regard the difference between these two approaches as differences between two paradigms. The differences are visible concerning the overall goal, the concrete approach to bring about change at the school, the conceptualisation of the core content and the weight of different aspects of the content matter (Breiting et al. 2009).

Practices belonging to each of these paradigms are still widespread as evidenced in recent presentations at conferences and readings of 'best practice' publications. And, despite the often voiced need for an empowerment strategy to be applied to EE, it seems that 'former versions', i.e. different approaches to modify behaviour, might still be the most common in school systems in various parts of the world. Behind these approaches are well-meant intentions to do something good for the environment.

When we take experience from EE as a point of departure for ESD it is obvious that the thinking and practice of 'the new generation of EE', integrating the empowerment strategy rather than the campaigning idea, is much closer to the intentions and conceptualisation of ESD than former versions. It can be argued that the difference between the two paradigms 'former versions of EE' and 'the new generation of EE' is more profound than the difference between ESD and 'the new generation of EE'. From this point of view, the development of ESD from 'the new generation of EE' should be seen as an evolution that strengthens the focus on inter-generational perspectives and aspects of social justice, democratic participation and inequality.

It is often mentioned that in reality the most common practice is a mixture of the two approaches 'the new generation of EE' and 'former versions of EE'. But our point here is that it is not easy to combine the two approaches because they form separate pedagogical paradigms. 'The new generation of EE' has an internal consistency that might be missed if mixed with former versions of EE. 'The new generation of EE' regards environmental problems as issues in the community related to the use of natural resources and considers that it is human beings who are the only ones able to identify them as unwanted. The competence to face that implies to be able to understand the inherited conflicting interests in the community. People's activities are the root cause of the issues and only people together (for example in politics, organisations or communities) and individually can solve the problems in accordance with their wishes and intentions. But even such insights have been difficult to accept by central advocates for EE during past decades (see Smyth 1995).

As part of the empowerment perspective of ESD there is a need for an extended interest of learners in such issues of development with their complexity and controversial nature (Mogensen and Schnack 2010). But learners might not have an immediate interest in these (world) issues. Therefore mechanisms to enhance their engagement are important and pedagogical approaches need to pay attention to them. It has proved fruitful to understand these mechanisms as mechanisms that develop a feeling of ownership among the learners of the issues in question (Breiting 2008). A feeling of ownership, or 'mental ownership', is a conception of how people are engaged and feel responsibility for things, issues, ideas, solutions etc. The mechanisms that enhance mental ownership are listed in Table 3.1.

Table 3.1 These aspects will enhance the development of a feeling of ownership (mental ownership) among learners to issues, ideas, innovations, and concrete changes

Aspects to take into account to support the development of mental ownership
All involved participate in the goal setting or strategy formulation
All concerned are regarded as 'equal' partners in the process
All have a direct interest in the changes
All involved give input to the process
All can find their 'fingerprint' in the final outcome
All receive some form of recognition for their contribution to the process
All feel they really understand the issue

Based on Breiting (2008)

It is remarkable how these mechanisms to enhance ownership fit into the daily life experiences of many adults, e.g. linked to bringing up children or to our own routes into feeling responsibility for something. With a high level of mental ownership to specific changes people will be very persistent in their concern for the quality and success of the changes. They will be willing to engage in further actions to sustain the changes and 'feel good' when the changes are well-functioning. They will be happy communicating about the 'progress' and defending the ideas and the outcome. People will react in an opposite way if they have a very low level of mental ownership for changes. From school project innovations it is well-known that most teachers lose their interest in innovations when the money runs out. Only a few key people who have been very much involved tend to keep involved. When 'movers' leave the school the ones left behind often feel that the responsibility for the innovations left with these people. Despite the obvious importance of the above mechanisms to enhance mental ownership of innovations, in reality they are not taken enough into account in many project initiatives and innovations. When developing and writing *Quality Criteria for ESD Schools* the authors aimed at making use of these mechanisms in the quest for inspiring schools to take ESD seriously and to avoid many of the traps that easily result from external initiatives. By integrating mechanisms with a track record of enhancing local ownership on all levels at schools, the inspiration from the QC booklet should imbed ESD in the local school culture and function as a catalyst for pedagogical change based on local ideas and joint engagement. But these mechanisms will only come into play if schools get on board trying to make use of the *Quality Criteria for ESD Schools*.

3.6   The Future for Quality Criteria for ESD Development

We can expect increasing demands on schools and other educational institutions to document their performance and to prove they are competitive in a globalised world. The philosophy of NPM seems to have an overwhelming appeal to politicians and civil servants in many parts of the world and seems to generate a self-enhancing mushrooming effect. And, of course, it is difficult to argue against tools to enhance the quality of education and at the same time to make the use of resources more efficient. From this trend it seems obvious that a need for good quality indicators, such as the ones proposed in the *Quality Criteria for ESD Schools* will also be in demand in the future. But from the previous sections it should be clear that to be successful with the development of ESD at school level it is mandatory to understand the mechanisms that generate a feeling of ownership to issues and innovations and that the same mechanisms are essential for the development of students' action competence related to development and SD issues.

The quality of school development, the quality of education and the quality and success of ESD seem to be strongly interlinked, but need a more open and genuine participatory approach than what is provided by the mechanisms of NPM with its focus on documentation, indicators and control measures.

References

Bateson, G. (2000). *Steps to an ecology of mind: Collected essays in anthropology, psychiatry, evolution, and epistemology.* Chicago: University of Chicago Press (Original 1972).

Breiting, S. (2008). Mental ownership and participation for innovation in environmental education and education for sustainable development. In A. Reid, B. B. Jensen, J. Nikel, & V. Simovska (Eds.), *Participation and learning perspectives on education and the environment, health and sustainability* (pp. 159–180). Houten: Springer Netherlands.

Breiting, S., & Mogensen, F. (1999). Action competence and environmental education. *Cambridge Journal of Education, 29*(3), 349–353.

Breiting, S., Mayer, M., & Mogensen, F. (2005). *Quality criteria for ESD schools. Guidelines to enhance the quality of education for sustainable development.* Vienna: ENSI/SEED & Austrian Federal Ministry of Education, Science & Culture. http://www.ensi.org/media-global/downloads/Publications/208/QC-GB.pdf. Accessed 8 May 2014. Available in 20 languages at http://www.ensi.org/Publications/Publications-references/. Accessed 5 May 2014.

Breiting, S., Hedegaard, K., Mogensen, F., Nielsen, K., & Schnack, K. (2009). *Action competence, conflicting interests and environmental education—The MUVIN Project.* Copenhagen: Research Programme for Environmental and Health Education, DPU, Aarhus University. https://www.ucviden.dk/portal/files/10041047/Engelst_udgave_af_MUVIN_bogen.pdf. Accessed 8 May 2014.

Centre for Educational Research and Innovation (CERI)-OECD (Organisation for Economic Co-operation and Development). (1991). *Environment, school and active learning.* Paris: OECD. http://www.bmukk.gv.at/medienpool/24211/envschool.pdf. Accessed 11 May 2014.

CERI-OECD. (1995). *Environmental learning for the 21st century.* Paris: OECD. http://www.ensi.org/media-global/downloads/Publications/224/OECD_environmental_Learning1.pdf. Accessed 11 May 2014.

Eco-Schools. (2014). http://www.eco-schools.org/. Accessed 8 May 2014.

Espinet, M. (2012). Critical perspectives on promoting quality criteria for ESD: Engaging schools into reflection on the quality of ESD. In M. Réti & J. Tschapka (Eds.), *Creating learning environments for the future. Research and practice on sharing knowledge on ESD* (pp. 97–100). Kessel_Lo: ENSI (Environment and School Initiatives). http://www.ensi.org/media-global/downloads/Publications/341/Creating%20learning%20environments.pdf. Accessed 8 May 2014.

Jensen, B. B., & Schnack, K. (1997). The action competence approach in environmental education. *Environmental Education Research, 3*(2), 163–178.

Liriakou, G., & Flogaitis, E. (2000). Quelle évaluation pour quelle Education relative à l'Environnement? *Education relative à l'Environnement, Regard Recherche Réflexions, 2,* 13–30.

Mayer, M., & Vare, P. (2012). Using quality criteria as a roadmap for ESD schools. In M. Réti & J. Tschapka (Eds.), *Creating learning environments for the future. Research and practice on sharing knowledge on ESD* (pp. 55–62). Kessel_Lo: ENSI (Environment and School Initiatives). http://www.ensi.org/media-global/downloads/Publications/341/Creating%20learning%20environments.pdf. Accessed 8 May 2014.

McKeown, R., & Hopkins, C. (2007). Moving beyond the EE and ESD disciplinary debate in formal education. *Journal of Education for Sustainable Development, 1*(1), 17–26. doi:10.1177/097340820700100107.

Mogensen, F. (2012). Preliminary descriptive analysis of the QC survey—A working paper. In M. Réti & J. Tschapka (Eds.), *Creating learning environments for the future. Research and practice on sharing knowledge on ESD* (pp. 69–84). Kessel_Lo: ENSI (Environment and School Initiatives). http://www.ensi.org/media-global/downloads/Publications/341/Creating%20learning%20environments.pdf. Accessed 8 May 2014.

Mogensen, F., & Mayer, M. (Eds.). (2005). *ECO-schools: Trends and divergences.* Vienna: BMBWK. http://www.ensi.org/media-global/downloads/Publications/173/ComparativeStudy1.pdf. Accessed 8 May 2014.

Mogensen, F., & Schnack, K. (2010). The action competence approach and the 'new' discourses of education for sustainable development, competence and quality criteria. *Environmental Education Research, 16*(1), 59–74.

OECD. (2009). *TALIS, creating effective teaching and learning environments, first results from TALIS.* http://www.oecd.org/dataoecd/17/51/43023606.pdf. Accessed 8 May 2014.

Pirsig, R. M. (1992). *Lila: An inquiry into morals.* New York: Bantam Books.

Rauch, F., & Steiner, R. (2006). School development through education for sustainable development in Austria. *Environmental Education Research, 12*(1), 115–127.

Réti, M., & Tschapka, J. (Eds.). (2012). *Creating learning environments for the future. Research and practice on sharing knowledge on ESD.* Kessel_Lo: ENSI (Environment and School Initiatives). http://www.ensi.org/media-global/downloads/Publications/341/Creating%20learning%20environments.pdf. Accessed 8 May 2014.

Robottom, J., & Hart, P. (1993). *Research in environmental education. Engaging the debate.* Geelong: Deakin University.

Schnack, K. (1998). Why focus on conflicting interests in environmental education? In M. Åhlberg & W. L. Filho (Eds.), *Environmental education for sustainability: Good environment, good life* (pp. 83–96). Bern/Frankfurt/M: Peter Lang.

Schwartz, P. (1991). *The art of the long view.* London: Century Business.

Seth, J. H., Lo, J., Vavreck, L., & Zaller, J. (2013). How quickly we forget: The duration of persuasion effects from mass communication. *Political Communication, 30*(4), 521–547. doi:10.1080/10584609.2013.828143.

Smyth, J. C. (1995). Environment and education: A view of a changing scene. *Environmental Education Research, 1*(1), 3–20.

Stables, A., & Scott, W. (2002). The quest for holism in education for sustainable development. *Environmental Education Research, 8*(1), 53–60.

Sterling, S. (2001). *Sustainable education: Re-visioning learning and change* (Schumacher Society Briefing No. 6). Dartington: Green Books.

Stevenson, R. B. (2006). Tensions and transitions in policy discourse: Recontextualizing a decontextualized EE/ESD debate. *Environmental Education Research, 12*(3–4), 277–290.

Tilbury, D. (2011). *Education for sustainable development. An expert review of processes and learning.* Paris: UNESCO. http://unesdoc.unesco.org/images/0019/001914/191442e.pdf. Accessed 8 May 2014.

UNECE. (2011). *Learning from each other: Achievements, challenges and ways forward—second evaluation report of the implementation of the UNECE ESD strategy.* Geneva: UNECE. http://www.unece.org/fileadmin/DAM/env/esd/6thMeetSC/Informal%20Documents/PhaseII ProgressReport_IP.8.pdf. Accessed 8 May 2014.

UNESCO. (2009). In A. E. J. Wals (Ed.), *Review of contexts and structures for ESD.* Paris: UNESCO. http://www.unesco.org/education/justpublished_desd2009.pdf. Accessed 9 May 2014.

UNESCO. (2012). In A. E. J. Wals (Ed.), *Shaping the education of tomorrow: 2012 full-length report on the UN Decade of Education for Sustainable Development.* Paris: UNESCO. http://unesdoc.unesco.org/images/0021/002164/216472e.pdf. Accessed 9 May 2014.

Vare, P., & Scott, W. (2007). Learning for a change: Exploring the relationship between education and sustainable development. *Journal for Education for Sustainable Development, 1*(2), 191–219.

World Commission on Environment and Development. (1987). *Our common future.* Oxford: Oxford University Press. http://www.un-documents.net/our-common-future.pdf. Accessed 9 May 2014.

Chapter 4
Education for Sustainable Development (ESD): A Critical Review of Concept, Potential and Risk

William Scott

4.1 In the Beginning . . . Sustainable Development

The last 50 years have shown how human socio-economic development continues to compromise the biosphere's ability to support life on Earth through phenomena such as climate change, widespread chemical pollution, ocean acidification, stratospheric ozone depletion, habitat destruction, biodiversity and species loss, freshwater depletion, disruption of material cycles, desertification, and the like (Ehrlich and Ehrlich 2013; Worldwatch 2013; WWF 2012). At the same time, there has been increased understanding that the continuing scourges of poverty, malnutrition, disease, illiteracy, discrimination, misogyny, racism, and so forth, comprise "growing inequalities between people across the world in terms of access to resources and achieving well-being" that are, as Vare and Scott (in press) put it, "both an affront to human dignity and a source of international and intercultural instability".

In response, United Nations' commissions, conferences, and Earth summits have resulted in ideas around sustainable development (SD), and much international activity on global socio-economic and environmental goals—most recently the UN's millennium development goals, agreed in 2000 covering issues such as poverty, child mortality, gender inequality and environmental degradation (UN no date).

This idea of sustainable development embodies the conjoined objectives of human well-being and a well-functioning biosphere in order to make widespread and enduring human fulfillment a possibility. It gained prominence through the *World Conservation Strategy* and the Brundtland Report (*Our Common Future*), which saw sustainable development as a socio-economic process in which

W. Scott (✉)
University of Bath, Bath BA2 7AY, UK
e-mail: w.a.h.scott@bath.ac.uk

© Springer International Publishing Switzerland 2015
R. Jucker, R. Mathar (eds.), *Schooling for Sustainable Development in Europe*,
Schooling for Sustainable Development 6, DOI 10.1007/978-3-319-09549-3_4

the exploitation of resources, the direction of investments, the orientation of technological development and institutional change, are made consistent with future as well as present needs. (World Commission on Environment and Development (WCED) 1987: 17)

Our Common Future describes sustainable development in this way:

(...) development that meets the needs of the present without compromising the ability of future generations to meet their own needs. It contains within it two key concepts:

- the concept of "needs", in particular the essential needs of the world's poor, to which overriding priority should be given; and
- the idea of limitations imposed by the state of technology and social organization on the environment's ability to meet present and future needs (WCED 1987: 41).

In this sense, a different way of socio-economic development is being sought to enable everyone to live well, and within the Earth's ability to support us—now and in the future—where the idea of sustainable development embodies an ethical commitment to the well-being of all humanity and the biosphere. Hamm and Muttagi clarify a crucial point:

Sustainable development is essentially not about the environment, but rather about the capacity of human society to enact permanent reform in order to safeguard the delicate balance between humans and their natural life-support system. (1998: 2)

However, as these views aim to reconcile environmental protection with economic growth, Bonnet wrote for the many who saw this as trying to square the circle, when he viewed sustainable development with suspicion because of its "(...) highly anthropocentric and economic motives that lead to nature being seen essentially as a resource." (2007: 170).

Oxfam has recently captured the substance of these goals in a striking fashion with a model of sustainable development that combines the concepts of *planetary* and *social boundaries*:

Achieving sustainable development means ensuring that all people have the resources needed—such as food, water, health care, and energy—to fulfil their human rights. And it means ensuring that humanity's use of natural resources does not stress critical Earth-system processes (...). (Oxfam 2012: 4)

These goals are set in a bounded framework (see Fig. 4.1 below) where:

The social foundation forms an inner boundary, below which are many dimensions of human deprivation. The environmental ceiling forms an outer boundary, beyond which are many dimensions of environmental degradation. Between the two boundaries lies an area—shaped like a doughnut—which represents an environmentally safe and socially just space for humanity to thrive in. It is also the space in which inclusive and sustainable economic development takes place. (ibid.: 4)

This is a compelling image where

the 11 dimensions of the social foundation are illustrative and are based on governments' priorities for Rio+20. The nine dimensions of the environmental ceiling are based on the planetary boundaries set out by Rockström et al. (2009b). (ibid.: 4)

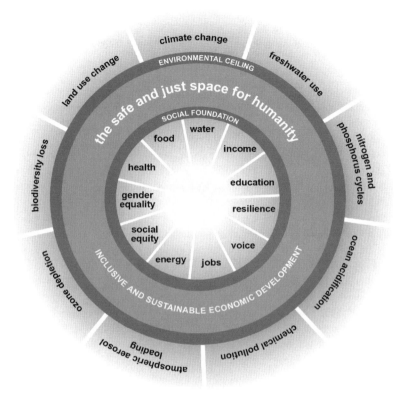

Fig. 4.1 A safe and just space for humanity to thrive in: a first illustration (From Oxfam 2012, reproduced with the permission of Oxfam GB, Oxfam House, John Smith Drive, Cowley, Oxford OX4 2JY, UK www.oxfam.org.uk. Oxfam GB does not necessarily endorse any text or activities that accompany the materials)

There are, however, problems with this analysis that illustrate some of the difficulties of thinking about, and operationalising, sustainable development. Most fundamentally these are about the way that Oxfam uses the idea of *boundaries*. It does this in two ways: first as socially-constructed desired minimum floors, and secondly as upper thresholds beyond which very significant environmental consequences are likely. But these social and environmental dimensions are not equivalent. It is uncomfortable, too much so for Oxfam perhaps, but one (the environmental) is likely to be more absolute than relative, and not amenable to social construction in the same way that the social one clearly is. For example, were income poverty (currently defined as <$1.25/day) ever to be eradicated, it would immediately be redefined as, say, <$2/day. Indeed, this would happen long before the $1.25 level was exceeded. In this sense, poverty levels will be re-defined such that poverty, relatively at least, endures.

Conversely, we cannot define for ourselves what the critical natural thresholds are for ocean pH, atmospheric carbon, etc., though we may come to learn what these are in time, if we are unfortunate. These are not socially constructed, except in the narrow sense that we create limits for ourselves in the policy process in order to increase our chances of staying within those limits—whatever they turn out to be. Think of blowing up a balloon. We may caution not to go beyond a 30 cm diameter, but there will be a limit set by the material-air system (not our values, wishes or thinking) at which the material will fail and the balloon will burst. This 30 cm diameter, just like a 350 ppm limit for atmospheric CO_2 (Hansen et al. 2008), is just our best guesstimate at staying well below the critical failure limit.

These two boundaries are obviously both important, but one *is* much more fundamental than the other, and we do ourselves no favours by asserting otherwise. Oxfam may well have felt it had little choice but to do this, given the political pressure it obviously feels from supporters and funders to ensure that enhancing social justice, viewed broadly, is at the forefront of its thinking and actions. Similar pressures are felt, as I shall explore in what follows, by many of those who espouse educational interventions around sustainability, especially if they come to this from development education backgrounds where people's interests are always placed first. Such pressures also suffuse the UN's post-Rio processes around moving from *millennium* to *sustainable development* goals (UN 2013a).

All that said, the approach by Oxfam is to be welcomed as a contribution to a setting out of the issues. Agreeing on meaning remains difficult, however, and relativism only gets you so far. In research carried out for the Higher Education Funding Council for England (Hefce 2008) it was clear that there was no one view of sustainable development that could command consensus across the sector. Although the researchers began by defining teaching and research activity relating to sustainable development as that containing . . .

> a significant element related to either or both of the natural environment and natural resources, PLUS a significant element related to either or both of economic or social issues

. . . it was impossible to maintain the conceptual tightness of this framing whilst collecting the data that academics in the institutions wanted the researchers to collect. This contrasted sharply with similar data collection in Wales through the STAUNCH initiative (Lozano 2010) where a much looser framing was allowed which did not specify or attempt to demand the significance of the natural environment, and so was more permissive. This clearly led to an over-counting of incidences of sustainable development as a focus of academic activity. Whilst sustainable development may be a socially constructed idea, this does not mean you can construct it any way you like.

All this matters because your conceptual framing of sustainable development influences how (and if) you think about ESD, and will be a key factor in determining your framing of ESD. This may go some way to explaining why there are such diverse, and often polarized, views on how to think about ESD—or whether it is worth thinking about at all; most teachers and academics, despite the Decade, do not think about ESD at all.

4.2   Sustainable Development, Learning, and ESD

If sustainable development is concerned with building our capacity to live well within the Earth's ability to support us, this will inevitably involve *learning* to do this, given where we are starting from. A popular view of this is as a process through which we shall need to learn to live in tune (or in harmony) with the environment.

So, what is this learning? Is it just the usual sort of thing? Can we view it just as the outcomes of what teachers, trainers, and work-based professional developers get up to in educational settings? If so, then it will be about re-visioning goals, curriculum re-design, new and pre-specified learning outcomes, re-oriented approaches to professional accreditation and training, and changes to examination and quality assurance systems. In other words, it will be about changes to what is learned and to how this is done, and why. If you are thinking about these issues from within formal education, or thinking about schools (and colleges/universities) as institutions, then it might seem obvious that this must be the case, to some degree at least. But if you are outside such systems, you might see all this as necessary, but far from sufficient, particularly if you see sustainable development itself as a social learning process that will not be taking place at all if learning is not happening.

John Foster (2008) argues that sustainable development makes no sense other than as a social learning process of improving the human condition that can be continued indefinitely without undermining itself. He argues that sustainable development does not depend on learning; rather, it is inherently a learning process of making the emergent future ecologically sound and humanly habitable, *as* it emerges through the continuous responsive learning which, Foster says, is the human species' most characteristic endowment. Foster neatly captures the idea of learning as a collaborative and reflective process, the extension of this into an inter-generational dimension, and the idea of environmental limits. In this view, the learning that takes place in schools, colleges and universities is a small part of what we shall need to do. The UN's focus on public awareness is an acknowledgement of this breadth, and community-based NGOs understand this, though not always to the extent that they understand that learning needs to be systemic.

Over the same 50 year timescale that we saw earlier, education has come to be seen by some as a crucial social strategy if new ways of socio-economic development are to emerge that will enable everyone to live well, and keep within the Earth's ability to support life. This has resulted in the idea of education for sustainable development (ESD), and a UN Decade (DESD: 2005–2014) of global activity focused on this, which is now drawing to a close.

ESD can be thought of as the bringing together of a wide variety of educational strategies aimed at addressing the existential problems of human socio-economic development. But, as we near the end of the Decade, what can we say about how ESD is conceptualised and interpreted; about its coherence and usefulness as an idea; about how well it fits within education systems and schools; about its potential as a strategy to change educational experiences across the globe; and about the uncertainties and ambiguities at its heart? This idea of ESD is not novel. Its origins lie within environmental education (EE) whose own genesis was in nature study, outdoor education and conservation education (Disinger 1983/1997 18; Roth 1978). Environmental education emerged in the 1960s as a result of the growing awareness of the environmental and social challenges humanity faced, and was first formalised by the World Conservation Union (IUCN) in 1970 as

> a process of recognising values and classifying concepts in order to develop skills and attitudes necessary to understand and appreciate the inter-relatedness among man, his culture and his biophysical surroundings. Environmental Education also entails practice in decision-making and self-formulating of a code of behaviour about issues concerning environmental quality. (1970)

These intertwined social and environmental goals were further developed through the seminal UN conferences of the 1970s, culminating in the *Tbilisi Declaration* (UNESCO-UNEP 1978), and were well summed up by Stapp et al. echoing the work of Harvey (1977):

> (...) the evolving goal of EE is to foster an environmentally literate global citizenry that will work together in building an acceptable quality of life for all people. (1979: 92)

During the 1980s, environmental education was promoted most vigorously by non-governmental organisations (NGOs), particularly, though not exclusively, in economically developed societies, as the global reach of, say, WWF, illustrates. In broad terms, NGOs' policy proposals and educational resources attempted to shift mainstream education practice towards the Tbilisi goals. Whilst there was some modest influence on curriculum and teacher professional development, this was not ultimately significant and made little lasting impact on national education systems. Smyth (1995) suggested that the adjective *environmental* had been a significant barrier as it signalled that environmental education was something separate from established disciplines, thereby outside mainstream educational debates and practice. Much the same can be said today of ESD.

The idea of ESD evolved in the 1990s stimulated through *Caring for the Earth: a strategy for sustainable living* (IUCN, UNEP & WWF 1991), the 1992 Rio *Earth Summit*, and *Agenda 21* which set out to be a comprehensive plan of action by governments, NGOs, and networks (globally, nationally and locally) to reduce human impact on the environment. Agenda 21 gave rise to much activity, but though there was a chapter (36) on education, training and public awareness, there was no mention of ESD other than, obliquely in a different chapter: "Demographic and sustainable development education should be coordinated and integrated in both the formal and non-formal education sectors." (Agenda 21 1992: 27).

There are, however, numerous references to both environmental education and development education in Agenda 21, especially in Chapter 36. Drawing on Agenda 21, the UN identified four overarching goals for "all Decade stakeholders":

- *Promote and improve the quality of education*:
     The aim is to refocus lifelong education on the acquisition of knowledge, skills and values needed by citizens to improve their quality of life;

- *Reorient the curricula*:

    From pre-school to university, education must be rethought and reformed to be a vehicle of knowledge, thought patterns and values needed to build a sustainable world;
- *Raise public awareness and understanding of the concept of SD*:

    This will make it possible to develop enlightened, active and responsible citizenship locally, nationally and internationally;
- *Train the workforce*:

    Continuing technical and vocational education of directors and workers, particularly those in trade and industry, will be enriched to enable them to adopt sustainable modes of production and consumption. (UNESCO 2005a: 5)

It is clear that this is a reference, not just to all provision of education, training and professional development, but also to the everyday business of living together in society. In this, it looks back to the 1977 Tbilisi Conference for its fundamental principles. There is no sense here of creating something separate from the education that already exists; rather, the idea was to improve, and sharpen the focus of, that education.

> Education, including formal education, public awareness and training, (. . .) is critical for promoting sustainable development and improving the capacity of the people to address environment and development issues. While basic education provides the underpinning for any environmental and development education, the latter needs to be incorporated as an essential part of learning. Both formal and non-formal education are indispensable to changing people's attitudes so that they have the capacity to assess and address their sustainable development concerns. It is also critical for achieving environmental and ethical awareness, values and attitudes, skills and behaviour consistent with sustainable development and for effective public participation in decision-making. To be effective, environment and development education should deal with the dynamics of both the physical/biological and socio-economic environment and human (which may include spiritual) development, should be integrated in all disciplines, and should employ formal and non-formal methods and effective means of communication. (Agenda 21 1992: #36.3: 320)

Following the Johannesburg *World Summit for Sustainable Development* in 2002, the UN General Assembly adopted resolution 57/254 to launch the UN Decade (2005–2014) (UN 2002). This invited Governments to consider the inclusion of measures to implement the Decade in their respective educational strategies and action plans by 2005, taking into account the international implementation scheme to be prepared by UNESCO (UN 2004). Later, Resolution 59/237 (2004) invited governments to promote public awareness of, and wider participation in, the Decade through cooperation with and initiatives engaging civil society and other relevant stakeholders. Looking back 10 years, this understandable strong focus on governments (it was the UN after all) looks odd, given just how much the Decade (and ESD) has proved so very non-governmental in its organisation.

UNESCO says that ESD:

- allows every human being to acquire the knowledge, skills, attitudes and values necessary to shape a sustainable future; and (. . .)
- touches every aspect of education including planning, policy development, programme implementation, finance, curricula, teaching, learning, assessment, administration, etc. (2013)

The difficulty of this sort of phrasing is that it positions ESD as having a separate existence, with this reification placing it outside the mainstream, with all the problems that Smyth noted. As UNESCO's report on the Decade puts it: "ESD [has] gained recognition internationally as an education relevant to addressing today's SD challenges" (2012: 6). What this loose phrasing actually means is that *education* is recognised as relevant to addressing sustainable development challenges. "*Recognised*" by whom, however, is never made clear, but the suspicion must be that this is only by a narrow insider grouping of committed activists and professionals. There is little evidence in the report of whole education systems being re-oriented, although it does say that "the need for ESD was well established in national policy frameworks" (2012: 5). Even this seems a generalisation too far. This reification, essentially seeing ESD as equivalent to a subject or discipline, inevitably leads to conclusions such as: "ESD is difficult to teach in traditional school settings where studies are divided and taught in a disciplinary framework." (McKeown 2002: 32), and to the compilation of examples "of ESD teaching" (Environmental Association of Universities and Colleges (EAUC) 2013).

4.3 Change, Continuity and Critique

For the UN (2004), the overall goal of the DESD was to integrate the values inherent in sustainable development into all aspects of learning to encourage changes in behaviour that would create a more sustainable future in terms of environmental integrity, economic viability, and a just society for present and future generations.

This emphasis on human behaviour change fits uneasily with the 1970s focus on values, cognition, skills and attitudes, but behaviour change as an educational goal was firmly established within environmental education. As Hungerford and Volk confidently asserted: "The ultimate aim of education is shaping human behaviour" (1990: 302) where

> responsible citizenship behaviour can be developed through environmental education. The strategies are known. The tools are available. The challenge lies in a willingness to do things differently than we have in the past. (Hungerford and Volk 1990: 317)

Put simply, this approach says that:

- if we can create a curriculum that takes sustainability issues seriously
- provide enough information about ecological concepts and environmental inter-relationships,
- provide carefully-designed opportunities for learners to acquire environmental sensitivity and a sense of empowerment,
- enable learners to acquire analytical and investigative skills, and citizenship action skills.

... then they will acquire understanding and both cognitive and social skills, their attitudes will shift, and then their behaviour will change in pro-sustainability ways.

This very influential model is rooted in a scientific-realist view of the world and draws on the notion of responsible environmental behaviours arising out of Ajzen and Fishbein's (1980) theory of planned behaviour. It sees behaviours as the interaction of the "desire to act" with "situational factors" and brings together issues associated with an understanding of scientific and ecological concepts and how these relate to our everyday lives, and the psychological influences *on* those lives. Hungerford and Volk elaborated two curricular strategies concerned with issue identification and action taking which found a reflection in work on action competence (Jensen and Schnack 1997; 2006) in Denmark, although the Danes did not make the mistake of thinking, as Hungerford and Volk (1990: 303) did, that the "major methods of citizenship action" could be divorced from the "investigation of issues". Nor did Jensen and Schnack ever think that education should set about developing citizens who will behave in desirable ways.

This continued emphasis on individual behaviour change pervades current thinking about the outcomes of ESD programmes. See, for example, Vare and Scott (2008) for a comment on the tendency within global learning and development education programmes in schools to promote (as opposed to critically appraise) fair trade schemes.

A serious problem with the Hungerford and Volk model lies in its separation of the desire to act from 'situational factors'; that is, from the social and economic context within which those acts will take place. This is a naïve notion of citizenship that assumes that the desire to act is volitional as opposed to existential, and it looks as if all non-psychological and rather awkward socio-economic issues were dumped into a box labelled 'situational factors'. A distinct benefit of using a sustainability discourse (as opposed to just an *environmental* one) is that such conveniences are ruled out on conceptual grounds. Sustainability's framing embraces economic and social issues together with environmental ones, the first two cannot just be wished away by focusing only on pro-*environment* behaviours, to the exclusion of social justice, the elimination of poverty, and the like. Our experience of living and working requires us all to navigate our way through these incommensurate ideals both at global, national, community and family levels.

It is understandable why all this invited the sort of criticism which soon came from critical realists (Robottom and Hart 1995) within environmental education in opposition to what they saw as a *behaviourist* emphasis and a complete failure to critique socio-political circumstances within which all such behaviours were to be embedded. Their purposes, rooted in emancipatory action research (Fien 1993), were to help teachers and students work towards social transformation. Theirs was an alternative model which was grounded in a desire to bolster social and ecological justice and through this reduce socio-economic disparities. It came to be associated with development education, and an opposition to neo-liberal approaches of all kinds. This model sets out to effect social rather than behaviour change and has cognitive and affective elements. Its use has largely been with teachers and teacher trainers (see Huckle 2006), and is associated with socially-critical theory and its focus on the economic forces that direct and buffet our lives. The purposes of this perspective on ESD is to show teachers and students how to analyse the values behind their socially-learned behaviour patterns and how to resist such forces and work towards social transformation. A key focus was helping students and teachers to ask appropriate socially-critical questions, typically of the *cui bono?* form.

This model, put simply, says:

- if we can influence opinion-formers (e.g. teachers), and through them, influence learners,
- raising their awareness and consciousness (and countering false-consciousness) of the issues that prevent a sustainable society,
- then their under-pinning values will be changed,

... and they will argue, work, vote and agitate for (pro-sustainability) social change.

This perspective also opposed what it saw as liberal education's tendency not to ask critical questions of society because of its focus on the individual as a learner where an education was, to a significant degree, seen as *for* itself—i.e., the outcome was an educated individual whose knowledge, understanding, skills and other attributes were well grounded within the prevailing culture, and whose literacy had critical dimensions as well as functional and cultural (Stables 2010). A critique of this liberal tradition would be that there is too *little* emphasis on behaviour modification and insufficient focus on social inequities and the need for change. The liberal educators' response to this is that encouraging critical questions of society, and looking for the need for change, is at its heart, it is just that the answers are never pre-specified as they tend to be in socially-critical or behavioural approaches. And it is, they argue, no business of educators to persuade learners to change behaviours or society in pre-specified ways. Rather, this is the business of politicians and socio-political activists, social marketers and the advertising industry, but not of educators. Of course, a pertinent response to this might be that it illustrates a liberal education blind spot, as its own preferred approach to education contributes towards the perpetuation of particular social models. See Huckle (2013) for an up-to-date consideration of these approaches in the context of ESD and the Eco-Schools movement.

What seems common ground is that ESD can helpfully be seen as an education in citizenship: a responsive social learning process which is a preparation for informed, open-minded, social engagement with the main existential issues of the day that occur in the family, the community and workplace—in all aspects of a lifelong learning. Clearly, being socially critical, and actively considering changes in entrenched behaviours, are each citizenly qualities that are necessary if societies are to actively re-create themselves, and a way has to be found to bring these together in schools and other institutions if building an acceptable quality of life for all people is to be possible. Schools, colleges and universities, as institutions, are an acknowledged, integral part *of* any learning society, with the key role of supporting young people in the *early* stages of their acquiring the wide-ranging understandings, skills and capabilities that they will need to continue to develop for successful and fulfilling engagement with, and living in, the world. In terms of sustainability, then, the purpose of schools, broadly speaking, might be seen as stimulating young people's development of awareness and interest in relation to living sustainably, with the hope (but not certainty) that this will give rise to social participation that can contribute, for example to the goals of greater social justice and human well-being, and the bolstering of the resilience of ecological systems. Further and higher education allow such ideas to be explored in much greater depth and sophistication and will likely have an emphasis on the resonance of these ideas in contemporary society and the workplace.

As we have seen, schools have been addressing such issues for over 40 years, in the main through the curriculum with some integrated work across subjects, and growing use of personal, social, health education and citizenship courses, and partnerships with external groups, all designed to enable students to develop more rounded and fuller sets of understandings and skills. There has also been a growth of more activist developments, particularly through clubs and eco/green councils, with a remit to effect change in (and sometimes beyond) the school in relation to management practice; for example, reducing energy and resource use, and increasing recycling and composting. The last few years have seen a growth, internationally, of similar developments in further and higher education, especially focused on students' campus experiences more generally, and has led to an increase in behaviour-change projects in relation to this.

But all this exposes a central question for those involved: at heart, are you really interested in educational or social outcomes? In what learners learn (broadly viewed), or what they *do*? This is a curriculum question, although not a particularly new one, that needs to be asked at a time when there is considerable social impetus to change individual behaviour, and the conscription of education to that end. Stables, in emphasising the role of institutions in "preparing people to make difficult decisions", privileges the "development of skills of critical thinking, dialogue and debate" above "content", stressing the iterative nature of learning, participation and decision-making through the life-span (Stables 2010: 594). However, education is most successful, perhaps, when it combines these elements. Vare and Scott have argued that it is helpful to think of two complementary approaches:

> ESD 1—Providing guidance about behaviours, shifts in habit, and ways of thinking about how we live now. This tends to be heavily content-focused, information-based, and grounded in everyday practice.
> ESD 2—Building students' capacity to think critically and develop abilities to make sound choices in the face of the inherent complexity and uncertainty of the future. This is much more dialogue and debate-oriented, and focused on controversial issues. (2007: 196; 2008)

ESD 1 promotes informed, skilled behaviours and ways of thinking where the need for this is deemed important by experts. This is about doing things *differently*. It is about greater efficiency: *level 1 learning*.

ESD 2 is building capacity to think critically about and beyond what experts say, and test out sustainable development ideas. This is about doing different *things*. It is about more effectiveness: *level 2 learning*.

Examples of ESD 1 include actions to be more efficient/less wasteful; (e.g. less greenhouse gas). All this is 'learning to be more sustainable' (or, usually, less *un*sustainable). This needs information and communication strategies, and is exemplified by approaches such as *social marketing* where things are explained to people. But people do not always make rational decisions.

Examples of ESD 2 include thinking about how what 'being more sustainable' means. This may well need *information* and *communication*; but it also needs something more sophisticated through which people are able to get to grips with conflicting ideas and values.

ESD 1 fits with the received view of sustainable development as being expert-knowledge-driven; the role of the non-expert is to do as guided with as much grace as can be mustered. This is UNESCO's view—by and large. It is what is driving the Decade. ESD 2 embodies a different view of what sustainable development *is*. In this view, sustainable development does not only depend on learning; it is inherently a *learning process*. This leads to radically different definitions of sustainable development, such as that of Foster: a social learning process of improving the human condition which can be continued indefinitely without undermining itself.

ESD 2 fits this view of sustainable development, recognising that

- many problems lack precise specification
- what can be known in the present is not always adequate, and desired 'end-states' cannot be specified with confidence
- there are competing problem definitions, and participants have incompatible value-sets
- its meaning remains provisional—it has to be as much 'worked out' as 'carried out'.
- the complexity and uncertainty we face cannot be wished, legislated, or educated away
- learning's a choice, but change is sure.

Vare and Scott (2007) argue that ESD 1 and ESD 2 approaches are complementary because people need to have relevant subject matter to debate and critically examine in their own contexts, and because ESD 2, although open-ended, cannot exist in a vacuum. This, if well constructed, could also bring the behavioural and socially critical together in the context of a liberal approach.

All of the foregoing relates to student learning, but there is another dimension to these considerations with the idea that the school itself, as an *institution*, has to become sustainable. This example from England illustrates its radical nature:

> Sustainable development will not just be a subject in the classroom: it will be in its bricks and mortar and the way the school uses and even generates its own power. Our students won't just be told about sustainable development, they will see and work within a school that is a living, learning place in which to explore what a sustainable lifestyle means. (Department for Education and Skills (DfES) no date)

In this, a key sustainable schools' focus is that of the institution's becoming a model for activity in the community:

> Schools (…) are invited to become models of sustainable development for their communities … turning issues like climate change, global justice and local quality of life into engaging learning opportunities for pupils—and a focus for action among the whole school community. (DfES 2006:3),

From a policy perspective, this duality and complementarity are not *necessarily* problematic, though they can be in practice. ESD1 is often straightforward for policy-makers to support, because it deals with concrete issues where outcomes (recycling, energy savings, fair trade adoption, reduced resource use, etc.) can be achieved quickly, with local NGOs ready uncritically to support all this. This is especially so where these map onto existing policy initiatives around, for example, welfare, saving energy, sustainable transport, waste, and international development. It is also easy to see that enthusiasm for this emphasis might, in an already busy curriculum, crowd out the other (i.e. ESD2) emphases, particularly as sustainable development can be hard to explain and there is a temptation to simplify the message to the point of completely diffusing it in order to get it across to busy professionals already heavily engaged with other priority aspects of government policy. Both are necessary, however, if education's contribution to sustainability is to be optimised. Martin et al.'s UK case study for UNESCO also explores these competing priorities.

4.4   Transformation, or Incremental Change?

Within the broad goals established by the UN General Assembly (2005), the sub-goals for the DESD at the national level were to:

- provide an opportunity for refining and promoting the vision of and transition to sustainable development through all forms of education, public awareness and training.
- give an enhanced profile to the important role of education and learning in sustainable development.

Its objectives were to:

  (i)   facilitate networking, linkages, exchange and interaction among stakeholders in ESD;
 (ii)   foster an increased quality of teaching and learning in education for sustainable development;
(iii)   help countries make progress towards and attain the millennium development goals through ESD efforts;
 (iv)   provide countries with new opportunities to incorporate ESD into education reform efforts. (UNESCO 2005b: 6)

One of the strengths of ESD is the variation that is found from one educational context to another which has arisen from local interpretations and developments as the concept is shaped to fit, more or less comfortably, with existing policy and practice. Inevitably, this all involves accommodations with preferred ideological and epistemological dispositions. Equally inevitably, all interpretations of ESD rest on understandings of what sustainable development itself *is*. How could this be otherwise, even if the conceptual links are loose, or talked about in hushed tones between consenting adults. This diversity of ESD, which is clear to see from a look at emerging practice, or any reading of the increasing number of journals that now cater for interested academics, is also a considerable weakness as it betrays a lack of shared understandings which, in turn, inhibit communication and collaboration. An aspect of this is that not all ESD is described *as* ESD, with a plethora of alternatives (EfS—education for sustainability, for example), some of which are supported by particular groups, sometimes to distance themselves from ESD which they see, variously, as too neo-liberal/pro-growth/conservative/capitalistic/'Western'/ etc., according to taste. As UNESCO notes:

> ESD is called by many names in national and local contexts. In some places, Environmental Education (EE) and other related "educations" (e.g. global education and climate change education) are defined and practiced to include socio-cultural and economic aspects alongside environmental aspects. (2013)

A number of dilemmas emerge from this confusion of language and goals. A particularly significant one is whether ESD should set out to have a transformative aim, or (merely) be focused on becoming a key component within the educational mainstream, attempting to change things between the margins and somewhere nearer the heart of things. If transformative, then it seems clear that it is institutions themselves that need to be transformed, and not just the educational opportunities they provide, otherwise developments will be at the mercy of leadership diktat or passing educational fashion. A question that immediately follows from this is whether educational systems need to be transformed in order that their embedded institutions can themselves have a chance of significant reforming. A further question then has to be: what chance is there of educational systems being really transformative if the national (and international) socio-political system within which *they* are embedded is not?

Put like that, it follows that a systems perspective is needed which acknowledges both embeddedness and interconnectivity. This is unsurprising given that we are considering sustainable development where systems thinking ought to be at its heart. A further question is whether a focus on ESD will be robust enough to make any of this more likely. Although for the UN, the overall goal of the DESD was to integrate the values inherent in sustainable development into all aspects of learning, there is nothing in this that suggests that the UN thinks that educational systems themselves should set out to be transformative in nature.

Sterling's (2001) outline of "sustainable education", underpinned by an ecological paradigm, is something that calls for such transformation in (and of) education. In relation to universities, Sterling illustrates the difference between these goals:

> (...) the effect of patterns of unsustainability on our current and future prospects is so pressing that the response of higher education should not be predicated only on the integration of sustainability into higher education, because this invites a limited, adaptive, response. Rather, (...) we need to see the relationship the other way round—that is, the necessary transformation of higher education towards the integrative and more whole state implied by a systemic view of sustainability in education and society, however difficult this may be to realise. (2004: 49–70)

In a later work, Sterling (2012) takes a broader view and invokes the arguments of Rosen et al. (2010) and Clark (1989) who see that nothing less than "a change of cultural worldview" (Rosen), and "conscious social change" (Clark) will do if we are to escape unsustainability (see also Sterling et al. (2013) for a much fuller treatment of these issues). Such transformative foci provide unparalleled contexts for useful learning experiences, which many would like to see as transformative in themselves, at least for the individual learner. Webster and Johnson (2009) make a similar set of arguments in relation to schools, and look to their becoming, through their institutional practice, restorative of both natural and social capital (Daly 1973; Meadows 1998). This has been elaborated upon by Scott (2013) in terms of stages of institutional development for a school that bring together student learning, leadership, and the enhancement of social and natural capital.

But can institutions be reformed in this way in the absence of supporting frameworks at the system and whole-society level? Mary Clark (1989) argues that in Western history there have only been two major periods where societies deliberately critiqued (and re-educated) themselves, creating new worldviews. The first was in the Greek city states (500–400 BC) when philosophers pursued new lines of thought, and social action emerged. The second was the Renaissance and Enlightenment when Western culture, through its natural and social philosophers, subjected itself to critical thought and renewal. Here, the result was the modern worldview that many (for example, Chet Bowers 2013) believe is implicated in the global socio-environmental crisis. Clark (1989: 235) argues that we need once again to "collectively create a new worldview that curbs ecological and social exploitation, and recreates social meaning", seeing this process as a society-wide phenomenon, and not something that can be entrusted to schools or to further, and higher education. This is a distinction and emphasis that Stables (op. cit.) also makes in distinguishing between formal education and "the learning society" in the context of needing to address these existential issues.

From a policy perspective, of course, a transformative stance is difficult to mandate directly for two main reasons: (i) there is evidence (Webster and Johnson) that suggests that institutions have to develop transformative volition and capability within themselves which may involve progressive stages of transformation; (ii) it is hard to justify changing educational policy in the absence of congruent changes to other policies, which implies at least some degree of collective agreement at the level of society and government; this, in turn, amounts, to some degree, to a transformation *of* society and government. The recent Martin et al. (2013) case study illustrates something of the difficult dynamics of this, across the different jurisdictions within the UK.

Sterling (2004) identified four possible responses to the challenge of sustainable development that can be:

A. null—no response needed
B. bolt-on—adding to what is done, at the margins
C. build-in—integrating, more centrally, into what is done currently
D. whole-system redesign—changing what is done to create a new system based on different principles (2004: 58)

... and it was this which informed the recent Webster and Johnson (2009) and Scott (2013) analyses. The latter sees the necessity of a progression of sorts between what are seen as developmental stages where [B] and [C] are more akin to emergent emphases than distinct positions, and are very much easier to establish than [D], although it is difficult to see how [D] could be achieved without progressing through [C] which itself probably needs [B] as a stimulus.

As noted, we might posit that these might be attempted at three principal levels:

1. *Institutional*—a school, college or university and everything that happens in it
2. *System*—all institutions of one type (e.g. schools) in a country
3. *National*—the government's view of sustainable development informs its entire thinking and action in relation to sub-systems, such as universities.

In terms of a whole institution approach, there is much that can be done even if it is not commensurate with shifts at the system or national levels, and the stimuli that can go upwards within a nested system to influence change at the next level should not be discounted. Indeed, real change probably depends on it. In the absence of directive, downward pressures, however, there will be significant limits on what is possible because of the mismatch between vision, purposes, etc., not to mention financial incentives. For example, no matter how 'sustainable' a school manages to be in restorative terms, if national examinations (a system level phenomena not much influenced *by* schools) are unreformed, then these will act as a considerable brake on what's possible in terms of student learning. Similarly, national curricula or legislative aims can be shaded by school-interpretation but not coloured in completely. And the financing of an individual school within a system level approach would likely be more permissive and persuasive of change than where this is missing.

A real example of this is in England where the Higher Education Funding Council (Hefce) has had funding schemes in place for almost 10 years which promote and reward institutional foci on sustainable development (particularly carbon-reduction). However, owing to traditions to do with academic freedom, it does not do this in relation to curricula. Instead it franchises this responsibility to other organisations that do not share the Council's vision or understanding, and which focus on ESD rather than on sustainable development. However, an ESD focus is not the same as the sustainable development one, which Sterling, Clark, and Webster and Johnson are writing about.

UNESCO's own analysis of progress through the Decade points to a number of examples of all this in relation to ESD. For example, in Bhutan ...

> ESD/GNH (Gross National Happiness) has been adopted as a national priority. [It] is an integral part of the performance management system that draws a lot of inputs from the school self-assessment. The school self-assessment tools have been oriented to take in GNH/ESD values and process. All schools make GNH/ESD plans and review these plans bi-annually. (2012: 73)

... although what results from all this is quite unclear.

Eco-Schools is cited as an example of a *whole-school approach to sustainability*, but Eco-Schools is only working with individual schools to help them shift focus at their own pace, and at the margins of their activity, and the award of a green flag has nothing necessarily to do with whole-school re-design in relation to sustainable development. In a similar vein, the *Sustainability and Education Academy* at York University, Canada, is cited by UNESCO as a *whole-system approach to sustainability*. But this is only an example of a leadership development programme that supports the creation of a culture of sustainable development. Like all such programmes round the world, its existence is not evidence *of* the operation *of* that culture.

More encouragingly, prima facie at least, Manitoba's education system is highlighted as somewhere where

> the philosophy of sustainability and employment of processes contribution (*sic*) to student engagement have been embedded in [the school division] for over 20 years even before the term ESD was coined. The larger scale ESD movements and the work of Manitoba Education has provided division staff with new evaluative frameworks to work with, professional development opportunities and resources. (UNESCO 2012: 74)

. . . but, again, new frames and approaches and resources are not the same as changed practice or re-orientated learning. Echoing the *Expert Review of Literature on Processes and Learning for Sustainable Development* (Tilbury 2011), the 2012 UNESCO report highlights:

> (. . .) a number of case studies of "whole-system engagement" (. . .) a number of concrete interactive methods and tools (e.g. values clarification techniques, critical incidents, debates, reflexive account, asking critically reflexive questions) that have not surfaced in this report's empirical review but have found their way in[to] ESD-related activities. Clearly, part of the support of professional development for whole-system engagement will consist in facilitating professional networks and providing tools and methods congruent with the proposed (or suggested) paradigm shift. (2012: 76)

Indeed it will, but "whole-system engagement", which is as loose a term as you would want to find, is not the same as "whole-system redesign" in Sterling's sense, or a new worldview (in Clark's). Each of these examples may well be some sort of indicator of the development of a focus on sustainability, but that is all. It seems unfortunate to claim more for them than they warrant.

Actually, it is more than unfortunate as it suggests that there are no compelling examples to be had. Indeed, that does seem to be the case. My own first choice of an educational system that is developing a whole-school-system approach (which might prove to be a re-design) would be Scotland (Martin et al. 2013, give a good account of this), and my first choice for a university would be the University of British Columbia (UBC) which sets out four graduate attributes whereby the graduate: demonstrates (i) holistic systems thinking and (ii) sustainability knowledge; (iii) is aware of, and integrates across, intellectual constructs; and (iv) acts to create positive change (UBC 2011).

One step on the road to re-orientation might well be the adoption of high-level aims that express the importance of sustainable development. For example, according to an expert panel established by the English Education ministry to advise on curriculum reform (HMG 2011), the 2004 school national core curriculum in Finland includes a set of underlying values of education. These are:

> human rights, equality, democracy, natural diversity, preservation of environmental viability, and the endorsement of multiculturalism. (2011: 63)

Similarly, New Zealand expresses its vision for schools in terms of:

> young people who will seize the opportunities offered by new knowledge and technologies
> to secure a sustainable social, cultural, economic, and environmental future for our country.
> (2011: 15)

The expert panel's report noted this:

> (…) many of the jurisdictions that we have considered that have recently conducted
> reviews of their curricula have introduced a high-level reference to sustainability. With
> this in mind and in the light of the Government's adoption of ambitious carbon reduction
> targets to 2027 we suggest the Government considers a recommendation that the school
> curriculum should also contribute to environmental "stewardship". (2011: 15–16)

Despite considerable dissatisfaction with the limitations of "stewardship", much
NGO and practitioner effort then went into lobbying government to adopt the
recommendations of the panel which included the promotion of "understanding
of sustainability in the stewardship of resources locally, nationally and globally."
(HMG 2011: 17) Alas, even such a modest step proved too great an effort for
government, and those (many) schools interested in a transformation to address
sustainability are left to struggle on without high-level encouragement or support.
Pragmatically, this could be all that is possible for most places. More heretically, of
course, it could be all that is really necessary for significant change to occur,
provided you are determined to lead it.

4.5 End Words

At the heart of the arguments of this chapter are three particular ideas. These
are summarized here:

1. If sustainable development only makes sense *as* learning, then effective ESD
   must always be a contribution *to* sustainable development, and our understanding of sustainable development will determine how we think about ESD, and, as
   Sterling (Pers. Com.) reminds us in a paper for UNESCO's celebratory end-of-
   Decade conference, about education itself. It follows that, for ESD to have
   meaning, and therefore effect, it needs to be grounded within a conceptual
   framing of sustainable development itself. There are, of course, different conceptual framings of sustainable development, and so more than one approach to
   ESD will endure, and even UNESCO acknowledges that some of these will
   continue to resemble environmental and development education. This is as it
   needs to be in free societies as we struggle to make sense of what we have done,
   and keep on doing, to the biosphere's systems, flows, cycles and sinks.

   A good educationally-critical sort of question to ask a teacher, trainer or
   lecturer who says they are involved in ESD is how what they are doing relates,
   and *contributes*, to sustainable development. If they cannot provide a convincing
   answer, then scepticism is in order about whether they know what they are

doing, and whether learners will benefit as much as they might expect, or at all. Another question would be to ask whether they think of ESD as a process or something to be taught, with appropriate conclusions being drawn if the response is the second of these.

2. Educational institutions need to prioritise student learning over institutional, behaviour or social change whilst making use of any such change to support and broaden that learning. In this sense it is fine for a school, college or university to encourage its students to save energy, create less waste, or get involved with initiatives such as fair trade (or *Fairtrade*), provided that these are developed with student learning in mind, including an umbilical link to their actual studies. To do otherwise is to forget why educational institutions exist. Being restorative of social or natural capital is laudable, but not if it neglects or negates the development of appropriate human capital, i.e. student learning. Doing all this in collaboration with students, and with the communities within which institutions are socially, economically and environmentally embedded, will aid everyone's learning, and perhaps even sustainable development.

3. Being socio-economically transformative remains an ideal, with being restorative of natural and social capital examples of would-be welcome outcomes. There is, however, little sign of such transformation's being achieved any time soon, or, indeed, that UNESCO is particularly convinced that it's a necessary goal for ESD. This is, perhaps, just as well as the evidence that ESD could lead transformation is not convincing. Indeed, why should it be, when it is a focus on sustainable development that is needed for a transformative effect, not a process of education such as ESD. It does seem persuasive, however, that a focus on transformation, per se, is not necessary to make progress towards that goal, and that it is small-scale, on-the-ground developments that are needed to create the conditions *for* transformation. The ground-breaking work of the Ellen MacArthur Foundation (EMF 2012), with its *circular economy* focus, is an example of such an initiative. Although not couched in the language of sustainable development, this *is* transformative in nature, and it is setting about its educational business by working *within* business and educational organisations.

All these seem important as the ESD Decade morphs into post-Decade activity, and MDGs (Millennium Development Goals) become SDGs (Sustainability Development Goals) although they are not obviously significant to the United Nations. UN Secretary-General Ban Ki-moon has initiated a number of processes to help devise the SDGs so as to maximise benefit for humanity during the years 2015–2030. One of these, the sustainable development solutions network (SDSN) has identified ten priority challenges of sustainable development:

1. End extreme poverty including hunger
2. Achieve development within planetary boundaries
3. Ensure Effective Learning for All Children and Youth for Life and Livelihood
4. Achieve Gender Equality, Social Inclusion, and Human Rights for All
5. Achieve Health and Wellbeing at All Ages
6. Improve Agriculture Systems and Raise Rural Prosperity

7. Empower Inclusive, Productive, and Resilient Cities
8. Curb Human-Induced Climate Change and Ensure Clean Energy for All
9. Secure Ecosystem Services and Biodiversity, Ensure Good Management of Water and Other Natural Resources
10. Transform Governance for Sustainable Development (UN 2013b: ix).

... which, the report says, could form the basis for the SDGs that would apply to all countries up to 2030.

This report has nothing to say about ESD, although there is a passing reference to "education in sustainable development" (see below). There are, however, numerous references to learning, including the idea that children everywhere should actually learn the SDGs to help them understand the challenges that they will confront as adults. Section 3 of the report, where reference to ESD might have been anticipated, is really an updating of the UN's *Education for All* goals. But if you want to see how little the UN understands about sustainable development, then turn to Annex 2 which sets out educational statements, disaggregated across the "four dimensions of sustainable development" [sic]. These are ...

	Economic development and eradication of poverty	Social inclusion	Environmental sustainability	Governance including peace and security
GOAL 3 Ensure effective learning for all children and youth for life and livelihood	Effective learning is critical for creating job opportunities and livelihoods for people at all ages, which in turn drives economic development	Effective learning is critical for creating job opportunities and livelihoods for people at all ages, which in turn promotes social inclusion	Improved education and awareness, including education in sustainable development, will generate innovation and leadership for environmental sustainability	Educated and informed citizens will contribute to and uphold good governance and lower the risk of conflict and insecurity

It is hard to know what to make of such an unsophisticated confection, save that the UN takes no notice of UNESCO, or the Decade, and has an astoundingly naïve view of sustainable development which contrasts poorly with what we had from Oxfam at the start of this chapter. All this is hugely disappointing, but instructive for those activists who promote ESD in that it is not ESD that is important to the UN; rather it is sustainable development and what it terms effective learning. It follows that promoting an interest in learning our way into the future in the post-Decade decade will be better done if the focus is on what students, academics and teachers are themselves interested in, and not what ESD orthodoxies tell them they really ought to be focused on. Then there might be more young, and not so young, people whose learning engages with existential issues—such as the future of life on the planet. Without this, there is the risk that we shall all continue to be ignored by those whose job it is to run mainstream education institutions and systems. They have, after all, had considerable practice at doing just this. Happily, however, there is emerging evidence that young people do take seriously the existential dilemma that we face. See, for example, Butters (2012), Gayford (2009), Hope (2013), International Institute for Sustainable Development (IISD 2013), and United Nations Environment Programme (UNEP 2011), and the on-going surveys of students entering higher education in the UK, where 67 % of them (Drayson et al. 2012) said that sustainability should be covered by their university through a re-framing of curriculum. This seems a suitably positive note on which to end.

References

Agenda 21. (1992). *United Nations Conference on Environment & Development Rio de Janerio, Brazil*, 3 to 14 June 1992. Conference documents: http://www.unep.org/Documents.Multilin gual/Default.asp?DocumentID=52. Fulltext of Agenda 21: http://sustainabledevelopment.un. org/content/documents/Agenda21.pdf. Accessed 27 Sept 2013.

Ajzen, I., & Fishbein, M. (1980). *Understanding attitudes and predicting social behaviour*. Eaglewood Cliffs: Prentice Hall.

Bonnett, M. (2007). Environmental education and the issue of nature. *Journal of Curriculum Studies, 39*(6), 707–721.

Bowers, C. (2013). *Personal webpage*. http://www.cabowers.net. Accessed 20 Aug 2013.

Butters, C. (2012). *Nordic success stories in sustainability*. Oslo: Stiftelsen Idébanken. http:// thebalancingact.info/riobok/NORDICSuccessstoriesinSustainability.pdf. Accessed 27 Sept 2013.

Clark, M. E. (1989). *Ariadne's thread*. New York: St. Martin's Press.

Daly, H. E. (1973). *Toward a steady-state economy*. San Francisco: WH Freeman & Co.

DfES (Department for Education and Skills). (2006). *Sustainable schools for pupils, communities and the environment delivering UK sustainable development strategy: A consultation paper*. London: Department for Education and Skills.

DfES. (no date). *Department for education and skills resource material: Our commitment*. http:// webarchive.nationalarchives.gov.uk/20070205113517/dfes.gov.uk/aboutus/sd. Accessed 20 Aug 2013.

Disinger, J. F. (1983/1997). *What research says: Environmental education's definitional problem*. ERIC Clearinghouse for Science, Mathematics and Environmental Education Information Bulletin No. 2, 1983. http://eric.ed.gov/?id=EJ312591. Accessed 20 Aug 2013.

Drayson, R., Bone, E., & Agombar, J. (2012). *Student attitudes towards and skills for sustainable development*. London: National Union of Students/Higher Education Academy. http://www. heacademy.ac.uk/assets/documents/esd/Student_attitudes_towards_and_skills_for_sustain able_development.pdf. Accessed 27 Sept 2013.

Ehrlich, P. R., & Ehrlich, A. H. (2013). Can a collapse of global civilization be avoided? *Proceedings of the Royal Society of Biological Sciences, 280*, 1754–1763.

Environmental Association for Universities and Colleges (EAUC). (2013). http://www.eauc.org. uk/sorted/home. Accessed 20 Aug 2013.

Fien, J. (1993). *Education for the environment: Critical curriculum theorizing and environmental education*. Geelong: Deakin University Press.

Foster, J. (2008). *The sustainability mirage*. London: Earthscan.

Gayford, C. (2009). *Learning for sustainability: from the pupils' perspective*. A report of a 3-year longitudinal study of 15 schools from June 2005 to June 2008. London: WWF. http://assets. wwf.org.uk/downloads/wwf_report_final_web.pdf. Accessed 27 Sept 2013.

Hamm, B., & Muttagi, P. K. (1998). *Sustainable development and the future of cities*. London: Intermediate Technology Publications.

Hansen, J., Sato, M., Kharecha, P., Beerling, D., Masson-Delmotte, V., Pagani, M., Raymo, M., Royer, D. L., & Zachos, J. C. (2008). *Target atmospheric CO_2: Where should humanity aim?* http://arxiv.org/pdf/0804.1126.pdf. Accessed 20 Aug 2013.

Harvey, G. (1977). *Environmental education: A delineation of substantive structure.* PhD thesis at Southern Illinois University at Carbondale. Dissertation Abstracts International 38 611-A, 1977.

Her Majesty's Government (HMG). (2011). *The framework for the national curriculum. A report by the Expert Panel for the National Curriculum review.* London: UK Government. https://www.gov.uk/government/publications/framework-for-the-national-curriculum-a-report-by-the-expert-panel-for-the-national-curriculum-review. Accessed 20 Aug 2013.

Higher Education Funding Council for England (Hefce). (2008). *Strategic review of sustainable development in higher education in England.* Report to Hefce by the Policy Studies Institute, PA Consulting Group and the Centre for Research in Education and the Environment, University of Bath. London: Higher Education Funding Council for England. http://www.hefce.ac.uk/data/year/2008/hefcestrategicreviewofsustainabledevelopmentinhighereducationinengland. Accessed 20 Aug 2013.

Hope. (2013). *Examples of young people's involvement in sustainable development.* http://www.activecitizensfe.org.uk/sustainability.html. Accessed 20 Aug 2013.

Huckle, J. (2006). *Education for sustainable development: A briefing paper for the Teacher Training Resource Bank (TDA).* http://john.huckle.org.uk/download/2708/Education%20for%20Sustainable%20Development,%20a%20briefing%20paper%20for%20the%20Teacher%20Training%20Agency.doc. Accessed 27 Sept 2013.

Huckle, J. (2013). Eco-schooling and sustainability citizenship: Exploring issues raised by corporate sponsorship. *The Curriculum Journal, 24*(2), 206–223.

Hungerford, H. R., & Volk, T. L. (1990). Changing learner behaviour through environmental education. *Journal of Environmental Education, 21*(3), 8–21.

IISD (International Institute for Sustainable Development). (2013). *Sustainability Leadership Innovation Centre.* http://www.iisd.org/slic/. Accessed 20 Aug 2013.

IUCN (International Union for Conservation of Nature). (1970). *Environmental education workshop at the conference of the International Union for the conservation of nature and natural resources.* Nevada: USA.

IUCN, UNEP & WWF. (1991). *Caring for the earth: A strategy for sustainable living.* Gland: IUCN, UNEP & WWF.

Jensen, B. B., & Schnack, K. (1997). The action competence approach in environmental education. *Environmental Education Research, 3*(2), 163–178.

Jensen, B. B., & Schnack, K. (2006). The action competence approach in environmental education. *Environmental Education Research, 12*(3/4), 471–486.

Lozano, R. (2010). Diffusion of sustainable development in universities' curricula: An empirical example from Cardiff University. *Journal of Cleaner Production, 18*, 637–644.

Martin, S., Dillon, J., Higgins, P., Peters, C., & Scott, W. (2013). Divergent evolution in education for sustainable development policy in the United Kingdom: Current status, best practice, and opportunities for the future. *Sustainability, 5*(4), 1522–1544.

McKeown, R. (2002). *ESD toolkit.* http://www.esdtoolkit.org. Accessed 20 Aug 2013.

Meadows, D. (1998). *Indicators and information systems for sustainable development.* A report to the Balaton group. Hartland VT: The Sustainability Institute. http://www.donellameadows.org/wp-content/userfiles/IndicatorsInformation.pdf. Accessed 2 Nov 2013

Oxfam (Raworth, Kate). (2012). *A safe and just space for humanity: Can we live within the doughnut?* Oxford: Oxfam. http://www.oxfamtrailwalker.org.nz/sites/default/files/reports/dp-a-safe-and-just-space-for-humanity-130212-en.pdf. Accessed 20 Aug 2013.

Robottom, I., & Hart, P. (1995). Behaviourist EE research: Environmentalism as individualism. *Journal of Environmental Education, 26*(2), 5–9.

Rockström, J., Steffen, W., Noone, K., Persson, Å., Chapin, F. S., III, Lambin, E., Lenton, T. M., Scheffer, M., Folke, C., Schellnhuber, H., Nykvist, B., De Wit, C. A., Hughes, T., van der

Leeuw, S., Rodhe, H., Sörlin, S., Snyder, P. K., Costanza, R., Svedin, U., Falkenmark, M., Karlberg, L., Corell, R. W., Fabry, V. J., Hansen, J., Walker, B., Liverman, D., Richardson, K., Crutzen, P., & Foley, J. (2009). Planetary boundaries: Exploring the safe operating space for humanity. *Ecology and Society, 14*(2), 32. http://www.ecologyandsociety.org/vol14/iss2/art32. Accessed 20 Aug 2013.

Rosen, R., Electris, C., & Raskin, P. (2010). *Global scenarios for the century ahead: Searching for sustainability.* Boston: Tellus Institute.

Roth, C. E. (1978). Off the merry-go-round and on to the escalator. In W. B. Stapp (Ed.), *From ought to action in environmental education* (pp. 12–23). Columbus: SMEAC Information Reference Centre.

Scott, W. (2013). Developing the sustainable school: Thinking the issues through. *The Curriculum Journal, 24*(2), 181–205.

Smyth, J. (1995). Environment and education: A view from a changing scene. *Environmental Education Research, 1*(1), 1–20.

Stables, A. W. G. (2010). New worlds rising. *Policy Futures in Education, 8*(5), 593–601.

Stapp, W. B., et al. (1979). Towards [a] national strategy for environmental education. In A. B. Sacks & C. B. Davis (Eds.), *Current issues in EE and environmental studies* (pp. 92–125). Columbus: ERIC/SMEAC.

Sterling, S. R. (2001). *Sustainable education, re-visioning learning and change.* Dartington: Green Books.

Sterling, S. (2004). Higher education, sustainability, and the role of systemic learning. In P. B. Corcoran & A. E. J. Wals (Eds.), *Higher education and the challenge of sustainability: Problematics, promise, and practice* (pp. 49–70). Dordrecht: Kluwer Academic Press.

Sterling, S. R. (2012). Afterword: Let's face the music and dance? In P. B. Corcoran & A. E. J. Wals (Eds.), *Learning for sustainability in times of accelerating change* (pp. 511–516). Wageningen: Wageningen Academic Press.

Sterling, S. R., Maxey, L., & Luna, H. (2013). *The sustainable university: Progress and prospects.* London: Routledge.

The Ellen MacArthur Foundation (EMF). (2012). *Towards a circular economy: Economic and business rationale for an accelerated transition.* Report Vol. 1. http://www.ellenmacarthurfoundation.org/business/reports/ce2012#. Accessed 20 Aug 2013.

Tilbury, D. (2011). Education *for sustainable development: An expert review on processes and learning for ESD.* Paris: UNESCO. http://unesdoc.unesco.org/images/0019/001914/191442e.pdf. Accessed 20 Aug 2013.

UBC. (2011). *Transforming sustainability education at UBC: Desired student attributes and pathways for implementation.* University of British Colombia position paper. http://sustain.ubc.ca/sites/sustain.ubc.ca/files/uploads/images/teaching_learning/transforming-sustainability-education-at-ubc-desired-student-attributes-and-pathways-for-implementat.pdf. Accessed 20 Aug 2013.

UN. (2002). *General assembly resolution 57/254.* New York: United Nations. http://www.un-documents.net/a57r254.htm. Accessed 20 Aug 2013.

UN. (2004). *UNESCO framework for the UN DESD—International implementation scheme.* New York: United Nations. http://unesdoc.unesco.org/images/0014/001486/148650e.pdf. Accessed 27 Sept 2013.

UN. (2013a). *Sustainable development goals: An action agenda for sustainable development: Network issues report outlining priority challenges.* http://unsdsn.org/2013/06/06/action-agenda-sustainable-development-report. Accessed 20 Aug 2013.

UN. (2013b). *An action agenda for sustainable development.* New York: United Nations. http://unsdsn.org/files/2013/10/An-Action-Agenda-for-Sustainable-Development.pdf. Accessed 20 Aug 2013.

UN. (no date). *Millennium development goals: We can end poverty.* New York: United Nations. http://www.un.org/millenniumgoals. Accessed 20 Aug 2013.

UNEP (United Nations Environment Programme) News Centre. (2011, October 17). *Young environmental leaders from developing countries showcase sustainable solutions in Germany.*

http://www.unep.org/newscentre/Default.aspx?DocumentID=2656&ArticleID=8902. Accessed 27 Sept 2013.

UNESCO. (2005a). *Promotion of a global partnership for the UN decade of education for sustainable development: The International implementation scheme for the decade in brief*. Paris: UNESCO. http://unesdoc.unesco.org/images/0014/001473/147361e.pdf. Accessed 31 Oct 2013.

UNESCO. (2005b). *United Nations decade of education for sustainable development (2005–2014): International implementation scheme*. Paris: UNESCO. http://unesdoc.unesco.org/images/0014/001486/148654e.pdf. Accessed 4 Aug 2014.

UNESCO. (2012). *Shaping the education of tomorrow: Full-length report on the UN Decade of education for sustainable Development*. Paris: UNESCO DESD Monitoring & Evaluation. http://unesdoc.unesco.org/images/0021/002164/216472e.pdf. Accessed 27 Sept 2013.

UNESCO. (2013). *DESD final report*—UNESCO: Email about an online consultation.

UNESCO-UNEP. (1978). *Inter-governmental conference on environmental education*, 14–26 Oct 1977, Tbilisi. Paris: UNESCO-UNEP. http://www.gdrc.org/uem/ee/EE-Tbilisi_1977.pdf. Accessed 27 Sept 2013.

Vare, P., & Scott, W. A. H. (2007). Learning for a change: Exploring the relationship between education and sustainable development. *Journal of Education for Sustainable Development, 1*(2), 191–198.

Vare, P., & Scott, W. A. H. (2008). *Two sides and an edge*. London: Development Education Association.

Vare, P., & Scott, W. A. H. (in press). From environmental education to ESD: Evolving policy and practice. In: C. Russell, J. Dillon, & M. Breunig (Eds.), *Environmental education reader*, Bern: Peter Lang.

Webster, K., & Johnson, C. (2009). *Sense and sustainability: Educating for a low carbon world*. Skipton, UK: TerraPreta.

World Commission on Environment and Development (WCED). (1987). *Our common future*. Oxford: Oxford University Press. UN Document at: http://www.un-documents.net/our-common-future.pdf. Accessed 27 Oct 2013.

Worldwatch. (2013). *State of the world 2012: Is sustainability still possible?* Washington, DC: Worldwatch Institute. http://www.worldwatch.org/bookstore/publication/state-world-2013-sustainability-still-possible. Accessed 20 Aug 2013.

WWF. (2012). *Living planet report*. Gland: World Wildlife Fund. http://www.wwf.org.uk/what_we_do/about_us/living_planet_report_2012. Accessed 20 Aug 2013.

Chapter 5
Perspectives on ESD from a European Member of UNESCO's High-Level Panel, with Particular Reference to Sweden

Carl Lindberg

"The UN Decade is the golden opportunity we cannot afford to miss" (Vilnius, March 2005, European Launch of the Decade) quickly became a mantra among many educators following the decision at the Johannesburg Summit in 2002. Ever since Agenda 21 was formulated 10 years earlier at the Earth Summit in Rio de Janeiro many of them had worked to highlight education as an important means to promote sustainable development (SD). At the Johannesburg summit the Japanese Prime Minister Junichiro Koizumi suggested that the UN should proclaim a decade to promote Education for Sustainable Development (ESD). The proposal came from a strong Japanese NGO in the environmental field under the chair of Professor Osamu Abe and was sent to the Japanese government. At the preparatory meeting in Bali before the Summit many wanted to delete the proposal, but Japan stood up for the idea and later the UN General Assembly proclaimed 2005–2014 as "The Decade for Education for Sustainable Development" (DESD).

The Swedish Prime Minister Göran Persson, together with other Government representatives in Johannesburg, had to offer the international community something new which would contribute to SD. He was well prepared by his advisers, but in his speech at the Summit he went from what his advisers thought would be the Swedish proposal and invited—to their surprise—participants to an international conference on ESD, and thereby gave strong support to the UN Decade.

5.1 Experiences of ESD in Sweden: Before Johannesburg

Sweden, a sparsely populated country in Northern Europe, had for many decades been active in environmental issues, as well as in social matters, by developing what is known as 'the Nordic welfare model'.

C. Lindberg (✉)
Swedish National Commission for UNESCO in ESD, Uppsala, Stockholm, Sweden
e-mail: carl.g.lindberg@telia.com

© Springer International Publishing Switzerland 2015
R. Jucker, R. Mathar (eds.), *Schooling for Sustainable Development in Europe*,
Schooling for Sustainable Development 6, DOI 10.1007/978-3-319-09549-3_5

Sweden was one of the initiators of the UN Conference "Only One World" on the Human Environment held in Stockholm in June 1972. The idea was created already back in 1967 by the Swedish UN delegation where Alva Myrdal, a Government minister, played a crucial role. Alva Myrdal had great interest in educational matters and initiated the formation of OMEP (World Organization for Early Childhood Education) in 1948. She was to become Director-General of UNESCO and was awarded the Nobel Peace Prize in 1982. Right from the start of the organisation of the 1972 UN conference environmental education was seen as one of the goals of the conference.

Sweden's strong commitment to the UN and its specialised agencies dates back to the Cold War when Sweden, a small neutral country, had to rely for its safety on international conventions drafted by the United Nations system. When the Swede Dag Hammarkjöld was elected UN Secretary General, it greatly contributed to the strong support for the United Nations among Swedes.

There are many explanations for the interest in environmental issues, in a broad sense, among the Swedish population. One, often cited, is the customary law that gives the public free access to nature, provided that they do no damage. This gives many Swedes access and closeness to nature.

In the early 1960s, Rachel Carson's book *Silent Spring* was an eye-opener also in Sweden. Swedish scientists contributed reports in the same spirit. Before and during the 1960s, environmental activities focused on protection and preservation of nature. Later on, efforts expanded to protect human health and well-being. In 1967 the relevant state agency, the Swedish Environmental Protection Agency, was established.

When the UN Conference was held in Stockholm in 1972, a state committee published a report called *Choosing the Future* which stated

> It is important and necessary that we ourselves are studying the future (. . .) In this way, the small states create public opinion for other possible options of how the future world should be constituted and the political bodies have to represent the interests of future generations. (Sverige. Justitiedepartementet. Arbetsgruppen för framtidsforskning 1972)

In the 1962 compulsory school curriculum there is no mention of the environment, conservation or environmental education outside of biology as a school subject. The curriculum of 1969 mentions the environment for the first time. On the other hand, since the period after the Second World War democracy education had established a very strong position in the Swedish curriculum. It has often been said that the school also has a 'democracy objective' alongside its 'knowledge objective'. In the curriculum of 1980, there was also support for the school to address issues of peace and conflict. Peace education was in the early 1980s as controversial as ESD is now, so there is a clear link here.

The Swedish Prime Minister's personal support for ESD in Johannesburg can certainly be explained by his earlier position as Minister for Schools. Already in 1990 he initiated the introduction in the Preamble of the Education Act that everybody working in schools has to respect our common environment. In an interview he explained that with this Preamble we'd conquer a new frontier of the school and education. Environmental issues are in a dominant position in both the short and long term, both in a global and individual perspective. The school system could not miss this chance—to take the leap—to lift the school in this context, by emphasizing this in the important opening paragraph in the act. (Wickenberg Per 1999: 140)

In September 1992 a committee presented a new draft curriculum for primary schools. Inspired by various sources including the Summit in Rio de Janeiro it proposed a new subject: Technology and Environment. Unfortunately this was not included when later the government proposed its version of the curriculum, nor was the then-current term 'sustainable development' used as one of the foundations for school—as it should have been given the wording of the Education Act.

Although governments in the 1990s showed a strong commitment to environmental issues and the follow-up of Agenda 21, it would not be correct to say that environmental education in schools and in tertiary education assumed a central role in environmental policy. In 1999, however, the Swedish National Agency for Education founded the Green School Award which increased support for environmental education. That same year, universities were obliged to establish environmental management systems to minimize their own environmental impact.

Baltic 21 was a political process to create an Agenda 21 for the Baltic Sea and its surrounding states. It had been initiated by the Swedish Environment Minister Anna Lindh. A number of reports had been produced in areas such as forestry, agriculture, industry and tourism which all called for support through education. The Minister for Education and the Minister for Schools decided to invite their colleagues from the 11 countries in the Baltic 21 cooperation to a ministerial meeting to create a Plan of Action on ESD.

The Ministerial meeting took place in Stockholm in March 2000 and the Baltic 21 Education Project was launched with Sweden and Lithuania as lead countries. The late Lithuanian vice-Minister, Vaiva Vebraite, showed serious commitment and played a crucial role. The work was done in three working groups (schools, higher education and non-formal education) with meetings in Vilnius, Karlskrona and Gdansk.

Sweden had the privilege to get access to research-based knowledge on which to build. The scientists Leif Östman and Johan Öhman conducted studies on how environmental education in Swedish schools related to ESD. They found three traditions in environmental education. The goal of the first, the 'fact-based tradition', was to foster well-informed students with scientific knowledge of environmental problems and their causes. In the second, 'the normative' tradition, the teacher emphasises the moral character of the environmental problems. Special emphasis is placed on the fact that students develop ethical positions based on knowledge-based arguments. The third tradition, 'the pluralistic Environmental Education', aims to give students the ability to critically consider different options and participate in the democratic debate on environmental and related social issues.

The Baltic 21 Education Action Plan was adopted by a Ministerial Conference in January 2002. It was distributed to all Swedish schools and higher education institutions. It has, in many dissertations and theses, been described as an important part of the introduction and promotion of ESD in Sweden and other Baltic Sea countries.

At the European Council meeting in Gothenburg in June 2001 the EU strategy for SD was adopted. In connection with this the Swedish Government formulated its vision that Sweden should continue to be at the forefront of the transition to a sustainable society.

5.2   Experiences Relevant to ESD in Sweden: After Johannesburg 2002

The Prime Minister's proposals in Johannesburg for an international ESD conference in Sweden gave a new impetus and legitimacy to promoting ESD. Less than a month later the first national conference was held. Enthusiasm was high among the participants. Information was spread among many stakeholders, such as state agencies. It was quite self-evident that the venue of the conference would be Gothenburg because of the long-standing interest shown by the University of Gothenburg and the Chalmers University of Technology in SD.

After Johannesburg, the Swedish Parliament decided that the outcome of the summit was a "political and moral" obligation and noted that ESD was strengthened by the summit. The Parliament also noted that "ESD should be present in all subjects and as an integral part of the education system" and that "all forms and levels of education in Sweden should include all three aspects of sustainable development" (Riksdagens betänkande 2002/03: 93). ESD was thus strongly supported by the parliament and by the prime minister personally.

A few weeks after Johannesburg Sweden was consulted by UNECE (United Nations Economic Commission for Europe, the UN regional organization of Europe and North America) and asked to develop, in cooperation with Russia, a strategy for ESD for all 56 countries of UNECE. Sweden accepted the proposal and in May 2003 the UNECE's Environment ministers met in Kiev, Ukraine, where they adopted a Statement on ESD. The statement formed the basis of the strategy for ESD, later developed by UNECE with the help of skilled experts from many member countries (UNECE 2005a). This strategy has become an important tool in the promotion process of ESD.

At a major conference called "Sweden after Johannesburg" in December 2002, which brought together many leading Swedish scientists and university representatives with interests in various areas of SD, only one participant talked about education—all the others talked about research. The exception was the Vice-Chancellor Christina Ullenius, Karlstad University, who discussed the normative aspect of the term ESD in a way which I share:

> The necessary paradigm shift in society cannot be achieved solely through research. Through education universities can bring out the knowledge created by research (. . .) The question then becomes whether education can be morally normative for these issues? Our education should be value-neutral—but prescriptive in certain areas, such as democracy, equality and diversity. Maybe it's just as important that education is prescriptive in terms of substantive issues in sustainable development? (Efter Johannesburg 2002: 34)

When the Nordic Council of Ministers convened a ESD conference in Karlskrona, Sweden, in June 2003, 140 participants from Luleå in the North to Warsaw in the South, from Greenland in the West to Moscow in the East attended. The conference worked in innovative ways where pre-school teachers and university researchers collaborated with representatives of NGOs, all animated by a desire to learn from each other in order to promote ESD.

At some Swedish higher education institutions the decision about the Decade took effect relatively fast. A conference on ESD at Lund University, to which Vice-Chancellor Göran Bexell, Professor of Ethics, invited in November 2003, is an example. Vice-Chancellor, deans and heads of departments discussed the university's responsibility to promote SD. A comment from one participant was: "It was the first time that researchers with such diverse backgrounds as historians and mathematicians discussed common concerns."

"Learning to Change our World" was the international conference in Gothenburg to which Prime Minister Göran Persson invited in Johannesburg. It took place in May 2004 with 350 participants from more than 75 countries at Chalmers University of Technology. The participants also made field trips to educational institutions in the surrounding municipalities.

The Swedish Government attached great importance to the conference. The Prime Minister and three ministers (education, environment and international development) addressed the Conference. The UN Secretary General Kofi Annan participated through video conferencing.

Already at the opening of the conference the Prime Minister announced a proposal for an amendment to the Higher Education Act, stating that all Swedish universities have to promote SD in all their activities.

Learning to Change our World?, an anthology written by 27 scientists from 13 Swedish higher education institutions, was presented and distributed to all participants. Its authors belonged to the national research network "Education and Sustainable Development" which was built in close cooperation with European colleagues. In the introduction to the anthology the researchers explained that the title without this question mark would be provocative from a purely scientific point of view. They argue that their research would thus be linked to a "normative statement" contrary to the idea that research should be value neutral (Wickenberg et al. 2004: 15).

They also argued that one can imagine other moral questions such as "Isn't it a problem that we do not see any substantial global progress regarding the efforts for sustainable development?" (Wickenberg et al. 2004: 15). They add later:

> In this anthology on education and sustainable development we are convinced that it is time for researchers to take a personal clear stand as well as professionally to clarify fatal issues of global sustainability for our common future. (Wickenberg et al. (Eds.), 2004: 15)

At a number of meetings of Education Ministers in the EU and the Council of Europe in 2003 and 2004 Sweden highlighted the importance of ESD and distributed information on the conference in Gothenburg. Knowledge of and interest in ESD and the UN Decade by ministers and their staff appeared extremely limited. To me this was extremely disappointing given that SD is an overall objective for the EU and the Union was a considerable driving force behind the decisions taken in Johannesburg. One reaction at a ministerial meeting was that "ESD is only a matter for UNESCO" (said in a debate at a meeting with Ministers of Education organised by Council of Europe in Athens, November 2003, by one of the Swiss delegates).

The committee that organized the conference in Gothenburg in May 2004 had been asked to suggest what steps the government should take to promote ESD nationally and internationally. It submitted its report in November 2004 (Committee for ESD 2004). Among the many proposals were the Prime Minister's own suggestion that ESD would be included in the Higher Education Act, but also added to the Education Act and that Sweden should establish a UNESCO institute for international aid to promote ESD. The proposal for the amendment of the Higher Education Act was well received by the overwhelming majority of the universities and was accepted by the parliament. This meant that the Swedish universities had to promote SD in all their activities from February 1 2006 onwards. The Higher Education Act states that: "In their course of operations, higher education institutions shall promote sustainable development to ensure for present and future generations a sound and healthy environment, economic and social welfare, and justice." (Swedish Ministry of Education and Research 2006: Chapter 1 Initial provisions Section 5).

5.3 Experiences Relevant to ESD in Sweden: Promotion of ESD in Sweden During the UN Decade

• The Swedish National Commission for UNESCO has made ESD a priority. Via well-established networks of contacts with Swedish agencies and NGOs, this body has exploited opportunities for promoting ESD throughout the UN DESD. It has also stressed the importance of ESD in international cooperation and at UNESCO's General Conferences, and at the Bonn meeting in 2009 where Sweden was well represented. Via the Commission the Swedish Government has supported UNESCO's activities on ESD, for example a successful ESD Side Event at the Rio +20 meeting in 2012.
• In 2008 the Swedish Government established SWEDESD (The Swedish International Centre of ESD) at the University College of Gotland with the objective to promote ESD in international development activities. The international conference "The Power of ESD" in October 2012 brought together leading ESD representatives from around the world and helped to provide a current and relevant picture of the successes and challenges they face in their activities.
• The Swedish National Graduate School in Education and Sustainable Development (GRESD) stems from a research partnership between eight Swedish universities working in this field. All in all, more than 15 PhD students have been part of the work in creating a research environment on ESD of international importance. The initiative for GRESD was taken by The Institute for Research in Education and Sustainable Development (IRESD) which gathers approximately 20 researchers and doctoral students from different higher education institutions in Sweden.

- The Global School, financed by SIDA (Swedish International Development Cooperation Agency), is closely involved in the promotion of ESD within the Swedish school system.
- The Baltic University Program is an international network with support from the Swedish Government bringing together 225 universities in the Baltic Sea region into an educational partnership based on SD.
- CEMUS (the Student Centre for Environment and Development Studies) is a unique institution largely run by students at Uppsala University which bases its activities on SD and ESD.
- The new curricula for Early Childhood Education and Primary School and the one for Secondary School, valid as from 2011, are permeated by the values of SD. Early childhood education is recognized as the starting point for ESD. The Swedish branch of the OMEP (the World Organization for Early Childhood Education) is working very hard on promoting ESD.
- The Swedish National Agency for Education encourages ESD through its triennial accolade "The Sustainable School Award", established in 2005, replacing "The Green School Award". To be awarded, a school must organize its work so that all pupils and staff are given the opportunity to take an active part in formulating ESD goals and in the planning, implementation and evaluation processes of these.
- Since 2007 the RCE Skåne (the Regional Centre of Expertise on ESD in the Skane region) has been a part of the International Network of RCEs. In 2013 RCE West Sweden has been established and Uppsala is to become an RCE shortly.
- The Keep Sweden Tidy Foundation (an NGO) is the Swedish branch of the international organization of Eco Schools and has some 2,000 affiliated schools and pre-schools. Swedish authorities financially support their activities.
- The Swedish section of the WWF, World Wide Fund for Nature, which is one of the country's largest environmental and nature conservation organisations, strongly promotes ESD as part of its operations focusing on schools and universities.
- In April 2012 the Conference "Stockholm +40" was held as a reminder of the first UN Conference on the Human Environment, held in Stockholm in 1972. One of the sessions was dedicated to ESD and turned out to be the most popular of all the sessions at the Conference.

5.3.1 ESD Activities in Sweden Have to Be Significantly Strengthened

Although Sweden has high potential to be a leading country in ESD, there is much to be done to improve the ability of the Swedish educational system—from pre-primary education to university level—to be permeated by the ideas and values of ESD. That the concept of ESD meets resistance in various countries is well known, and this also applies to Sweden, even at high levels of the education system.

The former Chair of the Swedish National Commission for UNESCO, Görel Thurdin, has drawn attention to this twice: In the report from a committee about a new Upper Secondary school "Path to the future—a reformed school" she wrote in 2009:

> It is with great surprise that we note that the coupling education and sustainable development generally is not included in the report. Only in one place in the nearly 700 page report we can find the term 'sustainable development'. The Swedish National Commission for UNESCO finds this very remarkable, especially given the Government's ambitions in the climate field. A letter from the chairman of the Swedish Commission for UNESCO on a government committee's proposal to a reformed upper secondary school (2009)

In the report on a new teacher education, "Sustainable Teacher Education", also published in 2009, Görel Thurdin wrote:

> Again, we note that an important report concerning education lacks serious discussion and proposals on ESD. How can an investigation into the future of teacher education lack insightful discussion on ESD? Furthermore, the report's emphasis of subject knowledge and scientific knowledge is not sufficient to deal with complex issues and problems from students' reality, global challenges and social development. (A letter from the chairman of the Swedish Commission for UNESCO on a government committee's proposal to a reformed teacher education, 2008)

This reluctance to promote ESD, even within the Swedish educational system, underlies the importance that the Ministry of Education and Research and its agencies clearly show that Sweden endorses the decisions of the EU, UNECE, UNESCO and the UN on the importance of education to promote SD. This was recently manifested in the decision *The Future We Want* at the Summit Rio +20 where the leaders stated with regard to ESD: "We resolve to improve the capacity of our education systems to prepare people to pursue sustainable development, including through enhanced teacher training, the development of sustainability curricula, . . ." (UN 2012: 40, Paragraph 230)

5.4   ESD Experiences as the European Member of UNESCO's High-Level Group on ESD

In July 2004 the UNESCO Director-General Koïchiro Matsuura appointed a High-Level Panel for ESD as part of UNESCO's responsibility for the UN Decade. Following the appointment of the High Level Panel, I was invited to Paris. together with the President of the African Union, the then President of Mali, Alpha Omar Konaré, Japan's former Minister of Education, Science, Culture and Sport, Professor Akito Arima, and Professor Steven Rockefeller, Chairman of the Rockefeller Brother Fund, United States. On location in Paris I was informed that I was appointed to be the fourth member of this group, representing Europe. The group was expanded a few years later with Mary Joy Pigozzi, Senior Vice-President, Academy for Education Development, USA, and Rosiska Darcy de Oliveira, former Environment Minister of Brazil.

UNESCO's major education conference in Geneva in September 2004 was the occasion to present the UN Decade and ESD. Mary Joy Pigozzi, Director for the Division for the Promotion of Quality Education at UNESCO, and I had been commissioned to present it to Education Ministers and their staff. Approximately 200 participants attended this session, which was an extra session beside the regular conference program. I was surprised by this arrangement, because I had thought it would be natural if we were to devote the entire Conference to discuss and prepare for the UN Decade, which would begin less than 4 months later. Discussions on UNESCO's educational programs Education for All (EFA) and ESD and the relationship between them came to attract particular interest. I participated in the UNESCO Conference in Dakar in April 2000 as head of the Swedish delegation. In Dakar, the EFA programme got new energy through the adoption of the six Dakar goals, so I knew this process relatively well. My response was the same as I have argued over the years since then: EFA should have SD as a guiding principle and ESD must be based on basic education as a human right.

The Norwegian Nobel Prize Committee announced in October 2004 that it awarded the Peace Prize in the Memory of Alfred Nobel to Kenya's Deputy Environment Minister, the late Wangari Maathai. This meant significant support for the concept of SD. She received the award for her work on SD, democracy and peace. Her Nobel Lecture meant strong support for ESD. In her speech she talked about the achievements of her Green Belt Movement:

> Thousands of ordinary citizens were mobilized and empowered to take action and effect change. They learned to overcome fear and a sense of helplessness and moved to defend democratic rights. We developed a citizen education program during which people identify their problems, the causes and possible solutions. They then make connections between their own personal actions and the problems they witness in the environment and in society. Although initially the Green Belt Movement's tree planting activities did not address issues of democracy and peace, it soon became clear that responsible governance of the environment was impossible without democratic space. (Nobel Peace Prize Speech by Wangari Maathai, Dec 10, 2004)

As a member of the High-Level Panel since 2005 I have had many opportunities to promote ESD internationally, so I present some of my experiences from those international ESD conferences which were held during the first years of the Decade.

UNESCO's biggest regional office is in Bangkok, Thailand. In February 2005 it was the site of a major international conference where ESD was discussed on the basis of the very different conditions that exist between the many countries that make up the Asia-Pacific region. The conference provided ample opportunity for me to develop great respect for the skills and dedication that representatives for nations outside Europe exhibited in their willingness to spread ESD.

At the UN headquarters in New York the International launch of the Decade took place on March 1, 2005 at a well-attended meeting, chaired by Professor Steven Rockefeller, USA. UNESCO's General Director Matsuura presented the purpose of the Decade and UNESCO's role as lead agency for the process.

Vilnius, capital of Lithuania, was on 18 March 2005 the site of the European launch of the UN Decade during UNECE's high-level meeting of Environment and

Education Ministers (UNECE 2005b). The Lithuanian President had taken this initiative inspired by the late Vaiva Vebraite, presidential adviser and former deputy Minister and one of the driving forces behind the Baltic 21 Education Project. Russia and Sweden shared the presidency when the UNECE meeting adopted its elaborate Strategy on ESD which, regrettably, was not supported by the United States. Unfortunately, this is still the case despite President Barack Obama's stated position in favour of ESD in a letter to the participants of the Bonn meeting in April 2009.

In Ottawa in April 2005 I was privileged to have been invited to a Conference on the results of Professor Charles Hopkins committed and long-term efforts to persuade people to embrace the rationale for ESD, and this time in his native country Canada.

"Sustaining the Future—Globalization and Education for Sustainable Development" was a major conference held in Nagoya, Japan, at the end of June 2005, in conjunction with the World Expo, Aichi, organised jointly by the UN University (UNU) and UNESCO. The conference was also the Asia-Pacific launch of the decade. "For UNESCO ESD is a vital aspect of quality education" (UNESCO & UNU 2005: 17), the Director General, Matsuura said. He also noted that the situational analysis of ESD showed that politicians and stakeholders in the Asia-Pacific region perceived ESD as Environmental Education, so he concluded: "Moving from EE to ESD will be a key challenge for the Decade in the region" (UNESCO & UNU 2005: 18).

UNU's Vice-Chancellor Hans van Ginkel signposted a broad history of how ESD emerged within the international scientific community and the universities. In conjunction with the Johannesburg Summit 11 of the world's foremost educational and scientific organizations under the leadership of UNU-IAS (UNU's Institute for Advanced Studies) signed the Ubuntu-Declaration which brought together for the first time science, technology and ESD. I very much appreciated that he stressed that "Education for Sustainable Development means what it says. It is not just environmental education or even sustainable development education but education for sustainable development" (UNESCO & UNU 2005: 25). Therefore Hans van Ginkel always writes ESD as EfSD. He also suggested that "ESD gives orientation and meaning to Education for All (EFA)." (UNESCO & UNU 2005: 25).

In addition, Hans van Ginkel presented the ideas behind RCE (Regional Centers of Expertise) on ESD as proposed by the UNU. The program to promote the development of RCEs was launched at the conference and five initiatives were presented: Toronto (Canada), Heerlen (Netherlands), Sendai (Japan), Suva (Fiji) and Kumasi (Ghana).

The former Japanese Minister of Education, Professor Akito Arima, a member of the High-Level Panel, warned nations *not* to make the same mistakes that Japan had made in the 1960s and until the mid-1970s: "During this period of very rapid economic growth air, land and water became a danger to our health." (UNESCO & UNU 2005: 60). He hoped everyone would learn from those mistakes and stated that ESD is the key to this awareness.

The Asian-Pacific Centre on Education for International Understanding in Seoul, South Korea, organised an international conference in August 2005, which showed strong support for ESD among many committed representatives of South Korean society. The Secretary General of the Korean National Commission for UNESCO Samuel Lee and Professor Sun-Kyung Lee were among them.

In Luneburg in September 2005 the German Commission for UNESCO and the University of Luneburg organised the "Higher Education for Sustainable Development Conference" which was an early sign that Germany was a nation that really wanted to take their share of responsibility for the success of the UN Decade. At the Conference Professor Gerd Michelsen received congratulations as UNESCO had appointed him Chair in Higher Education for Sustainable Development. The German award "Dekade-Projekte" was awarded for the first time to several deserving organisations for dedicated promotion of ESD.

Beijing in October 2005 was the site for "The Second International Forum on ESD". It was organized in cooperation between the Chinese National Commission for UNESCO and the Beijing Academy of Educational Sciences. It clearly showed that there is strong support for ESD in China. Immediately before the opening of the conference, I was asked by the Vice-Minister for Education which country could show the best ESD activities. I had to admit that I did not know enough to make a fair assessment of this complex issue. In my speech, I chose to specifically highlight the need for pupils and students to participate in decisions on how they are to learn, as well as the need for critical thinking and problem solving in fostering a feeling of responsibility and the will to actively contribute to the development of a sustainable society.

After this initial year of the UN Decade five more years of participating at international conferences followed for me. In January 2011, UNESCO's High-Level Panel was dissolved by Director General Irina Bokova.

5.5 Experiences as the European Member of UNESCO's High Level Panel on the Decade of ESD from Asian Countries

My knowledge and experience of ESD is a result of speeches and presentations delivered at conferences and can of course reproduce only a very small part of what happens in practice, in the individual classroom in a primary school or lecture hall in a higher education institution. They do, however, give an indication of the difficulties and opportunities that exist in a particular country when one is getting feedback on the ideas and values that support ESD.

My experiences of ESD activities in Asian countries are against this background, naturally, quite superficial. They derive almost exclusively from what I heard and learned at conferences in Japan, Korea, China, India, Indonesia, Thailand and Mongolia.

In Japan, I have been privileged to return as a speaker at major ESD conferences several times during the decade. At a global conference in the UN building in Tokyo on the role of higher education in creating a sustainable future I found it very important that the Japanese Crown Prince participated. Japan's strong commitment to ESD, as demonstrated for example by several years of funding of ESD activities implemented by UNESCO's regional office in Bangkok, was also shown at an international conference in December 2008, during which they carefully prepared the World Conference in Bonn, which took place less than 6 months later. Already in the subsequent fall 2009, yet another international ESD conference was organized in the UN building.

India has, at the highest level, manifested a strong commitment to ESD. At the Ahmadabad-conference Tbilisi +30 I met hundreds of Indian ESD activists among more than 1,000 participants from around the world. They gathered at CEE (Centre for Environmental Education) in November 2007. The Centre's impressive activities for ESD in different parts of India under its dynamic founder and leader Kartikeya Sarabhai can serve as a model for activities in most countries. Furthermore, the Indian Minister for Human Resources, Kapil Sibal, reconfirmed this commitment in 2010 at the preparatory conference in New Delhi, with the purpose of setting up a so called Category II-institute with working relations to UNESCO. This Mahatma Gandhi Institute of Education for Peace and Sustainability—so its provisional title—aims to "bring ESD into every classroom across India" (Kapil Sibal 2010).

In China, the government representatives showed strong commitment to ESD. At the International Forum on ESD in 2005, 2009 and 2011, organized by the Chinese National Commission for UNESCO, Deputy Education Ministers participated and spoke in favour of ESD. Vice-Minister of Education, Chen Xiaoya, who led the Chinese delegation at the 2009 Bonn conference, emphasized there that "China has imposed ESD into a national public educational policy in the 'National Outline for Education', the midterm and long-term plan (2010–2020) for school education" (Speech by Dr SHI Gendong, Uppsala, May 18, 2011). There is a special Working Group on ESD in the Chinese National Commission for UNESCO.

For historical reasons, the Korean National Commission for UNESCO has a strong position. Dr. Samuel Lee showed, during his time as the Commission's Secretary General, a strong and internationally respected commitment to ESD. The same applies to Professor Sun-Kyung Lee, especially in her role in ENSI (Environment and School Initiatives) which helps to link ESD activities in Europe and Asia.

Mongolia is hit hard by climate change, which is one reason for the strong commitment to SD in this large but sparsely populated country shown by its Government and President. President Tsakhiagiin Elbegdorj's personal commitment to sustainability was honoured by UNEP with the award Champion of the Earth 2012. To further illustrate Mongolia's commitment to SD UNEP decided to hold the international celebration of World Environment Day in June 2013 in the capital Ulaan Bataar. Mongolia's Government has, for a number of years, shown great interest in how education can contribute to SD.

5.6   ESD—As I Encountered It in Some European Countries

Germany has played a very prominent role in the promotion of ESD. The unanimous decision of the Bundestag on 1 July 2004 to support the UN Decade forms a strong and legitimate basis for the work. This has been evident in all the conferences I have been privileged to witness: Luneburg, Bonn, Berlin and Bremen have been places for big and well-organized conferences. The Bonn conference in April 2009 which resulted in the Bonn Declaration was of course particularly important.

In the Netherlands, there was already before the Johannesburg conference a strong support for ESD. Heerlen, Maastricht, Amsterdam and Leyden are some of the sites of rewarding ESD conferences where I have encountered strong and inspiring commitment.

ESD in the UK has been described to me at conferences almost exclusively outside the UK by its dedicated representatives, or in book form. My only visit there was during midsummer 2006 at the University of Bath in response to an invitation to a conference from Professors William Scott and Stephen Gough. The conference was primarily for researchers from the UK and Germany with a view to finding indicators to further evaluate the effects of ESD. Through one of their books (Gough and Scott 2007). I got a detailed picture of the opportunities and problems that the promotion of ESD encounters in higher education in the UK.

Russia and the former Soviet republics have many dedicated and knowledgeable ESD representatives that I have listened to at various international conferences, including at the Leningrad State University Pushkin, in St. Petersburg. The Vice-Chancellor of this university—a scientist in the environmental field, with a strong ESD commitment—wanted to make the university available for the conference. A reason to do so was that he was disappointed over how environmental issues were completely ignored during Soviet times, and also the difficulties encountered by citizens who tried to point this out.

In the Baltic Sea states—Estonia, Latvia and Lithuania—I have met many university representatives dedicated to ESD and many have been involved ever since they participated in the drafting of the Baltic 21 Education Project.

Many teachers and researchers from Norway attended the Baltic 21 Education work, even though Norway does not border the Baltic Sea. The Norwegian teacher network Miljölaere has been well developed for decades. The NGO Idea Bank in Oslo conducts much innovative work in ESD in other Nordic countries as well. The international conference in Bergen in September 2010, which was the conclusion of the great SUPPORT Project (http://support-edu.org/), was a strong manifestation of ESD. Professor Victoria Thoresen's impressive work to develop education for sustainable consumption has been widely appreciated in many countries.

The Danish University of Education was involved early in the ESD field by Professors Bjarne Bruun Jensen, Søren Breiting, Karsten Snack and Jeppe Lessoe. Also the environmentalist network Eco-net has for many years been a strong promoter of ESD. Denmark has the privilege to get support in its ESD work from one of the country's most famous artists, Jens Galschiøt. He has made many sculptures on a special theme—the Balance Act—in support of the UN Decade, expressing mankind's vulnerable situation. Many Danish government officials have shown strong support for ESD. One of the Danish princesses has in her role as patron of the Danish National Commission for UNESCO participated in a Nordic ESD conference in Odense in 2010.

Finland was the first country which most determinedly took note of the recommendations of the Baltic 21 Education Project. After just a few years they developed a far-reaching action plan for ESD for all levels of the education system. The Finnish President Tarja Halonen's strong commitment to SD was evident before and during the Rio +20 conference where she expressed direct support for ESD as well.

5.6.1 On an Overall European Level

The Decision by the EU Council of Ministers on 18 November 2010 in support of ESD is very important. Unfortunately, it is likely to be a relatively unknown decision among ESD promoters in many countries. In the decision the 27 Ministers of Education invited "THE MEMBER STATES AND THE COMMISSION, WITHIN THE LIMITS OF THEIR respective COMPETENCES, TO Support ESD and Promote These Council Conclusions." (3046th EDUCATION, YOUTH, CULTURE and SPORT Council meeting Brussels, 18 and 19 November 2010). This decision on ESD should be referred to by ESD advocates in their activities. It ought to be a responsibility for EU Ministers for Education to enable this decision to become well known and translated into practical policy.

The networks ENSI (Environment and School Initiatives; http://www.ensi.org/) and SEED (School Development through Environmental Education; http://seed. schule.at/) have over the years done a lot for ESD promotion in Europe.

UNECE's work for ESD in accordance with its organizational strategy is of great importance in Europe. In this work representatives of 55 countries in and outside Europe meet and exchange views. Statements by UNECE's preparatory conferences for the Rio +20 conference were very important for ESD to become so clearly reflected in the Rio + 20 final document *The Future We Want*.

5.7 Concluding Remarks and Recommendations

I have recounted my experiences in promoting the ideas and values of ESD over more than 14 years. I have visited 25 countries in Europe and Asia for this reason. Of course, I have also taken note of the deeper evaluation of the impact of the UN Decade which Daniella Tilbury and Arjen Wals have presented at various occasions.

For nearly seven of these years, I had the privilege to be a member of UNESCO's High-Level Panel on ESD. From November 2005, I have also had the opportunity to be a special advisor to the Swedish National Commission for UNESCO with the task of promoting ESD. Overall, I have spoken at more than 150 conferences, including about 50 outside Sweden. Here are my reflections and recommendations based on this experience:

(A) The political will and responsible leadership for tackling many of the major challenges facing humanity in all countries can only be created by a well-informed and educated public.

(B) The national education sector is very often the biggest and most important sector in countries. Therefore ESD ought to be not only a key means but **the Key Means** in the endeavour to establish a green economy and SD. Only by making full use of the enormous potential of reorienting the education sector is it possible to be successful in the struggle for SD.

(C) The ministers responsible for schools and education must be the ones responsible for pursuing the ESD process. Too many Ministers of Education have neither been given that task, or have been unwilling to take responsibility for the ESD process.

(D) ESD is particularly important in the wealthy part of the world with 'the western lifestyle' but of course also in other parts of the world where the middle and upper classes have adopted this lifestyle. By consuming most of the world's limited natural resources and producing most emissions of greenhouse gases, we are producing by far the largest ecological footprints.

(E) Education in general has not been able to create sufficient insight into the need to change production and consumption patterns. The higher the average levels of education, the greater the destructive impact on our planet.

(F) Top-down and bottom-up processes are of equal importance. The top-down process involves people with power and influence at various levels making use of their power and influence to promote ESD. The bottom-up process is based on individual citizens' knowledge of ESD and their willingness to assume responsibility to promote ESD even when there is no support from above. Citizens (teachers, scientists, parents etc.) as well as NGOs must use their knowledge and prestige to put pressure on the different political levels in our societies to promote ESD.

(G) To be able to achieve quality education in the ongoing EFA process it is necessary to use SD as an overarching goal.

References

Committee for Education for Sustainable Development. (2004). *Att lära för hållbar utveckling* [Teaching for sustainable development]. Stockholm: Statens Offentliga Utredningar. http://www.regeringen.se/content/1/c6/03/41/44/0fe2bc94.pdf. Accessed 13 Sept 2013.

Efter Johannesburg—utmaningar för forskarsamhället. Dokumentation från Konferensen den 5 December 2002 i Rosenbad, Stockholm (Skriftserie 2003:1) (pdf-format 44 sidor) http://www.sou.gov.se/content/1/c6/21/35/95/6dd40768.pdf. Accessed 26 Feb 2014.

Gough, S., & Scott, W. (2007). *Higher education and sustainable development; Paradox and possibility*. London: Routledge.

Letter from the chairman of the Swedish Commission for UNESCO on a government committee's proposal to a reformed teacher education. (2008).

Letter from the chairman of the Swedish Commission for UNESCO on a government committee's proposal to a reformed upper secondary school. (2009).

Riksdagens betänkande. (2002/03). *Johannesburg—FN:s världstoppmöte om hållbar utveckling*. http://www.riksdagen.se/sv/Dokument-Lagar/Utskottens-dokument/Betankanden/200203 Johannesburg————FNs–_GQ01UMJU1/?html=true. Accessed 13 Sept 2013.

Sverige. Justitiedepartementet. Arbetsgruppen för framtidsforskning. (1972). Att välja framtid: ett underlag för diskussion och överväganden om framtidsstudier i Sverige: betänkande. Stockholm: Allmänna förl. http://libris.kb.se/bib/7257473. Accessed 13 Sept 2013.

Swedish Ministry of Education and Research. (2006). *Higher education act*. http://www.lunduniversity.lu.se/upload/staff/higher_education_act.pdf. Accessed 13 Sept 2013.

UN. (2012). *The future we want. Outcome document of the Rio +20 conference on sustainable development, Rio de Janeiro, Brazil*, June 2012. http://www.uncsd2012.org/content/documents/727The%20Future%20We%20Want%2019%20June%201230pm.pdf. Accessed 11 Dec 2013.

UNECE. (2005a). *Strategy for education for sustainable development*. http://www.unece.org/fileadmin/DAM/env/documents/2005/cep/ac.13/cep.ac.13.2005.3.rev.1.e.pdf. Accessed 13 Sept 2013.

UNECE. (2005b). *Vilnius framework for the implementation of the UNECE strategy for education for sustainable development*. http://www.unece.org/fileadmin/DAM/env/documents/2005/cep/ac.13/cep.ac.13.2005.4.rev.1.e.pdf. Accessed 13 Sept 2013.

UNESCO & UNU. (2005). *Sustaining the future—Globalization and education for sustainable development*, 28–29 June 2005 at Nagoya University, Nagoya, Japan. http://archive.unu.edu/globalization/2005/. Accessed 13 Sept 2013.

Wangari Maathai. (2004). *Nobel peace prize speech*. Accessed 10 Dec 2004.

Wickenberg, Per. (1999). *Normstödjande strukturer*. Lund: Lunds universitet [=Lund Studies in Sociology of Law, Vol. 5]. http://www.lub.lu.se/luft/diss/soc_176/soc_176.pdf. Accessed 13 Sept 2013.

Wickenberg, P., et al. (Eds.). (2004). *Learning to change our world? Swedish research on education & sustainable development*. Lund: Studentlitteratur.

Chapter 6
Social Learning-Oriented Capacity-Building for Critical Transitions Towards Sustainability

Arjen E.J. Wals

6.1 Introduction: The Rise, Meaning and Challenges of Social Learning in the Context of the DESD[1]

Around the globe sustainable development and sustainability have moved from the periphery to the mainstream of policy, business development, governance, science and society. Sustainable development and sustainability tend to evoke a common understanding in that they generally refer to balancing multiple interests, being mindful of future generations and remaining within the carrying capacity of the Earth. At the same time there is also an increased consensus within society as a whole that the search for sustainability and sustainable development is critically urgent and that capacities need to be developed to enable multiple-actors and multiple levels to participate actively in that search. However, there is no consensus about what is happening to the world's ecosystems, what needs to be done, by whom, when and where.

Meanwhile—during the United Nations Decade on Education for Sustainable Development (DESD) 2005–2014—consensus has been building among scientists and policy-makers that the road towards sustainability and sustainable development is currently ill-structured, ill-defined and context dependent (Hopwood et al. 2005; White 2013). Essentially this means that even though there is a general

[1] I wish to acknowledge that this chapter is based on a report (*Social Learning-oriented ESD: meanings, challenges, practices and prospects for the post-DESD era*) that I was commissioned to write by the DESD section of UNESCO for the end of DESD conference in Aichi-Nagoya, 10–12 November 2014. I also wish to acknowledge Meng Yuan (Mong) Jen and Mutizwa Mukute for their contributions.

A.E.J. Wals (✉)
Education & Competence Studies, Wageningen University, The Netherlands

Department of Education, Gothenburg University, Sweden
e-mail: arjen.wals@wur.nl

© Springer International Publishing Switzerland 2015
R. Jucker, R. Mathar (eds.), *Schooling for Sustainable Development in Europe*,
Schooling for Sustainable Development 6, DOI 10.1007/978-3-319-09549-3_6

understanding of the terms, their meanings vary in time (what might be considered sustainable today may turn out to be unsustainable tomorrow) and space (what might be considered sustainable in Amsterdam might be considered unsustainable in Dodoma). Furthermore, there is an increased awareness that there is no one single perspective that can resolve or even improve such issues and that there are often competing claims by societal interest groups (e.g. private sector parties, non-governmental organisations, civil society organisations, government bodies, scientists) that cannot easily be reconciled as they sometimes represent conflicting values. It is no surprise that given these uncertainties and the inevitable lack of foolproof solutions that withstand the test of time and work no matter where you are, the meaning of sustainability shifts towards *the ability to continuously reflect on the impact of our current actions on people and planet here and elsewhere, now and in future times.* From such a learning perspective it becomes key to translate the lessons learnt into the fine tuning of current actions (optimisation and improved efficiency) or the re-thinking of those actions altogether (system re-design and transitions).

Transitions towards more sustainable lifestyles and structures, cultures and forms of governance to support such lifestyles are increasingly seen as a part of a social innovation project that requires multi-stakeholder interaction and learning (Fig. 6.1).

In this chapter Education for Sustainable Development (ESD) is viewed as a social innovation process that strengthens peoples' capacities and mind-sets to

Fig. 6.1 An integrated multi-stakeholder approach towards social innovation and sustainability (Source: http://www.beaming.com/2013/06/a-social-innovation-movement.html. Accessed 1 September 2013)

participate actively in the continuous search for a more sustainable world; and not as a tool that can be used to *teach* people how they should behave or live their life— which presupposes that we can confidently prescribe the right or the best way. A quote from a key stakeholder from Nigeria taken from the recent *Results from ESD UNESCO Questionnaire 2—Input from online survey for Member States, Key Stakeholders and UN Agencies* illustrates this: "ESD has helped in promoting transformative education and creating a new system and sustainability thinking as a drive for great social innovations" (direct quotation: Nigeria, Key Informant Survey, UNESCO DESD 2012 Review, unpublished). Sceptics might argue that the social innovation approach, just like sustainability itself, might easily be hijacked by business interests to help grow companies and profit which indeed may be the case when the approach is stripped from a moral-ethic underpinning that signifies a planetary concern.

Among the capacities or capabilities needed to help social innovation and transitions towards a more sustainable world we find: anticipatory thinking, systems thinking, inter-personal skills, critical thinking and mind-sets that like empathy, solidarity and empowerment (Wiek et al. 2011). Furthermore, dealing with insecurity, complexity and risk are considered critical capacities or competencies for moving people, organisations, communities and, ultimately society as a whole, towards sustainability (Wiek et al. 2011). Social learning is introduced here as a form of multi-stakeholder engagement that is increasingly seen as particularly promising in developing such capacities and mind-sets. In Box 6.1 below there is a description of social learning that is highly compatible with the perspective of ESD as social innovation towards sustainability, but stands in sharp contrast with the more conventional views of social learning as a tool to modify social behaviour (e.g. Bandura 1963). This chapter seeks to generate a better understanding of ESD as social learning by contrasting or connecting it with other forms of learning, providing an organising framework for conceptualising social learning, highlighting exemplary European practices and discussing ways to evaluate or assess social learning-oriented ESD. The chapter ends with recommendations for the post-DESD period.

Advocates of ESD have long recognised that traditional discipline-based forms of learning and the one-way transfer of knowledge from a 'more knowledgeable other' or sender to a more or less passive receiver are insufficient and even inappropriate for dealing with sustainability challenges and engaging people meaningfully in transitions towards a more sustainable school, community, company, city, region and so on. Hence, it is no surprise that forms of learning that cultivate pluralism as well as joint sense and meaning making are receiving increased attention within ESD. Social learning in particular has resonated well with ESD researchers, practitioners and ESD policy-makers, especially in settings where classic distinctions between formal, informal and non-formal education, between sectoral and disciplinary boundaries appear to be fading. This observation was made in the 2012 UNESCO report on the processes of learning unfolding in the context of ESD (Wals 2012), but also in the first DESD global monitoring report by UNESCO in 2009 (Wals 2009a) where it was stated that a whole range of interactive methods and new forms of so-called 'knowledge co-creation' involving a wide range of societal actors, was emerging.

6.2 ESD and Multi-stakeholder Social Learning

In the second DESD Monitoring and Evaluation report, social learning was described as: "bringing together people from various backgrounds with different values, perspectives, knowledge and experiences (both from inside and outside the group or organisation initiating the learning process) to creatively find answers to questions lacking ready-made solutions" (Wals 2012: 27–28). The report refers to Peters and Wals (2013) who add to this that multi-stakeholder social learning:

- involves learning from one another together;
- assumes that we can learn more from one another if we do not all think or act alike, in other words, people learn more in heterogeneous groups than they do in homogenous groups;
- requires the creation of trust and social cohesion, precisely in order to become more accepting and to make use of the different ways in which people view the world;
- cultivates 'ownership' with respect to both the learning process as well as the solutions that are found, which increases the chance that things will actually take place;
- ideally results in collective meaning making, sense making and change.

There is not one single definition or description of social learning that adequately captures all its potential meanings. Rodela (2011) did a review of social learning in the context of natural resource management which distinguished three levels of social learning (Table 6.1). The table provides a typology of social learning which can help understand social learning within ESD.

Table 6.1 shows that social learning can operate at different levels (individual, organisation/network, whole system) and that at each of these levels the learning outcomes, processes and operational measures vary. Whereas the socio-ecological system tends to be featured in natural resource management contexts, it is the education system that often is central in ESD. Whole school approaches, for instance, tend to require social learning between teachers, pupils, administrators, parents and members of community organisations as the entire system seeks a transformation towards sustainability. At the same time the school is an organisation itself in need of becoming more responsive to community relationships and more capable of linking the school's operations and environmental management to the curriculum. Finally, the whole system of education, learning and community engagement needs to respond, which means, for example, that all the interactions between all actors, between school and community, between curriculum and school greening need to be reconfigured (Hargreaves 2008).

The utilisation of 'diversity' seems to be a common thread in emerging social learning practices within ESD. As quoted above, social learning brings together people from various backgrounds with different values, perspectives, knowledge and experiences (both from inside and outside the group or organisation initiating the learning process) and challenges one-dimensional or piece-meal solutions while

Table 6.1 Three research approaches to social learning: main characteristics (Rodela 2011: 30)

		Individual-centric	Network-centric	Systems-centric
Characterising features	Learning process	Transformative: learning as a transformative process that occurs during a participatory activity and involves the individual	Experiential: learning as a process embedded in past experience, and/or observation of other practitioners	Emergent: learning as an emergent property of the socio-ecological system
	Learning outcomes	A change of participants' internal-reflective processes; a change of participant's behaviour	A change in established resource use or management practices	Shift of the socio-ecological system on a more sustainable path
Level of analysis	Unit of observation	The individual	The individual, network, multi-stakeholder platform	The individual, ecosystems, institutions
	Unit of analysis	The participant	Networks, multi-stakeholder platforms	The socio-ecological system
	Learning agent of interest	The individual who participates in a participatory workshop	The practitioner, member of a community of practice and/or network of practitioners	The stakeholder, community member or practitioner who is involved in resource management
Operationalisation	Operational measures	Moral dimension (civil virtues), cognitive dimension (improved understanding of problem domain), relational dimension (relational base), trust (trust towards participants, process)	Change in how things are done. Improved relationships	Change of institutions and management practices at higher levels (e.g. policy), with interest for ecosystem responses

at the same time co-creating more holistic ones. As such, social learning also creates a change in understanding through social interactions between actors within social networks (Reed et al. 2010). A recent study by Sol et al. however, shows that bringing together people with different backgrounds, perspectives, values and so on does not automatically lead to social learning but requires the cultivation of commitment and trust between all involved (Sol et al. 2013). Social cohesion is considered conditional for allowing diversity to play its generative role in finding innovative and co-owned solutions to sustainability challenges. Not surprisingly social learning literature stresses the importance of investing in relationships, 'de-formalising' communication, co-creation of future scenarios and joint fact-finding (e.g. Johnson et al. 2012; Lotz-Sisitka 2012).

6.3 Contexts for Social Learning-Oriented ESD

Four contexts appear to be particularly relevant for social learning-oriented ESD: the interface between school and community, the interface between science and society, local and regional development arenas; and policy-making and gover-nance. In the next section these four contexts will be briefly described and illus-trated. The examples are chosen somewhat randomly from the recent UNESCO commissioned paper on social learning-based ESD (Wals et al. 2014). Some are well-connected to or can easily be linked to the DESD, although it should be acknowledged that several of them existed well before the DESD commenced. All cases have in common that they bring together people/groups/organisations/networks from various backgrounds in order to create innovative solutions.

6.3.1 ESD as Social Learning at the Interface of School and Community: CoDeS

Nowadays, in many parts of the world—albeit with great variation in intension and intensity—attention is paid at all levels of formal education to sustainability-related topics that affect a community, country or region. In some parts of the world, this coincides with a call for educational innovation and strengthening of school-community linkages. Social learning at the interface of school and community potentially deepens the learning in school and makes knowledge from books and other forms of media more significant and relevant.

CoDeS is a Comenius multilateral Network, consisting of 29 partners and funded by the Lifelong Learning Programme of the European Union, that focuses on school community collaboration addressing sustainability. The activities of the network aim at providing a European perspective on the processes of learning, models, values and tools for successful collaboration. Based on the partners' wide range of experience and background in ESD, the network produces, publishes and dissem-inates a range of products useful for school and community stakeholders to help them engage in successful practice. Products include: case study reports, a tool box, a travelling guide, and different types of workshops.

CoDeS provides a platform for school-community innovation using sustainabil-ity as both a means and an end. The network investigates ways of maintaining collaborative structures and involving isolated communities. Its resources are made available to the public electronically. Key assumptions or drivers of CoDeS are the ideas that:

- through *collaboration* development of competencies relevant for science learning and for social and environmental responsibility can be achieved;
- *open inquiry* will result in motivating students and broadening their scientific knowledge;
- inquiries about models provide a basis for designing *learning arenas at the school-community interface*;
- providing a set of indicators, can help create the right conditions for *process-orientated planning* of collaboration and the development of the teachers' competencies needed for such work;
- introducing a toolbox can provide direct support for teachers;
- meeting inspiring exemplars fosters *gender mainstreaming* but also helps in recognising problem areas;
- placing emphasis on *inclusive design* and *participative planning* support *balancing inequities* in the field of science learning and also promote *intergenerational discourse*;
- establishing *indicators for success* and survival of partnership structures will help assure sustainability;
- organising *innovative ways of dissemination* and training will provide teachers with the genuine learning experiences needed for successful implementation;
- fostering and facilitating a *dialogue* between representatives of *different levels and sectors of stakeholder groups* guarantee better understanding and opening a European perspective on the area.

In the above list all concepts that suggest a strong social learning orientation are italicised. CoDeS is a capacity-building network of reflective practitioners, researchers and educational policy-makers identifying exemplar cases, exploring the role of social media in communities of practice and sharing collaborative approaches.

The whole-school approach referred to earlier—linking ESD with the everyday curricular work undertaken by schools as well as enabling and encouraging closer links between schools and the communities around them—fits well with the CoDeS philosophy (Fig. 6.2). Scott (2013) describes how schools working in this vein might make such a contribution to sustainable development and discusses how we might come to know how effectively this is progressing.

Figure 6.2 suggests a number of processes that require social learning between multiple actors in enabling an environment of continuous improvement towards sustainability in curricula, school grounds, community engagement, environmental management and participation and decision-making. Such actors include teachers, students, parents, local businesses like bike repair shops and garden centres, and representatives of local government and NGOs focusing on, for instance, health, nature conservation and energy efficiency and carbon neutrality. Organising and facilitating the interactions between all these stakeholders as a social learning process is key to the success of a whole school approach to sustainability (Hargreaves 2008).

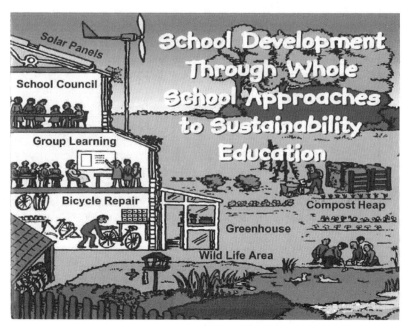

Fig. 6.2 A whole school approach to sustainability (Source: Comenius SEEPS, cover of the project CD)

6.3.2 ESD as Social Learning at the Interface of Science and Society: GUPES and PERARES

The challenge of (un)sustainability is becoming a focus in research and education across the planet. There is a recognition that pursuing sustainability is not simply or only a scientific and technical project but one that comprises complex ethical, philosophical, and political dimensions. To respond to these societal needs and challenges of sustainable development, universities have to break out of traditional academic boundaries by engaging with multiple stakeholders and connecting with contemporary sustainability challenges such as loss of food security, the prevalence of micro-plastics in water and soils, climate change and biodiversity loss. Dealing with such challenges requires new methodologies and modes of inquiry and knowledge creation. As a result there is an emerging re-orientation toward different forms of education, learning, research and community engagement in higher education. Based on the responses to the Global Monitoring and Evaluation Survey (GMES) used for the second review of the DESD (Wals 2012) and to the DESD key informant survey of the same review, it can be concluded that there are many higher education initiatives (e.g. new degree programmes, courses, modules and alternative approaches to learning) sprouting across the globe that emphasise the societal relevance of higher education and a "science as community" perspective (Peters and Wals 2013). Research and education are considered essential partners in

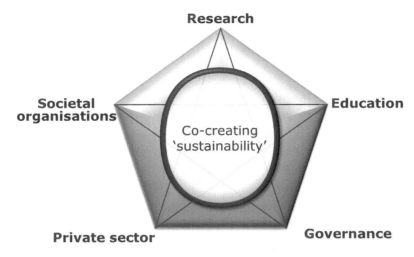

Fig. 6.3 The Big Five—co-creating sustainability at the cross-roads

what is increasingly referred to as co-creating sustainability. Co-creating sustainability entails the linking of multiple forms of knowledge: indigenous knowledge, local knowledge and scientific knowledge in a joint learning effort between partners representing the big five (Fig. 6.3).

A particular form of co-creation is citizen science which involves members of the public in collecting data about the state of the (local) environment, the health of a river, changing weather patterns, changes in biodiversity and so-on, alongside scientists (Shirk et al. 2012). Especially in the digital age where people have access to accurate monitoring devices connected to their smart phones and to apps that can help monitor environmental change, citizen science and crowd-sourcing data is promising in engaging people in joint learning about local sustainability issues (Dickinson et al. 2012; Silvertown 2009). The potential of ICT-mediated citizen science as a component of social learning has been little explored in the DESD but some of the cases highlighted make reference to it.

The Global Universities Network for Environment and Sustainability (GUPES) is one example of universities seeking to find ways to address sustainability issues by linking education, research to community engagement and outreach (http://www.guninetwork.org/guni.hednews/hednews/global-universities-partnership-on-environment-and-sustainability-gupes, accessed 19 April 2014). Based on the successful experience of Mainstreaming Environment and Sustainability in African Universities (MESA), GUPES transforms the mainstreaming of environment and sustainability concerns into teaching, research, community engagement and management of universities globally, as well as to enhance student engagement and participation in sustainability activities both within and beyond universities. This is done in accordance with the DESD 2005–2014.

In GUPES, a new kind of teaching and research that benefits communities has emerged. Increasingly university sustainability initiatives integrate sustainability issues into their operations and maintenance, teaching and research, and community engagement. This feature seems to permeate all disciplines involved (e.g. law, engineering, science, education, journalism). Evidence of this 'new kind of teaching and research' can be found in the way that participating universities are:

- Enhancing participation in research design and in the conduct of research that benefits communities, and in paying attention to the way that research outcomes are used for community benefit;
- Engaging students in service learning and problem solving projects in 'real life' contexts;
- Forging stronger partnerships with local communities and development groups to identify priorities for research and development work.

In the European context 'the living knowledge' network (http://www.livingknowledge.org/livingknowledge/, accessed 19 April 2014) and the European Union supported PERARES network of universities that have so-called science shops to link science and society, work in the same vein as GUPES:

Box 6.1: Mission and Vision of the PERARES Project: http://www.livingknowledge.org/livingknowledge/perares, Accessed 19 April 2014

The PERARES project aims to strengthen public engagement in research (PER) by involving researchers and Civil Society Organisations (CSOs) in the formulation of research agendas and the research process.

It uses various debates (or dialogues) on science to actively articulate research requests of civil society. These are forwarded to research institutes and results are used in the next phase of the debate. Thus, theses debates move 'upstream' into agenda settings. For this, partners link existing debate formats—such as science café's, science festivals, online-forums—with the Science Shop network—already linking civil society and research institutes. To be able to respond to research requests, it is necessary to enlarge and strengthen the network of research bodies doing research for/with CSOs. Thus, ten new Science Shop like facilities throughout Europe are started, mentored by experienced partners. Science Shop-like work is advanced by adding studies on good practices to the available knowledge base and organising workshops. Guidelines to evaluate the impact of engagement activities are developed and tested.

6.3.3 ESD as Social Learning in the Context of Local and Regional Development: RCE Rhine-Meuse

There is another trend of social learning sprouting in regional/local development contexts. This kind of social learning is closer to the concerns of citizens. Some are promoted by specific organisation, such as governments, international organisations or NGOs; others grow more spontaneously and organically. Much of this type of social learning takes place at the interface of community and the public and/or private sector of the regional/local, and it is cross-boundary in nature.

In some parts of the world, both in Western and Non-Western contexts, we see multi-stakeholder partnerships emerge that use social learning to co-create their own pathways towards sustainability. The rapid rise of Regional Centres of Expertise (RCE's) across the globe—in early 2012 100 RCEs had been established—provide testimony to the potential of multi-stakeholder social learning. A Regional Centre of Expertise (RCE) is a network of existing formal, non-formal and informal education organisations, mobilised to co-create sustainability within local and regional communities. RCEs bring together regional/local institutions, build innovative platforms to share information and experiences and promote dialogue among regional/local stakeholders through partnerships for sustainable development. Hence, RCEs involve interactions between actors that ordinarily do not work together easily as they never before saw common challenges and the complementarities of their expertise (e.g. Dlouhá et al. 2013). The RCE Rhine-Meuse is one of the older RCEs located at the cross-roads of Belgium, Germany and the Netherlands:

> The RCE Rhine-Meuse is a Regional Centre of Expertise on Education for Sustainable Development. In the RCE schools, science, industry and governmental institutions work together in order to generate learning processes that contribute to a more sustainable future.
> A more sustainable future can best be achieved by [giving] young people the best opportunities to continuously learn about future relevant themes, granting them access to the best sources of knowledge and practice in today's society. Thus their environment should promote and welcome their inquiry-[based] and real life learning, challenge and support them to development competences a sustainable future calls for; in regional society as well as worldwide as tomorrow's citizens. (Vision and mission of RCE Rhine-Meuse: http://www.rcerm.eu/RCE_Rhine_Meuse_global.html, accessed 1. September 2014)

The RCE Rhine-Meuse is part of a global network of RCEs to ensure that the RCEs benefit from each other's experiences. In addition to promoting multi-stakeholder participation the RCE also seeks to develop a platform for dialogue between NGOs, school teachers and administrators, and between environmental and educational administrations. Whereas several peer reviewed papers describing and advocating RCEs have been published (Mochizuki and Fadeeva 2008; Fadeeva and Mochizuki 2010; Ofei-Manu and Shimano 2012), there have, unfortunately, not been any critical reviews of the way the RCEs work and the extent to which the rhetoric of social learning is matched by concrete actions on the ground.

6.3.4 ESD as Social Learning Within Policy Making and Governance: The Dutch Learning for Sustainable Development Policy (LfSD)

Clearly, social learning-oriented (E)SD tends to build bridges among different groups in society. Through social learning, people cross traditional boundaries, and linkages are established between the five corners of the big five (see Fig. 6.3). Increasingly policy-makers are looking at social learning as a policy-tool for social innovation and sustainable development. As Rodela (2011) indicated, social learning in the context of policy-making and governance tends to operate at the system-level. It supports the assumption that social learning is a process involving system-wide change processes. Here three cases that focus on the policy-context in which (E)SD takes place are described to illustrate this.

Consistent with the underlying philosophy of 'sustainability as learning' the Dutch LfSD's goals are rather process-oriented: focusing on things like capacity-building, connectivity, emergence and reflexivity. The Dutch case is well described in the first collection of National ESD Journeys published by UNESCO (Mulà and Tilbury 2011). In the past, policy programmes focusing on environment and sustainability sought to change specific behaviours and looked for 'evidence' that such change indeed occurred. In a way these different orientations to policy-making reflect the government's dilemma of wanting to create a more sustainable society but having no definitive answers or prescriptions for how to act in order to be sustainable. So the LfSD-programme seeks to be a catalyst for capacity-building and the creation of so-called vital coalitions to enable citizens, young and old, to determine for themselves what it takes to move from the current situation/practice to a more sustainable one (Sol and Wals 2014). The 'vital coalitions' refer to (temporary) configurations or arrangements between different groups in society that are in each other's vicinity but until they were challenged by a common sustainability issue saw no immediate reason to work together. A hybrid learning configuration then constitutes a vital coalition of multiple stakeholders engaged in a common challenge using a blend of learning processes in a rich context where the whole is more than the sum of its parts.

The policy programme specifically mentions multi-stakeholder social learning as a vehicle for taking advantage of each other's qualities and the sometimes divergent perspectives they bring to the sustainability table. In order to assure that a vital coalition consists of groups representing different vantage points and perspectives but also holds some key areas of expertise (e.g. topical, local and process-related) a coalition of actors can only get government funding when four kinds of parties are represented (Sol and Wals 2014):

- members of (local) government and governance (e.g. local water board, food and health board);
- providers of facilitation and tools that can improve the quality of the interaction (e.g. consultants, community-organisers, EE-centres);
- societal actors who actually wish to address a local sustainability challenge (e.g. schools, local businesses, NGOs);
- people representing relevant societal and educational trends (e.g. cradle-to-cradle and closed cycle design experts, environmental app designers, after school programme managers).

6.4   Assessing the Impact of Social Learning-Oriented ESD

From the literature on assessing the impact of social learning-based transitions towards sustainability two things can be concluded: first, the evidence-base of social learning-based transitions is still rather weak; and secondly, it is difficult and even undesirable to formulate the intended results of a social learning process in terms of changes in the actual behaviour of people, living communities and organisations beforehand. After all, one of the features of social learning is that an aspect of the process itself is to determine the desired goals and results and to recognise that these may shift or grow organically in the course of the project in accordance with emerging insights. This is not to say that one cannot formulate beforehand the intended results in terms of process (i.e. that it be equitable and inclusive also giving voice to marginalised groups or perspective) or capacities or competencies that participants need to develop in the process (i.e. systems thinking, reflexivity, empathic understanding, conflict resolution, open-mindedness).

An additional challenge here is that social learning processes are characterised by a high degree of dynamics and unpredictability. Learning occurs at various levels: at the level of the individual, at the group level, and in the networks and communities of which the individuals and groups are a part. Furthermore, the relationships between the actors involved are also constantly changing. All of these make it quite difficult to determine the extent to which a project actually contributes to sustainable development, but it may well be possible to show progress in terms of, for instance, an increase in social cohesion between actors. It may be possible later on to claim that there has been some change in, for instance, concrete sustainability behaviour (e.g. an increase in car pooling, less consumption of meat), but the degree to which the change was the result of a project based on social learning and whether or not this change can be considered positive or negative in sustainability terms is more difficult to assess. In reviewing the DESD-related cases highlighted in some of the key UNESCO ESD-related reports this type of research is not available.

Considerations with respect to monitoring and evaluation in fact start already in the very initial stages of a planned intervention and the decision whether or not to emphasise social learning. One topic of deliberation is the perceived level of certainty about the desired change and the effectiveness of the possible solutions that are already available (see Table 6.2, the change matrix, column A). Finding an answer to this question is an important part of the start of a project and defines where an ESD-intervention can be placed on the 'certainty axis'. If we know for sure what is good and what is bad for sustainability, because this has been demonstrated through scientific research or learned from every day practice (position at the top of column A), then we can more easily proceed to establish clear-cut goals. Imposing such goals with some level of authority is then the more obvious course to follow (by means of public relations, campaigns, but also in the form of, for example, regulations and subsidies).

Table 6.2 The change matrix (Adapted from Wals et al. 2009: 14)

Column A	Column B	Column C	Column D	Column E	Column F
Certainty axis	Type of goals	Type of intervention	Intervening role	Results-axis	Sample methods
Much certainty regarding solution and the way to get there	Closed/pre-determined/established	Transmission Instrumental Transfer	Instruction Training Coaching	'Hard' results Emphasis on concrete *products* and measurable changes in behaviour and/or environmental conditions-(SMART)	Lectures, instruction, public service announcements, persuasive messages, seminars, (Ted) talks, googleing and youtubing, teaching, modules, indicator frameworks and checklists, social marketing, games and simulations, role playing
Little certainty regarding direction of the solution	Open/to be determined in dialogue/flexible	Social learning Co-creation Transformation	Facilitation	'Soft' results Emphasis on *processes*, capabilities and quality of the learning environment	Open-spaces, world-cafe's, multi-stakeholder platforms, communities of practice

It should be pointed out that issues that appear to be clear now often prove to be more complex or different from what we first believed (e.g. scientific knowledge and insights also change and are not always univocal). If we (dare to) recognise the fact that we have no certainty (bottom part column A) and that we actually face a collective quest, then other goals (column B),interventions (column C), roles (column D), results (column E) and methods of interaction (column F) will be more suitable. In Table 6.2 it is recognised that there are different kinds of issues/problems and that for some, which are well defined, the top of the scale might well be appropriate, but for those that tend towards 'wicked' problems, clearly working at the bottom of the scale is more appropriate.

The result of these deliberations has consequences for the link between people involved, the process of change and its possible outcomes. If we consider a position at the top of the matrix, then the participants will for the most part undergo an intervention, without having much influence on the process of change. If we consider a position at the bottom of the matrix, then the active contribution and the capacities of involved actors come into play: they will have much influence on both the process itself as well as its direction. In other words: if we find ourselves at the lower end of the matrix, as well as in an uncertain process of change, then a social learning process is the more obvious choice. However, if there is certainty regarding both the direction towards a solution as well as the solution itself then one might be more inclined to rely on training and convincing people and teaching/coaching them towards a new behaviour. The positioning of a desired change in the matrix above also has consequences for the type of results that can be expected and the choice of a monitoring and evaluation strategy (column E).

Given the dynamics within social learning-based (E)SD and the difficulty of establishing causality between ends and means, it may be more fruitful to focus the monitoring and the evaluating of social learning on: the degree to which the capacity of individuals, organisations and networks is developed and utilised for the purpose of contributing to social learning processes within the context of sustainable development; and the degree to which the environment in which multiple-stakeholders are brought together is conducive to social learning in the first place. The importance of choosing adequate tools and methods of interaction cannot be emphasised enough (column F). Fortunately, there are a great number of on-line resources available such as www.mindtools.com (accessed 19 April 2014) or the knowledge co-creation portal at Wageningen University (http://www.wageningenportals.nl/msp/term/alternate-terminology/social-learning, accessed 1 September 2013) that can be used to help identify adequate methodologies and methods.

An important question to ask is: how can we know whether the capacities of individuals, organisations and networks for contributing to social learning processes that are aimed at sustainable development are actually being developed and utilised? The learning process of the actors, organisations and networks often generates all kinds of 'soft' results that, at first glance, appear to have little to do with sustainable development but may be essential to creating a sustainable society (consider: social cohesion, empathy, involvement, co-operation and the like)

(Wals 2007). Recent ESD research has been focusing on (E)SD competence and on finding better descriptors and indicators of such competence in order to be able to assess their development (Wiek et al. 2011; Barth et al. 2007). In *Learning for the future: Competences in Education for Sustainable Development* (UNECE 2011) UNECE provides a framework of core ESD competences for educators assembled into three categories: the holistic approach, envisioning change and achieving transformation. Such frameworks can be helpful in monitoring and evaluating capacity building for (E)SD.

The main evaluation insights from the limited research that has been done on ESD-oriented social learning (Mukute et al. 2012) suggest that:

• *Multi-stakeholder social learning*, underpinned by individual and collective reflexivity, facilitates the development of capabilities that include relational agency, workplace learning and working at the interface of policy and practice;
• *Building communities of practice* (*CoPs*) within the 'Big Five' (research, policy, society, business and education, see Fig. 6.3) around a joint sustainability challenge using a range of novel methods for interaction can be effective in engaging multiple stakeholders;
• *Joint learning and action* between policy-makers, practitioners and academics for sustainability helps to create a common language between them and provides a space for the creation of conceptual capital that is relevant to a wide range of actors; ICT-mediated citizen science can play a big role in such joint learning;
• *Continuous dialogue* and social learning help refine interactions, practices and create practice-based knowledge that expands agency. However, some forms of partnerships require formalisation in order to realise strategic and continuous benefits.

6.5 Conclusions

This review shows that in many parts of the world the boundaries between schools, universities, communities and the private sector are blurring as a result of a number of trends, including the call for lifelong learning, globalisation and ICT-mediated education, the call for relevance in higher education and education in general, and the increased interest of the private sector in human resource development and sustainability. The resulting 'boundary-crossing' is re-configuring formal, informal and non-formal learning, changing stakeholder roles, and public-private relationships. This new dynamic can provide a source of energy and creativity in education, teaching and learning which—when underpinned by the quest for a more sustainable and liveable world—provides a powerful entry point for ESD.

ESD increasingly occurs and needs to occur within these boundary-crossing contexts as it is clear that an adequate response to sustainability challenges cannot be limited to single perspectives, disciplines or ways of knowing. The cases provided tend to emphasise the possibilities and the excitement that comes with these hybrid forms of learning and the formation of new partnerships and coalitions, possibly downplaying the difficulties of organising such learning. ESD-oriented social learning can be instrumental in facilitating boundary-crossing, creativity and innovation. In the concluding chapter of the full report of the second DESD review (Wals 2012: 64–68) a number of premises of ESD were listed that all seem to resonate with social learning. These premises are:

- ESD implies a life pedagogy which recreates the model of the present society and presents a more sustainable civilisation project, with social justice and reduction of poverty;
- ESD implies a new idea of curriculum, based on meaningful subjects and interdisciplinary proficiency which contributes to build a feeling of belonging to the Planet;
- ESD implies cooperative, supportive, dialogic and democratic learning processes, which require the participation of all members in the planning, execution and evaluation of education;
- ESD implies new public policies that can articulate the educative potentialities present in schools, civil society, government and in the private sector aiming at activities, projects and plans that intermingle when in action.

In the post-DESD era it will be crucial to support and further develop ESD as a catalyst for a transition in education, teaching, learning and professional development towards more holistic, integrative and critical ways of tackling sustainability issues. Doing so will require the strengthening of multi-stakeholder social learning in hybrid learning environments and the creation of (sometimes temporary) vital coalitions of actors jointly seeking change, innovation and transitions towards sustainability. At the same time it will become crucial to find ways to assure the more equitable inclusion of marginalised or 'powerless' groups, peoples and perspectives that may not be mainstream, but could hold the key towards re-orienting society towards sustainability. The issue of power and inequity has hardly been touched upon within the DESD. It is often the prevalent and dominating ideas and routine ways of doing things 'as usual' that blind us from seeing their shortcomings and keep us from developing healthier alternatives. As such the inclusion of counter-hegemonic perspectives and giving voice to the marginalised can be justified both on moral grounds and on sustainability grounds.

Simultaneously mechanisms will need to be put in place to ensure the effective involvement of stakeholders from all levels and fields of society in the decision-making processes. Governments can support ESD educators by stimulating the creation of 'learning environments' at the societal level: creating spaces where ESD practitioners meet, learn from each other, join forces and strengthen their individual activities. One mechanism to be developed further is the role of social media, the internet and other ICTs in strengthening participation and engagement in transitions and transformations towards sustainability. There are several examples at the international policy-making and lobbying level that show that this mechanism can be powerful in mobilising groups and voices from around the globe. The process of the on-line involvement of multi-stakeholder groups in the Rio+20 and

Future We Want (post 2015 debate) is a good example of ICT-supported social learning in the context of SD (http://www.un.org/en/sustainablefuture/, accessed 19 April 2014).

Finally, the evidence-base of the impact of social learning-oriented (E)SD remains weak. In part because funding for ESD research has been scarce during the DESD, and in part because of the difficulty of establishing causality in emergent and dynamic change and transition processes. More research will be needed but, more importantly perhaps, innovative monitoring and evaluation approaches and research methodologies need to be developed and employed to be able to capture the learning taking place at the various levels and to get a better sense of whether this learning contributes to sustainability as agreed upon by those involved.

Appendix 1: Resources for the Cases

CoDeS—Collaboration of Schools and Communities for Sustainable Development

Comenius-CoDeS http://www.comenius-codes.eu/Project_CoDeS/Goals/. Accessed 1 Sept 2013.

Case: Global Universities Network for Environment and Sustainability

GUPES-UNEP. http://www.unep.org/training/programmes/gupes.asp. Accessed 1 Sept 2013.
GUPES http://gupes.org. Accessed 1 Sept 2013.
Global Universities Network for Environment and Sustainability (GUPES). GUPES meeting report: Proceedings of the High Level Planning, Consultative, Sharing and Learning Meeting for University Leaders. Santiago, Chile, 5–6 September 2011. http://www.pnuma.org/educamb/Red%20de%20Formacion%20Ambiental/Contenido%20GUPES/Docs%20GUPES%20-PNUMA%202012/1%20GUPES%20Meeting%20Report_Chile%202011.pdf. Accessed 22 Apr 2014.
Granados-Sanchez, J., Wals, A. E. J., Ferrer-Balas, D., Waas, T., Imaz, M., Nortier, S. Svanstrom, M., Van't Land, H., & Arriaga, G. (2011). Collaborative curriculum innovation as a key to sprouting transformative higher education for sustainability. In Global University Network for Innovation (GUNI) (Eds.), *Higher education's commitment to sustainability: From understanding to action* (pp. 193–209), Series on the Social Commitment of Universities, Higher Education in the World Part 4, Basingstoke, UK: Palgrave/Macmillan.

Case: RCE Rhine-Muese

Vision and mission of RCE Rhine-Meuse: 330 http://www.rcerm.eu/RCE_Rhine_
Meuse_global.html. Accessed 1 Sept 2014.

Jos Rikers, H. A. N., & Jos Hermans, H. C. L. M. (2008). Regional Centre of
Expertise (RCE) Rhine-Meuse: A cross-border network. *International Journal
of Sustainability in Higher Education, 9*(4), 441–449.

Case: The Dutch Learning for Sustainable Development Policy (LfSD)

Based on:

van der Waal, M. (2011). The Netherlands. In I. Mulà, & D. Tilbury (Eds.),
National Journeys towards Education for Sustainable Development (pp. 77–
102). Paris: UNESCO. http://unesdoc.unesco.org/images/0019/001921/
192183e.pdf. Accessed 17 Apr 2014.

Sol, J., & Wals, A. E. J. (2014) Strengthening ecological mindfulness through
hybrid learning in vital coalitions. Special issue on ecological mindfulness and
cross-hybrid. *Cultural Studies of Science Education* (in press).

References

Bandura, A. (1963). *Social learning and personality development*. New York: Holt, Rinehart,
and Winston.

Barth, M., Godemann, J., Rieckmann, M., & Stoltenberg, U. (2007). Developing key competencies
for sustainable development in higher education. *International Journal of Sustainability in
Higher Education, 8*(4), 416–430.

Dickinson, J. L., Shirk, J., Bonter, D., Bonney, R., Crain, R. L., Martin, J., Phillips, T., & Purce,
K. (2012). The current state of citizen science as a tool for ecological research and public
engagement. *Frontiers in Ecology and the Environment, 10*(4), 291–297.

Dlouhá, J., Barton, A., Janoušková, S., & Dlouhý, J. (2013). Social learning indicators
in sustainability-oriented regional learning networks. *Journal of Cleaner Production, 49*,
64–73. http://dx.doi.org/10.1016/j.jclepro.2012.07.023. Accessed 19 Apr 2014.

Fadeeva, Z., & Mochizuki, Y. (2010). Roles of regional centers of expertise on education for
sustainable development: Lessons learnt in the first half of the UNDESD. *Journal of Education
for Sustainable Development, 4*(1), 51–59.

Hargreaves, L. G. (2008). The whole-school approach to education for sustainable development:
From pilot projects to systemic change. *Policy & Practice: A Development Education Review,
6*, 69–74.

Hopwood, B., Mellor, M., & O'Brien, G. (2005). Sustainable development: Mapping different
approaches. *Sustainable Development, 13*(1), 38–52.

Johnson, K. A., Dana, G., Jordan, N. R., Draeger, K. J., Kapuscinski, A. R., Schmitt Olabisi, L. K., & Reich, P. B. (2012). Using participatory scenarios to stimulate social learning for collaborative sustainable development. *Ecology and Society, 17*(2), 9. doi:10.5751/ES-04780-170209. http://www.ecologyandsociety.org/vol17/iss2/art9/. Accessed 19 Apr 2014.

Lotz-Sisitka, H. (2012). *(Re)views of social learning: A monograph for social learning researchers in natural resources management and environmental education.* Grahamstown/Howick: Environmental Learning Research Centre, Rhodes University.

Mochizuki, Y., & Fadeeva, Z. (2008). Regional centres of expertise on education for sustainable development: An overview. *International Journal of Sustainability in Higher Education, 9*(4), 369–381.

Mukute, M., Wals, A., Jickling, B., & Chatiza, K. (2012). *Report on the evaluation of the southern African development community regional environmental education evaluation (SADC REEP): Contract C51369.* Nairobi: Sida.

Mulà, I., & Tilbury, D. (Eds.). (2011). *National journeys towards education for sustainable development.* Paris: UNESCO.

Ofei-Manu, P., & Shimano, S. (2012). In transition towards sustainability: Bridging the business and education sectors of Regional Centre of Expertise Greater Sendai using education for sustainable development-based social learning. *Sustainability, 4*(7), 1619–1644. doi:10.3390/su4071619. http://www.mdpi.com/2071-1050/4/7/1619. Accessed 19 Apr 2014.

Peters, S., & Wals, A. E. J. (2013). Learning and knowing in pursuit of sustainability: References for trans-disciplinary environmental research. In M. Krasny & J. Dillon (Eds.), *Trading zones in environmental education: Creating transdisciplinary dialogue* (pp. 79–104). New York: Peter Lang.

Reed, M. S., Evely, A. C., Cundill, G., Fazey, I., Glass, J., Laing, A., Newig, J., Parrish, B., Prell, C., Raymond, C., & Stringer, L. C. (2010). What is social learning? *Ecology and Society, 15* (4): response 1. http://www.ecologyandsociety.org/vol15/iss4/resp1/. Accessed 19 Apr 2014.

Rodela, R. (2011). Social learning and natural resource management: The emergence of three research perspectives. *Ecology and Society, 16*(4), 30–41.

Scott, W. (2013). Developing the sustainable school: Thinking the issues through. *Curriculum Journal, 24*(2), 181–205. doi:10.1080/09585176.2013.781375. http://dx.doi.org/10.1080/09585176.2013.781375. Accessed 19 Apr 2014.

Shirk, J. L., Ballard, H. L., Wilderman, C. C., Phillips, T., Wiggins, A., Jordan, R., McCallie, E., Minarchek, M., Lewenstein, B. V., Krasny, M. E., & Bonney, R. (2012). Public participation in scientific research: A framework for deliberate design. *Ecology and Society, 17*(2), 29. doi:10.5751/ES-04705-170229. http://www.ecologyandsociety.org/vol17/iss2/art29/. Accessed 19 Apr 2014.

Silvertown. (2009). A new dawn for citizen science. *Trends in Ecology & Evolution, 24*(3), 467–471.

Sol, J., & Wals, A. E. J. (2014). Strengthening ecological mindfulness through hybrid learning in vital coalitions. *Cultural Studies of Science Education* (in press). doi: 10.1007/s11422-014-9586-z.

Sol, J., Beers, P. J., & Wals, A. E. J. (2013). Social learning in regional innovation networks: Trust, commitment and reframing as emergent properties of interaction. *Journal of Cleaner Production, 49*, 35–43. doi:10.1016/j.jclepro.2012.07.041.%20Accessed%2019%20April%202014. http://dx.doi.org/10.1016/j.jclepro.2012.07.041. Accessed 19 Apr 2014.

United Nations Economic Commission for Europe (UNECE). (2011). *Learning for the future: Competences in education for sustainable development.* Geneva: UNECE.

Wals, A. E. J. (Ed.). (2007). *Social learning towards a sustainable world.* Wageningen: Wageningen Academic Publishers.

Wals, A. E. J. (2009a). *Review of contexts and structures for ESD.* Paris: UNESCO.

Wals, A. E. J. (2009b). A mid-decade review of the decade of education for sustainable development. *Journal of Education for Sustainable Development, 3*(2), 195–204.

Wals, A. E. J. (2012). *Shaping the education of tomorrow: 2012 full-length report on the UN Decade of Education for Sustainable Development*. Paris: UNESCO. http://sustainabledevelopment.un.org/content/documents/919unesco1.pdf. Accessed 19 Apr 2014.

Wals, A. E. J., van der Hoeven, N., & Blanken, H. (2009). *The acoustics of social learning* (p. 32). Wageningen: Wageningen Academic Publishers.

Wals, A. E. J., Meng Yuan, J., & Mukute, M. (2014). *Social learning-oriented ESD: Meanings, challenges, practices and prospects for the post-DESD era*. Paris: UNESCO.

White, M. A. (2013). Sustainability: I know it when I see it. *Ecological Economics, 86*, 213–217.

Wiek, A., Withycombe, L., & Redman, C. L. (2011). Key competencies in sustainability: A reference framework for academic program development. *Sustainability Science, 6*(2), 203–218.

Chapter 7
Education for Sustainable Development in a Cultural Ecological Frame

Patrick Dillon

7.1 Preamble

The problem of human impact on the environment first became real for me as a teenager in the 1960s. I grew up in central southern England where the River Thames cuts through the chalk and separates the North Wessex Downs from the Chiltern Hills. It was a landscape of diverse habitats—flood plain meadows, chalk grassland, woodland—with an abundance of wildlife. In Britain, in 1963, we had our first National Nature Week and a set of postage stamps was issued to commemorate it. Rachel Carson's *Silent Spring*, published a few months earlier, was much talked about. Nature conservation became one of the issues of the day and was to develop into a 'movement' during the 1970s. Environmental education started to take shape in school curricula. At university, I studied biology and after a period of teaching in schools I got my first job in higher education teaching ecology in courses on social biology and landscape studies. By now I was well aware that much of the change I saw in the landscape of rural England, and its diminishing wildlife, could be attributed to the intensification of farming. Elsewhere people were reporting similar concerns about urbanisation, industrialisation, pollution, etc.

The simple cause and effect models of environmental change were slowly giving way to more nuanced approaches which recognised that the environment and the activities of people are inextricably linked in a complex of constantly adjusting interrelationships. In 1955 some of the most prominent thinkers of the day had gathered at a symposium in Princeton, New Jersey, to share perspectives and attempt a synthesis of the then state of knowledge. The two volumes that emerged from the symposium, *Man's Role in Changing the Face of the Earth* (Thomas 1956), became a seminal work. The words of Lewis Mumford still resonate:

P. Dillon (✉)
Faculty of Philosophy, University of Eastern Finland, Joensuu, Finland
e-mail: P.J.Dillon@exeter.ac.uk

© Springer International Publishing Switzerland 2015
R. Jucker, R. Mathar (eds.), *Schooling for Sustainable Development in Europe*,
Schooling for Sustainable Development 6, DOI 10.1007/978-3-319-09549-3_7

> In passing from the past to the future, we pass from memory and reflection to observation and current practice and thence to anticipation and prediction... this is a movement from the known to the unknown, from the probable to the possible... But in fact these aspects of time and experience cannot be so neatly separated. Some part of the past is already present in the future; and some part of the future is already present in the past. (Mumford in Thomas 1956: 1141)

Closer to home, the interplay in the landscape between past and present, continuity and change, had been recorded by W.G. Hoskins in *The Making of the English Landscape* (1955). Hoskins's observations and ideas were brought to prominence in a television series in the 1970s. As I immersed myself in the works of Hoskins and the landscape itself, it became clear to me that in order to better understand what was happening in the environment I had to look beyond ecology and more deeply into its interface with human activities. In parallel with my teaching in the 1980s, I undertook PhD research on the interrelationships between land-use, agriculture and environment. Meanwhile, the 'nature conservation movement' had broadened into the 'environmental movement' and the notion of sustainable development had arrived. My work in education was now concerned with finding ways of educating about the interconnections between nature and culture. Out of this came a conceptual framework, cultural ecology, as a means of theorising the environment holistically. Associated with the conceptual frame were practically orientated pedagogies of connection and difference and tools for investigating the interconnections.

7.2   A Cultural Ecological Framework

My work on cultural ecology is a project in progress. The ideas have been developed and refined over a number of years and shaped by the influences set out in the previous section. The cultural ecological framework given below builds on work undertaken personally and in collaboration with colleagues and is thus a consolidation and refinement of ideas published in earlier papers, notably Dillon (2008a, 2012), Dillon and Loi (2008) and Vesisenaho and Dillon (2013).

Two fundamental premises underpin the framework: first, nature and culture, the environment and the activities of people, are inextricably linked and it is meaningless to consider one without the other. This means that polarised stances arising from ecocentric and technocentric approaches are unhelpful. Second, it follows that when we speak of 'the environment' we mean more than just physical surroundings, ecosystems and economic activities. Environment includes social relations and the collective capabilities of all the people who inhabit it—their lifestyles, beliefs, ideas and aspirations. Critics will say this is a human-centered perspective on the environment. My response is that to engage with issues of sustainability is to engage with the complex interactions that take place between people and their environments. The environments we inhabit today have been shaped by the activities of people over millennia and only by addressing the detail of the engagement will we be equipped to proactively influence the future. This perspective aligns broadly with the European tradition of environmental education.

Cultural ecology in its most basic form concerns the relationships between people and their environments, about how they interact and transform each other. At its most basic, cultural ecology reflects relationships between the way we think and the way we act, the way we conceptualise the environment and the way we behave in it. These are fundamentally *educational* matters. It is thus futile to argue about the fine detail of distinctions between sustainable development and education for sustainable development. In the final analysis they are one and the same thing.

Many of the issues around sustainability arise from tensions between continuity and change, tensions between, on the one hand, the needs of people to enjoy the security offered by tried and tested ways of knowing and doing and, on the other hand, the desire to find new ways of engaging with the environment.

Tried and tested ways of knowing and doing, leading to linear routes of continuity, are *relatively* 'safe', i.e. they are routine; moving in a new direction may carry greater risk. Sometimes the routine way of doing things is no longer fit for purpose and it is necessary to find new ways forward. In its simplest use, the term sustainability is applied to the level at which resources are used. Most people continue to think of resources in economic terms with a focus on the market conditions which determine the rate of use of a given resource. Economists argue that it is artificial to make a distinction between those practices which are sustainable and those which are not. Within absolute boundaries of supply, the utilisation of resources is relative to price mechanisms and demand. In these terms, what is sustainable in one set of economic circumstances may not be sustainable in another. However, because, in cultural ecological terms, the tensions between continuity and change are based on a complex of such matters as lifestyles choices, individual and collective capabilities, beliefs, ideas and aspirations, it is not sufficient to quantify resources in purely economic terms. Value is assigned in different ways by different people. And what is valued is a reflection of what is *known* and what is *acted upon*.

Knowledge is organised into domains, bodies of disciplined knowledge which can be acquired, practised, and advanced. Note that the term 'discipline' not only denotes 'a body of knowledge' but includes the sociocultural contexts in which that knowledge has meaning. Working with disciplined knowledge necessarily involves discerning what is important from what is not, and the skilful use of the tools and techniques that are available. In a broad sense, educational systems have co-evolved with the growth of modern knowledge disciplines and reflect the values embodied within them: the academic separation of 'pure' knowledge from 'applied'; the orientation of learning towards 'received wisdom', which is typically embodied in textbooks or handed down by teachers. This is *relational* education, i.e. education that relates to established ways of doing things. In cultural ecological terms, relational education is where situations are defined relative to each other (e.g. through separate disciplines), behaviour may be predicted or directed (e.g. by setting objectives) and understanding is part of a historical continuum which recognises a past, a present and a future. Relational education clearly makes an important contribution to human well-being but of itself it is not sufficient to equip people to engage with the complexities of modern life and their responsibility towards the environment.

Relational education is built on a Platonic view of knowledge as 'justified true belief'. But as Macfarlane (2013) points out, this over simplifies the relationship between knowledge and beliefs: "we are complex, irrational and emotional creatures constantly negotiating our own compromises between beliefs and knowledge" (Macfarlane (2013: 19). Research is now much more focused on consciousness and it has been shown that the human mind works in complex and dynamic interaction with the environment where meaning is fluid rather than fixed and information is integrated with experience in processes that involve the continual construction, deconstruction and reconstruction of knowledge. In this view knowledge is seen as a set of possibilities rather than a fixed entity to be transmitted in linear form. The 'fluid' conception of knowledge recognises that in addition to relational interactions with the environment, our day-to-day activities are shaped by more tentative 'in the moment', or *co-constitutional* engagements. In co-constitutional interactions, situations 'emerge', behaviours and environments co-construct each other, and things happen 'in the moment'. Knowledge is idiosyncratic and 'localised'; beliefs are strongly influenced by emotions and the focus of engagement. An education that recognises these types of interactions might be called *co-constitutional* education.

The cultural ecological interactions in most formal educational institutions like schools and universities are relational. They depend on the transfer of established constructs within defined structures. Similarly, managerial approaches to environmental matters and the policies that shape them are also relational. They are derived from systematic models and applied in generalised ways. In both educational and environmental management the things that matter most to people, the things that arise from their intimate day-to-day engagement with their surroundings, their 'lived knowledge' (see also Moll et al. 1993) are either ignored or compromised as they are accommodated within relational structures. The cultural ecological challenge to both sustainable development and education for sustainable development is in bringing together, on the one hand, policy and managerial interventions derived from specialist knowledge and a concern for the bigger picture and the longer term and, on the other hand, living traditions and improvisations arising from local knowledge and local concerns.

Ideally, therefore, educational interactions in schools and universities should be both relational and co-constitutional when they can facilitate both conceptual and perceptual understandings and the connections between them. The co-constitutional aspect of education would recognise personal 'in the moment' experiences. Taken together, the relational and the co-constitutional offer a vision of learning in a cultural ecological sense: meaningful ways of relating personal, 'in the moment' experiences to historically established practices and ways of knowing. This vision reflects the integral relationship between nature and culture: the environment is appreciated 'in the moment', for what it is, for how it affects us emotionally and spiritually, as well as being an object of study and reflection.

It follows that learning in a cultural ecological sense requires an integrational form of education which does not separate school from other facets of life—the home, leisure, the workplace. Distinctions between academic and vocational and formal learning and informal learning are meaningless. A cultural ecological approach to learning would provide people with frameworks in which they can locate themselves and their experiences. Some experiences will be in formal situations, others informal; some in the home, some in the workplace. Experiences overlap and connect in some way with other experiences. Each experience relates to a personal story. And, in turn, each personal story relates to the story of humanity.

This may sound idealistic but the Reggio Emilia system shows that it is not necessarily unattainable. Reggio Emilia is a town in northern Italy where an ecological approach to education was pioneered in the 1960s and has since been developed and refined into a system of international renown. In this approach, the reciprocal relationship between people and their environment is recognised. The focus is on the different ways that spaces are used, the connections between things, and the experiences offered by the spaces and the connections. Space is given shape and identity by the relationships created within it. Students inhabit the space by continuously constructing places, imaginary and real. Within these spaces, student's relationships and transactions with others, and with the environment itself, determine the possibilities and qualities of learning. Learning is situated, adapted, localised, and connected through a continual dialogue between the students and their environment (Ceppi and Zini 1998).

The technological means of achieving this vision are substantially in place. Educational niches can be formulated around all the learning resources, human and physical, institutional and virtual, in educational institutions, homes, libraries, workplaces, community and adult learning centres, science and art museums, and through television and public services, and so on. The image here is of individuals co-constructing learning journeys with families, communities, networks, and educational professionals, of education that reflects the particularity, subtlety, idiosyncrasy, and patina of locality at scales, at time frames, and through modes of organisation appropriate to those places and the enterprises within them. Unfortunately, the way we deal with knowledge and beliefs has not kept pace with the technological potential. Education is too reliant on narrowly defined courses. Learning it too often fragmented, superficial and repetitive.

7.3 Pedagogies of Connection and Difference

The notion of pedagogies of connection and difference is closely connected with the cultural ecological framework and thus shares some of its developmental history. What follows draws heavily on earlier papers (Dillon 2006, 2008b) and recent work undertaken with colleagues at the University of Eastern Finland (see Dillon et al. 2014a).

The argument so far: people think and act in their environments in two reciprocal ways: relationally and co-constitutionally, the first is a normative and regulated way of being, the second subjective and idiosyncratic. Policies and strategies for both the environment of formal education and the wider cultural environment are based on relational structures arising from relational forms of thinking. Co-constitutional interactions, which characterise people's day-to-day engagement with their environment and which are heavily mediated by immediate perceptions and emotional responses, are superficially accommodated with the relational. This reinforces patterns of compartmentalised thinking and acting; it dissociates the localised experiences of individuals from the wider concerns of humanity.

The problems facing humanity, the environmental issues of the day, are too complex to be addressed solely through one mode of thinking and acting. Holism, that is learning to make connections, to 'think outside the box', to 'see the bigger picture', is a mantra of sustainable development and is fundamentally an educational concern. Cultural ecology emphasises holism through promoting an understanding of the interactional dynamics of people and their environments, between continuity and change, routine and innovation, past and present. A cultural ecology approach to education therefore makes space for the co-constitutional in dialogue with the relational and seeks connection and integration whilst recognising difference.

An educational system can be influenced at many cultural ecological levels: policy generation, institutional management, curriculum development, teaching strategies, approaches to learning. In what follows I concentrate on pedagogy, the interactions between learner and teacher in the construction, deconstruction and reconstruction of knowledge. Specifically, I look at pedagogies of connection and difference and explore at a general level ways of crossing the boundaries that compartmentalise how we think and act.

Elsewhere, I have made a case for a *pedagogy of connection* for engaging with complex and multifaceted issues by setting out some overarching principles for establishing the contexts of connection and developing tools for making the connections (Dillon 2006, 2008b). But holism is not just about connecting everything and seeing it 'in the round'. It also involves recognising how difference fits into the picture: the uniqueness of human experience means that absolute agreement, unanimity of thinking and doing, is unlikely. In a later paper, arising from work with colleagues at the University of Eastern Finland on intercultural and multicultural education, I introduced the idea of *pedagogy of difference* (Dillon et al. 2014a). The argument in bringing together the two pedagogical approaches is to move towards a form of transdisciplinarity which consolidates connections and accommodates difference. The consolidated pedagogical framework is given in Fig. 7.1.

The core of the framework, which makes up the central part of the diagram, is the notion of a 'boundary' through which connections are made and differences recognised. A boundary is very often a conceptual entity arising out of our tendency to compartmentalise knowledge into disciplines or subject areas where meaning is assigned in very specific ways. But, especially in policy making and environmental

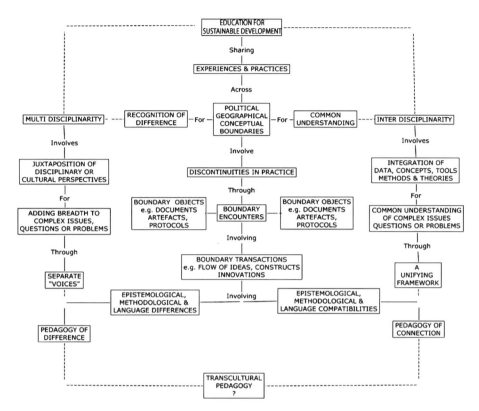

Fig. 7.1 The relationships between boundaries, multi- and interdisciplinarity, and pedagogies of connection and difference in education for sustainable development (Adapted from Dillon et al. 2014a)

and institutional management, boundaries may be political or economic, or ethical. Walker and Creanor (2005) define a boundary as a discontinuity in some form of practice, often determined by the limits of effective communication. Seminal work on boundaries was undertaken by Star and Griesemer (1989) as they brokered between different academic, professional, amateur and administrative interests in bringing to fruition a project on environmental and archival conservation. One now talks of 'boundary objects', including artefacts, documents and institutional and administrative protocols that have to be addressed by people from different communities if shared understandings are to be built. 'Boundary encounters' occur as people interact across boundaries, interpersonally and through the mediation of information and technologies. 'Boundary crossings' are the flow of ideas, constructs and innovations across the boundaries. In the framework (Fig. 7.1) a distinction is made between multidisciplinary and interdisciplinary approaches to working across boundaries.

Interdisciplinarity seeks integration, literally 'to integrate', being able to function in new environments, with people from different cultural or disciplinary backgrounds bringing with them new ideas and/or ways of working. It implies motivation and a willingness to work and learn together. It is based on understanding and cooperation between people in different settings and thus is as much about attitudes and beliefs as it is about knowledge. Interdisciplinarity cannot be reached by any individual working alone but rather it demands committed and determined long-term collaboration and networking between different individuals. The goal of interdisciplinarity is the integration of data, concepts, tools, methods and theories to gain a common understanding of a complex issue, problem or question.

Multidisciplinarity involves the juxtaposition of disciplinary or cultural perspectives to add breadth to understanding through making good use of available knowledge and methods but through the 'separate voices' of the contributing individuals. It recognises the multiplicity of meanings in different contexts but seeks to bring them together in order to broaden dialogue and understanding. Multidisciplinary approaches prepare individuals to face *others* and other points of view.

Both inter- and multidisciplinary boundary encounters offer opportunities for working with new ideas and approaches. In authentic boundary encounters, the protocols that maintain the boundaries of the contributing disciplines (for example the constructs and methods that distinguish science from, say, history, or the practices that distinguish working in one situation from those of another) are relaxed to allow free movement across the boundaries. The co-constitutional possibilities in these engagements may emerge spontaneously or they may need the stimulus of interventions such as those used by Daria Loi. Loi's workshops deploy "playful triggers", she introduces "eccentric objects", the participants share "odd experiences" (see Loi and Dillon 2006). These approaches have been employed in both educational and organisational situations where the emphasis has been on the transfer of ideas from one context to another with the expectation of a new synthesis to solve an existing problem or to find a new practical application. The goal is to 'free up the creative space' in order to allow 'in the moment' responses, no matter how bizarre, to be explored and either consolidated or left to drift away.

This may all sound highbrow but there is plenty that can be done with primary school children. In work conducted in schools in Finland and England colour and symbol associations were used as a stimulus for getting students to engage with their school grounds, describe their feelings about them, and produce a lexicon of 'emotion words'. They used the lexicon in explanatory writing, elaborating their associations between places, colours and symbols. The symbol, colour and word associations provided an insight into students' feelings about their school grounds, reflecting their 'in the moment' responses or emotional attachments (which we called "place attachments"). The reasons they gave for their feelings in their texts reflected their 'place meanings'. In cultural ecological terms, there is a reciprocal relationship between place attachments and place meanings, between perceptual and sensory experiences and conceptual understandings. Another way of looking at this is to equate emotions with 'in the moment' experiences and feelings with relational constructs. But feelings are more than conceptual constructs, rather they may be thought of as 'embodied knowledge'. In this study language captured in students' explanatory texts acted as the mediator between lived experiences and conceptual meanings and demonstrated strong links between emotion and cognition (Dillon et al. 2014b). Engaging young people in this type of activity is a good precursor to getting them to work holistically in their later school careers.

Freeing up creative spaces to stimulate co-constitutional activity has much in common with improvisation whose characteristics have been described by Barrett (1998). Barrett is concerned with improvisation in jazz but the characteristics he describes have wider applicability. For example: deliberate efforts to interrupt habitual patterns; embracing errors as sources of learning; minimising structures that constrain flexibility; continual negotiation and dialogue towards synchronisation; retrospective sense making.

The outcomes of this type of activity can be consolidated into new relational forms of knowledge or practice, a process that Sternberg and Lubart (1999) call combinational cross-disciplinarity. The participants bring with them formalised, discipline-based analytical thinking: they know the ideas, concepts and methods that are central to their disciplines and how to justify them. As they engage with others across boundaries, moving ideas between the contributing disciplines, their thinking becomes more integrative. The next stages involve synthesis and application: recognising the new patterns that emerge from the reconfigurations of ideas and concepts and developing new frameworks and practices to accommodate them.

There is some evidence of a set of general traits that can be applied to integrating and synthesising ideas regardless of expertise, e.g. fluency, flexibility and originality, things that are now called transferable skills. A pedagogy of connection is based on the premise that, in addition to any general traits, there are teaching and learning strategies that can be deployed to actively promote creative work across and between disciplines. Emphasis is placed on tools of connection, for example, comparison, association, analogy, metaphor, mapping and blending. Such tools build on long traditions of combinational cross-disciplinarity. Examples include Levi-Strauss's 'bricolage', the use of objects and tools immediately to hand to develop and assimilate ideas (Turkle 1996) and Koestler's (1964) 'bisociative thinking' for combining two previously separate associative streams to generate a new idea. Techniques have been developed to promote creativity (e.g. Sternberg and Lubart 1999), critical thinking (e.g. Bowell and Kemp 2002) and philosophical enquiry (e.g. Baggini and Fosl 2003). Many of these are intrinsically inter- and multidisciplinary. Bricolage has been re-purposed as a foundation for educational research (Kincheloe and Berry 2004).

One of the greatest challenges is finding ways to maintain dialogue between collectively agreed understandings about how things might be and the inertia arising from the tendency to revert to doing things the way they have always been done. There is also the challenge of misunderstanding and misconceptions that inevitably arise through differences in language and culture as ideas and practices move across boundaries. Most concise language is discipline specific and involves technical terminology and jargon. Boundaries between disciplines often generate considerable terminological difficulties. There are genuine language and associated conceptual difficulties arising from the fact that many constructs do not have an equivalence of meaning in different languages and/or cultures. Nor do some constructs transfer comfortably between one worldview and another. Misunderstandings may also arise because of a failure to recognize that something that works well in one context is not necessarily transferable to another. Generating discipline neutral language is difficult. Often, the language difficulties are overcome by finding unifying metaphors (Dillon et al. 2014a).

Leibniz looked for a transdisciplinary unity in knowledge, a theme taken up more recently by Capra, a systems theorist, Wilson, a socio-biologist, and, more controversially, by Sheldrake, a physiologist. For Capra (2002) it is the network that integrates the biological, cognitive and social dimensions of life, for Wilson (1998) "consilience" comes from the fundamental principles underlying the unity of all knowledge; and for Sheldrake (1988) it is the action of memory at a distance in both space and time.

Education that is both interdisciplinary and multidisciplinary offers routes to exploring matters such as: the structure of power; customary and alternative ways of taking decisions, settling disputes and allocating resources; the nature and basis of duties, responsibilities, and rights and the roles of custom and law in prescribing them; fairness, justice and moral responsibility; how communities reconcile the needs of individuals with the structure of society and responsibility towards the environment; and respect for different ways of life, beliefs, opinions and ideas. Such matters are at the heart of education for sustainable development. Whereas there are many guiding principles and theoretical and methodological perspectives on inter- and multidisciplinarity, there is no universal formula for them. Each community of practitioners and educators must build on the expertise available to it and set its own agenda relative to the challenges it faces.

7.4 Postscript

In the preamble, I explained how my interest in cultural ecology started. It seems appropriate to end with a short note about where this thinking has taken me and how this relates to education for sustainable development. My point of reference is the landscape and the ecology of central southern England which so influenced my love of the countryside as a child. People reading this will have very different personal histories, different points of reference, different places that are special to them, different priorities. What we all have in common is the experience of change. Much of the wildlife that I valued as a child has declined: some species have vanished; others are much reduced in status. New species have arrived and some have been reintroduced. What one grows up with is not necessarily the norm. One cannot take the landscape as a set of habitats with discrete plant and animal populations isolated from the human enterprises associated with them.

Better to think of them as cultural ecological environments where resources, human enterprise, ecologies and landscape co-construct each other in constantly shifting mosaics shaped by complex physical and socio-economic factors. In my part of the world, consumption of the countryside as a 'leisure resource' is now as significant as farming. Human mobility has opened up the environment in ways undreamed of 50 years ago, with all the associated costs and benefits. People's beliefs, ideas and aspirations can be fickle: what is valued in one generation is not necessarily valued in the next. Behaviours thought unthinkable in one generation become common-place in the next. There are no absolutes. For those of us who are passionate about it, sustainable development is an ideal but in practical terms it an uneasy compro-mise between the desirable, the possible and the expedient.

References

Baggini, J., & Fosl, P. S. (2003). *The philosopher's toolkit. A compendium of philosophical concepts and methods*. Oxford: Blackwell.

Barrett, F. J. (1998). Creativity and improvisation in jazz and organisations: Implications for organisational learning. *Organisation Science, 9*(5), 605–622.

Bowell, T., & Kemp, G. (2002). *Critical thinking. A concise guide*. London: Routledge.

Capra, F. (2002). *The hidden connections*. London: HarperCollins.

Carson, R. (1962). *Silent spring*. New York: Houghton Mifflin.

Ceppi, G., & Zini, M. (1998). *Children, spaces, relations: Metaproject for an environment for young children*. Reggio Emilia: Commune di Reggio Emilia and Ministero della Pubblica Istruzion.

Dillon, P. (2006). Creativity, integrativism and a pedagogy of connection. *International Journal Thinking Skills Creativity, 1*(2), 69–83.

Dillon, P. (2008a). Creativity, wisdom and trusteeship—niches of cultural production. In A. Craft, H. Gardner, & G. Claxton (Eds.), *Creativity and wisdom in education* (pp. 105–118). Thousand Oaks: Corwin Press.

Dillon, P. (2008b). A pedagogy of connection and boundary crossings. Methodological and epistemological transactions in working across and between disciplines. *Innovations in Teaching Education International, 45*(3), 255–262.

Dillon, P. (2012). Framing craft practice cultural ecologically: Tradition, change and emerging agendas. In M. Ferris (Ed.), *Making futures: The crafts as change-maker in sustainably aware cultures* (pp. 72–78). Plymouth: Plymouth College of Arts.

Dillon, P., & Loi, D. (2008). Adaptive educational environments: Theoretical developments and educational applications. *UNESCO Observatory Refereed E-Journal, 3*.

Dillon, P., Hacklin, S., Kantelinen, R., Kokko, S., Kröger, T., Simola, R., Valtonen, T., & Vesisenaho, M. (2014a). Developing a cross-disciplinary framework for collaborative research in multi- and intercultural education. In V. C. X. Wang (Ed.), *Encyclopedia of education and technology in a changing society* (pp. 630–643). Hershey: ICI Global

Dillon, P., Vesala, P., & Montero, C.S. (2014b). Young people's emotional engagement with their school grounds expressed through colour, symbol and lexical associations. A Finnish-British comparative study. *Children's Geographies*. DOI 10.1080/14733285.2014.894962.

Hoskins, W. G. (1955). *The making of the English landscape*. London: Hodder and Stoughton.

Kincheloe, J. L., & Berry, K. S. (2004). *Rigour and complexity in educational research. Conceptualizing the bricolage*. Maidenhead: Open University Press.

Koestler, A. (1964). *The act of creation*. London: Hutchinson & Co.

Loi, D., & Dillon, P. (2006). Adaptive educational environments as creative spaces. *Cambridge Journal of Education, 36*(3), 363–381.

Macfarlane, A. (2013). Information, knowledge and intelligence. *Philosophy Now, 98*, 18–20.

Moll, L. C., Tapia, J., & Whitmore, K. F. (1993). Living knowledge: The social distribution of cultural resources for thinking. In G. Salomon (Ed.), *Distributed cognitions: Psychological and educational consideration* (pp. 139–163). Cambridge: Cambridge University Press.

Sheldrake, R. (1988). *The presence of the past.* London: Collins.

Star, S. L., & Griesemer, J. R. (1989). Institutional ecology, translations and boundary objects. *Social Studies in Science, 19*, 387–420.

Sternberg, R. J., & Lubart, T. I. (1999). The concept of creativity: Prospects and paradigms. In R. J. Sternberg (Ed.), *Handbook of creativity* (pp. 3–15). London: Cambridge University Press.

Thomas, W. L. (Ed.). (1956). *Man's place in changing the face of the earth.* Chicago: The University of Chicago Press.

Turkle, S. (1996). *Life on the screen. Identity in the age of the Internet.* London: Weidenfeld & Nicolson.

Vesisenaho, M., & Dillon, P. (2013). Localizing and contextualizing information and communication technology in education: A cultural ecological framework. *Pedagogy Culture Society, 21*(2), 239–259.

Walker, S., & Creanor, L. (2005). Crossing complex boundaries: Transnational online education in European trade unions. *Journal of Computer Assisted Learning, 21*, 343–354.

Wilson, E. O. (1998). *Consilience: The unity of knowledge.* London: Little Brown.

Part II
Case Studies from Individual Countries or Regions

Chapter 8
Project Variety and Established Structures: Development and Actual Practice of ESD in Germany

Reiner Mathar

8.1 Environmental Education on the Way to Education for Sustainable Development

The discussions and the outcomes of the Brundlandt commission (WCED 1987) were highly noticed by environmental education (EE) practitioners in Germany, and the following Agenda 21 process reached the educational sector from the outset. In particular Chapter 36 of the report, on education and its contribution to realise sustainable development (SD), was the focus of discussions in Germany. In 1994, first studies on EE as a contribution to SD influenced the scene of environmental educators. In the school system, some states started eco-school networks with a wider focus on environment and future development. The federal government established a national committee on SD which included a wide section of stakeholders from across society. Starting in 1996, several case studies on how to realise education for sustainable development (ESD) were published. In 1997, the "Bund-Länder-Kommission für Bildungsplanung und Schulentwicklung" (a joint venture between the federal government and the German states to improve school development and research) published a baseline study which consisted of a cross-analysis of the case studies in the states (BLK 1997). Based on this 'Framework for ESD in Germany' several states established programmes on EE/ESD:

- Eco-Schools (six states in 1997),
- School development programmes with a focus on ESD (Hessen, North Rhine-Westphalia),
- Network of regional EE centres (including Bavaria, Lower-Saxony, Hessen, Thuringia and Saxony).

R. Mathar (✉)
Co-ordinating expert for Education for Sustainable Development,
Ministry of Education and Training, Hessen, Germany
e-mail: reiner.mathar@t-online.de

© Springer International Publishing Switzerland 2015   123
R. Jucker, R. Mathar (eds.), *Schooling for Sustainable Development in Europe*,
Schooling for Sustainable Development 6, DOI 10.1007/978-3-319-09549-3_8

All these activities influenced the discussion on the federal level about establishing a special school development programme with the main focus on ESD. This programme was designed by a research team at the Free University of Berlin, headed by Professor Gerd de Haan, one of the most influential figures in ESD in Germany.

In 1997, de Haan's team published a study on *Bildung für nachhaltige Entwicklung: Gutachten zum Schulentwicklungsprogramm* (*ESD: expertise on the school development programme*) (BLK 1999).

In July 1999, 15 of the 16 German states and the federal Ministry of Education signed off a joint programme on ESD for the years 1999–2004, with an investment of 12.5 million euros. The programme was called "BLK21", in reference to the Agenda 21 process from Rio 1992.

On the basis of this study, the programme was designed in three modules:

- Interdisciplinary knowledge (*interdisziplinäres Wissen*)
- Participatory learning (*Partizipatives Lernen*)
- Innovative structures (*Innovative Strukturen*).

As a result of other school development programmes in Germany the programme was clearly structured on the following framework:

- Defining the standards of qualification for involved schools
- Regional and state-wide networks as a basic structure of the programme
- Network coordinators in all states
- Cross-state cooperation and supporting structures
- Nationwide elements of the programme
- A nationwide steering committee.

More than 180 pilot schools were involved in the programme and they developed together:

- elements of curricula-based lessons;
- school plans and school programmes for ESD;
- the concept of 'Gestaltungskompetenz' (competencies to deal with SD issues);
- new elements of students' participation.

Teaching materials were developed and their implementation tested by the pilot schools. The schools identified themselves as laboratories, disseminating their experiences to other schools outside the programme even during the life of the programme.

Experience has shown that co-operation between social and natural science and language and arts subjects worked very well. The introduction of a spiral curriculum has proven to be very appropriate. It assigns SD issues in a compulsory form to each school year. Thus it is guaranteed that social, economic and ecological aspects were equally taught. One result of the first three programme years was that EE became the linking tool between the different thematic areas of SD.

8.2   Interdisciplinary Knowledge

Out of the module on interdisciplinary knowledge a recommendation was prepared by de Haan for a long-term integration of ESD into schools. This was made available to the Curriculum Commission of the Federal States (de Haan 2004). This recommendation has been integrated into the current definition of educational standards in the context of ESD for all subjects (as far as possible). Particular significance (an additional consequence of 'BLK21') was placed on the competencies for active involvement in future development, including:

- future-oriented thinking and knowledge about future scenarios and planning;
- ability for interdisciplinary work on solutions to problems, and innovation;
- systemic (connected and combined) thinking and planning competence;
- solidarity;
- ability to communicate and co-operate;
- ability to motivate oneself and others;
- ability to look critically at one's own culture and foreign cultures;

A considerable step forward was to assign these competencies to individual issues and subjects (de Haan 2004).

8.3   Participatory Learning

Within this module schools were developing new methods of participation for students and new ways of co-operating with partners outside the schools, and the following are examples of this:

- creation of energy consultancy by vocational schools for the local region;
- project on seawater desalination, in co-operation with Africa;
- pedagogical design of a house where water is experienced with all senses;
- research into fertilising methods (which are kind to water) conducted by high school students in co-operation with local farmers;
- opening of the school as a learning place for the neighbourhood;
- co-operation in the processes of the Local Agenda 21.

The experiences of the BLK21 schools demonstrate that the relevance of the schools within their social and cultural environment improves dramatically when the schools define their regional surroundings as a teaching subject and thus contribute to the creation of local knowledge.
Examples of this are:

- issue of brochures about local water supplies;
- portraits of the students' villages of origin;
- development of remediation concepts for regional waters (de Haan 2004).

8.4 Innovative Structures

In this module students endeavour to achieve a continuous integration of ESD into the school curriculum and the following examples illustrate this:

- system of energy management as a school project under the control of the students;
- eco audit: creation of a system of energy management in co-operation with external consultants;
- school as a learning area when integrating environment into the school profile.

The experiences of work within the programme demonstrate that the crucial aspect consists in the installation and integration of planning and steering groups. As part of middle management, these groups have been responsible for the planning, implementation and evaluation of the ESD projects in the schools. Equipped with the corresponding mandates of the school decision-making committees, they became the pillars for continuous integration.

The necessary paradigmatic changes cannot take place as a centrally steered conversion process through 'top-down' strategies. For SD, a crucial mental change is required enabling consideration for nature, standing up for social justice and a willingness to implement with a great richness of ideas the indispensable technical, economic and social innovations.

This richness of ideas is based on further competencies suggested by the German Programme of ESD. These are the competencies to co-create the future and to be actively involved in future society. Both are recognised as crucial criteria for quality ESD. They include the ability to change and preserve the future of the community in which one lives in the context of SD. The latest version of the Set of competencies was published at Transfer 21 2007)

At the end of the BLK21 programme period in 2004, the discussions about the results and the further development of ESD in Germany were combined with questions about desirable structures and practices for the UN-Decade on ESD (DESD). Finally, the federal government and the state ministries of education established a programme on mainstreaming the results of BLK21 which was re-named 'Transfer 21', emphasising that transfer was the key to mainstreaming. The programme guide formulates the main issues as:

- involving 10 % of the schools of Germany (of a total of around 43,000 schools);
- training of 100 multipliers for ESD;
- establishing local and regional structures to facilitate ESD in schools (Transfer 21 2005).

The multipliers training programme with more than 100 h of training and additional blended learning elements consists of ten modules:

- The basic concept of SD: what is sustainable, what is unsustainable?
- Methods and competencies of ESD
- ESD and the concept of interdisciplinary learning
- ESD orientated school programmes, support structures for schools
- ESD and students' participation

- ESD and sustainable students' enterprises
- Summer university for all participants
- ESD and quality assessment and quality development
- Consultancy and moderation competencies of multipliers
- Project management and ESD (Transfer 21 2005).

After the training, the group of around 100 people became one of the backbones of school development with a focus on ESD in some states. In the following years, Hessen, a state in the centre of Germany (around Frankfurt), established a network of regional centres of expertise to support ESD in schools as the main element of the DESD. The focus of the regional centres is on networking and consulting schools on their ESD development. As of 2014, 12 trained experts work there. Ten of those centres and coordination by the relevant state guarantee a quality standard for supporting schools in all parts of the states. By 2014 more than 10 % of all schools were regularly part of the network and had implemented ESD into their school plans and the everyday curriculum. Apart from providing on-going support, the regional centres have developed new elements of ESD in pilot phases, such as "School year for Sustainability" for primary schools or lessons on climate change— "Climate snack bar" and "Lifestyle for the climate"—for secondary and upper secondary level (Schule und Gesundheit Hessen 2014). The "School year for Sustainability" aims to implement ESD in the lessons of all children at grade 4, the last year of primary education in Germany. By working in this programme the students should get a basic understanding of the concept of SD and should be able to adopt elements of it to their own life. The programme contains of:

- In-service training for all teachers at the school.
- Ten to twelve extra activities for children of grade 4 run by partners from outside the school on topics such as climate change (no ice bear without ice), food and consumption (what does a climate friendly breakfast taste like?), waste and waste management, energy and resources, fair trade and mobility. All elements are directly connected to the everyday life of the children.
- Evaluation and implementation of the programme at the school for further years.

After a 1-year pilot phase and evaluation, the programme was spread over the whole state, starting with five more regional centres.

Another element of school support for ESD is the cooperation with partners from outside the school which can assist schools with their competencies in their special areas. To help schools find the best partners, the Ministry of Environment, the Ministry of Education and the Ministry of Social Affairs, which has the responsibility for elementary education, established a quality label for ESD: "Zertifizierter Bildungsträger BNE-Hessen" (certificated partner of ESD in Hessen) (Hessian Government 2012). To receive this certificate, organisations have to work on quality development for ESD in their organisation, using a self-evaluation instrument which includes the elements: concept of SD, ESD and the concept of competencies, ESD in curricula and standards of education.

In the future, the main focus of the collaboration between partners in local and regional educational settings will be the development of regional educational landscapes for ESD. Therefore the second element needed to move from project into long-term structure is the building of regional networks for ESD which include all existing regional partners in the areas of SD: social development, economic development, ecological development and cultural diversity. This means organising a process of exchange and cooperation between different stakeholders and educational authorities and educational NGOs on local and regional levels. In 2012 and 2013 a pilot phase on the way to a regional landscape of ESD was run in the eastern region of the state. Organised by the regional centre for ESD and more than 30 different partners, which brought together all aspects of SD, this resulted in building the regional network "Bildungsregion Nachhaltigkeit Ost-Hessen" (Educational landscape for sustainability in Eastern Hessen). Following an evaluation and the advice of the Free University of Berlin, the concept is now mainstreamed in the regions of five other regional centres of ESD and by 2015 it is anticipated that more than 50 % of the state will offer those networks for ESD to schools as well as to external partners.

8.5 Other Doorways to ESD—The Cross-Curricular Framework to Global Development Education in the Context of ESD

Apart from EE, which was very influential on the way ESD developed, education for global development or global responsibility is a long and successful tradition in Germany. As Agenda 21 tries to combine the main questions of environment and development, initially a parallel process to the development of EE took place in the 'world' of global educators. Following the proposal of the federal Ministry for Economic Cooperation, the Kultusministerkonferenz (KMK, Standing Conference of Ministries of Education of the German States) established a commission with the main aim to develop a cross-curricular framework for global development education (KMK 2007). This framework was developed by a high level commission of subject and didactic experts and contains a concept of competencies. The framework provides hints and examples of how to integrate ESD into a first set of subjects at lower secondary level: (science, geography, civic education, religion and ethics, economics) as well as some information for primary education and vocational training.

The framework (KMK 2007) is a conceptual paper for the development of syllabi and curricula, for designing lessons and extra-curricular activities as well as for setting and assessing requirements for specific subjects and learning areas. It offers inspiration for school profile and day school programme development, for cooperation with external partners and for teacher education. In addition, it offers concrete recommendations and suggestions for the interdisciplinary and cross-disciplinary organisation of instruction, and offers class-room materials to work out global development issues.

In recent years this framework has influenced the development of new curricula in most of the states in Germany and was introduced to curriculum commissions as a background paper. Most of these states have already integrated or are on the way to integrating the framework into state-based curricula, either as part of individual subjects and/or as part of cross-curricular standards of education at school.

Furthermore, the framework was adopted and used be a large number of NGOs in the field of development cooperation to link their offers, materials and educational practice to the needs of the schools, on the one hand, and, on the other hand, to the question of SD. In this way the framework has the function of bridging the gap between the educational offers of the NGOs and the situation and the possibilities at schools.

In 2011, the Ministry of International Cooperation made a second move to increase ESD at schools with a focus on development. Following its suggestion the KMK decided to develop a new version with contributions to all subjects of lower secondary education and this will be published in spring 2015. This second edition will contain proposals for all subjects, even art, music and sports and their contributions to ESD, illustrated by concrete examples of lessons for grades 8–10. Following discussions at UNESCO level, the whole institutional approach for ESD will also be part of the new edition.

Until 2014 these two concepts of ESD, with backgrounds from EE (Transfer 21) and Global Education (Cross-curricular framework), were discussed in Germany. It is now planned to present and publish a basic document which integrates both concepts at the final conference of the DESD in Germany towards the end of 2014. This document has been developed and discussed by the School Working Group of the Round Table for ESD in Germany.

8.6 UN Decade for ESD in Germany—Structure and Integration into Civil Society

Following a unanimous decision of the federal parliament in 2004 the German UNESCO Commission was tasked to develop a structure to implement the DESD in Germany. It was decided to:

- build a national committee;
- organise a round table to involve all stakeholders and civil society;
- develop a national action plan, with structures and areas of activities;
- develop a website for more public recognition of ESD in Germany (BNE-Portal 2014).

8.7 The National DESD Committee

The Committee represents federal and state ministries, the parliament, non-governmental organisations, the media, the private sector and the scientific community. It is chaired by the educational scientist Gerhard de Haan (Free

University of Berlin). The National Committee is supported by a secretariat in Bonn and an office in Berlin, set up with the financial assistance of the Federal Ministry of Education and Research.

8.8 The Round Table

Stakeholders in the field of SD from Germany meet once a year at a Round Table event to work on the implementation of the DESD in Germany. Representatives from federal, state and municipality levels are involved, as well as non-governmental organisations. Between sessions, the working groups set up by the Round Table contribute to the implementation of the DESD.

8.9 The Working Groups on ESD

The working groups develop concrete proposals to embed the guiding principles of SD into the different educational areas. Another important task of the working groups is to make the ESD activities in Germany visible. In this regard, the working groups support the German National Committee in selecting outstanding projects that are suitable as vehicles for and applications of ESD. The following working groups exist:

- Adult education and non-formal education
- Vocational education and training
- Biodiversity
- Elementary education
- School education
- Higher education
- ESD and community
- Media
- Cultural education
- Economics and consumption (UNESCO Germany 2005).

8.10 Strategy and Instruments to Implement the DESD in Germany

The DESD in Germany started with the main issue to raise the visibility of ESD in Germany. This was addressed by organising public activities in all sectors of education and for the wider public. From the outset the so-called "Dekade Projekte" (decade projects) were backbones of this awareness raising campaign. The idea offers the possibility to everyone and every organisation to develop a concrete project idea for ESD, integrating the three elements of SD (economics, ecology and social questions) and work on implementation in the specific area of education and/or interest.

The project team applies then for the nomination as an official DESD project. A nationwide jury meets twice a year. After its decision the awarded projects receive the award in a public ceremony combined with an official event in the educational and public sector, such as the *Didacta*, the leading fair for education in the German speaking countries (since 2006 a special day for ESD has been part of the 3 day fair). In this way the decade projects received public recognition and the project teams could meet educators and educational policy makers.

By 2014, more than 1,900 projects had been realised in the specific area of education and had been awarded the title "Dekade Projekt" (see http://www.bne-portal.de/index.php?id=13, accessed 21 April 2014). The following examples provide an insight into the wide range of topics and stakeholder involvement in Germany:

- "KarmaKonsum" is a young enterprise working in the fields of LOHAS (Life-styles of Health and Sustainability) and CSR (Corporate Social Responsibility) as a research office and consultant agency. Under the slogan "A new spirit in business" they promote the idea of a green and social responsible economy. As consultants, they advise businesses on how to develop a strategy for SD in their specific area. As part of their public engagement for SD "KarmaKonsum" orga-nises public events and film presentations for a new sustainable economy and sustainable lifestyles at the hot spot of economic development in Frankfurt/Main.
- The programme "Sustainability 2020" is a Masters programme which promotes the training of experts for SD: "Nachhaltigkeit in gesamtwirtschaftlichen Kreisläufen" (Questions of SD in macroeconomic cycles).
- "Virtuelle Akademie für Hochschulbildung für nachhaltige Entwicklung" (Virtual academy for higher education for SD) is a project of the "Deutschen Bundesstiftung Umwelt" (German foundation for the environment) at the University of Bremen and offers nationwide online courses for ESD and parallel workshops at different universities in Germany (VAN 2014).
- "Kreativwerkstatt für Kinder und Jugendliche": The young generation is to acquire the skills and competencies to deal with the challenges of the twenty-first century. Against this background the People's Bank Ruhr has launched an educational initiative in mid-2007. It stimulated and funded the development of a creative workshop for children and adolescents in Gelsenkirchen. Thanks to the construction of this creative workshop, ESD in Gelsenkirchen will be further cross-linked and expanded. Cooperation and joint projects with public and voluntary organisations in ESD are developed and carried out. These focus around topics such as encountering nature and natural sciences, media technol-ogy and artistic design. In addition, district-wide projects are promoted, such as a sustainable student company called "added value", the environmental diploma Gelsenkirchen, media editors "Spinxx" and the Gelsenkirchen dye gardens.

Following the focus on projects, the national committee and the round table developed a second phase of activities, called "Communities for ESD in Germany". This tries to attract local communities of all sizes to develop a strategy and practice for ESD within their community. By 2014, 21 cities and small villages had applied for the title and established their own working platforms. From big cities such as Frankfurt and Hamburg down to small villages such as Ahlheim and Hellenthal, all of them have worked on integrating ESD into their educational activities at all stages and in all areas of education. In 2014 the UNESCO National Committee published an overview of all these activities (UNESCO Germany 2014a).

As a third element beside decade projects and decade communities the decade "measures" are the ones that built the structure of the DESD. Those measures involved ministries in the states and larger national and international NGOs with the aim to establish lasting concepts and activities for ESD in Germany and beyond. Typical examples are Eco-School or ESD school networks, initiatives for curriculum implementation of ESD, international networks of experts in the field of ESD, international exchange programmes for students, networks of student enterprises for sustainable businesses, special ICT projects and structures for ESD, linking to international activities such as the Earth-Charta initiative. The main aim of these measures is to move from a single project to a long-term institutionalised structure for ESD and/or to integrate ESD into already existing structures such as the school system, the structures for adult education and vocational education and training (UNESCO Germany 2014b).

8.11 ESD in Germany Beyond 2015—First Discussions and Strategy

> One year before the end of the DESD (2005–2014) it has become clear that there has been a series of national and international successes, but that the necessary implementation of ESD in all areas of education still faces significant challenges. Together with the German Bundestag, the Standing Conference of the Ministers of Education and Cultural Affairs (KMK) and also the ESD stakeholders in Germany, the German National Committee is therefore pleading for follow-up activities to the current Decade under the aegis of the United Nations. The National Committee is entirely in agreement with its international partners in this respect, as demonstrated by the UNESCO's General Conference in November 2011 and the Rio+20 Conference in June 2012. (UNESCO Germany 2014b: 4)

The position paper quoted here, discussed and adopted by the German UNESCO Commission, has two main objectives: first, to serve as a common self-awareness and reference framework for ESD stakeholders in Germany, and second, it is intended to serve as a basis for positioning in the following years (UNESCO Germany 2014b: 5).

For the activities after 2014 the paper identifies some general objectives:

- "The strategic relevance of ESD for initiating and achieving SD needs to be made clear to political decision makers and the general public.

- The National Committee considers the structural implementation of ESD in all areas of formal and non-formal education to be a central task for all stakeholders. (...)
- In practice, the development and strengthening of regional educational landscapes which integrate ESD must be central" (UNESCO Germany 2014b: 9).

In general the discussion of how ESD should develop in Germany after 2015 has been mainly influenced by a broader discussion about education as the key to implementing SD in society in general and in the lifestyle of its citizens. In some regions of Germany the idea of ESD has already realised and other regions will follow. The main idea behind educational landscapes is to bring different stakeholders in local education into contact with each other, and develop links between the different topics and structures of education. In this way a local network of competencies in ESD will be developed for education at all stages. As an example, the state strategy for SD in the state of Hessen has developed a special initiative for ESD with four elements:

- regional networking for ESD and establishing regional landscapes for ESD
- School year for Sustainability for all grade 4 children in primary education with the main aim that these children have a basic idea of the concept of SD and are able to integrate elements of it into their own lifestyles.
- Renewable energy and renewable raw materials as a main element of an SD-road show for all schools in Hessen
- Climate change and lifestyle, a topic for upper secondary education (Hessen 2014).

To sum up, Germany will follow its experiences in ESD made in the period of the DESD and will sustain the structures and elements of ESD developed in the last 10 years. This opens up an extended field of activities including international cooperation (ESD expert-net 2014). The basic elements of ESD in the upcoming period should be this international exchange and a focus on different elements of ESD, the recognition of different perspectives on problems of SD and its challenges for education in different parts of the world, as well as the realisation of the importance of social and cultural diversity.

References

BLK. (1999). *Bildung für nachhaltige Entwicklung: Gutachten zum Schulentwicklungsprogramm* (Studien Heft, Vol. 72). Bonn: BLK.

BLK (Bund-Länder-Kommission für Bildungsplanung und Forschungsförderung/German commission for educational planing and research support). (1997). *Bildung für nachhaltige Entwicklung* (Studien Heft, Vol. 69). Bonn: BLK.

BNE-Portal. (2014). *Bildung für nachhaltige Entwicklung. Weltdekade der Vereinten Nationen 2005–2014.* http://www.bne-portal.de. Accessed 21 Apr 2014.

De Haan, G. (2004). *Box 21: Products, outcomes and sources for ESD from the programme 21.* Berlin: Freie Universität Berlin (in German).

ESD expert-net. (2014). *Education for sustainable development—A shared vision and common approaches from Germany, India, Mexico, and South Africa*. http://esd-expert.net/. Accessed 21 Apr 2014.

Hessen. (2014). *Nachhaltigkeitsstrategie Hessen. Frankfurt: Hessisches Ministerium für Umwelt, Klimaschutz, Landwirtschaft und Verbraucherschutz*. http://hessen-nachhaltig.de/. Accessed 21 Apr 2014.

Hessian Government. (2012). *Das Zertifikat: Geprüfter Bildungsträger BNE-Hessen*. Wiesbaden: Ministry of Education, Ministry of Environment and Ministry of Social affairs.

KMK. (2007). *Global development education: A cross-curricular framework in the context of education for sustainable development*. Bonn: KMK.

Schule und Gesundheit Hessen. (2014). *Umweltbildung & Bildung für eine nachhaltige Entwicklung*. http://www.schuleundgesundheit.hessen.de/themen/umweltbildung-bildung-fuer-eine-nachhaltige-entwicklung.html. Accessed 21 Apr 2014.

Transfer 21. (2005). *Multiplikatorenprogramm* (Vol. 21). Berlin: Freie Universität Berlin. http://www.transfer-21.de/index.php?p=230. Accessed 21 Apr 2014.

Transfer 21. (2007). *Developing quality criteria at ESD-schools, quality areas, principles and criteria*. Berlin: Freie Universität Berlin.

UNESCO Germany. (2005). *National action plan for the UN-Decade for ESD*. Bonn: German UNESCO Commission. http://www.bne-portal.de/fileadmin/unesco/de/Downloads/Dekade_Publikationen_national/Nationaler_Aktionsplan_fuer_Deutschland_engl.pdf. Accessed 21 Apr 2014.

UNESCO Germany. (2014a). *Vom Projekt zur Struktur: Kommunen der UN-Dekade "Bildung für nachhaltige Entwicklung"*. Bonn: German UNESCO Commission. http://www.bne-portal.de/fileadmin/unesco/de/Downloads/Kommunen/Kommunen.pdf. Accessed 21 Apr 2014.

UNESCO Germany. (2014b). *The German national committee for sustainable development: position paper 'strategy for ESD 2015+'*. Bonn: German UNESCO Commission. http://www.bne-portal.de/fileadmin/unesco/de/Downloads/Nationalkomitee/ESD-Position-paper-2015plus_english.pdf. Accessed 21 Apr 2014.

VAN. (2014). *Virtuelle Akademie Nachhaltigkeit*. http://www.va-bne.de. Accessed 21 Apr 2014.

World Commission on Environment and Development (WCED). (1987). *Report of the World Commission on Environment and Development: Our common future*. http://www.un-documents.net/our-common-future.pdf. Accessed 21 Apr 2014.

Chapter 9
Education for Sustainable Development Between Main-Streaming and Systemic Change: Switzerland as a Case Study

Rolf Jucker and Florence Nuoffer

> *True learning can only be the leisurely practice of free people. (Ivan Illich, quoted in Esteva 2004: 20)*

9.1 Introduction

Education for Sustainable Development (ESD) in Switzerland is problematic and this is mirrored elsewhere. In general, there seems to be ample agreement that the overarching goals of ESD include, at least, the following:

- "**Reorient the curricula**: From pre-school to university, education must be rethought and reformed to be a vehicle of knowledge, thought patterns and values needed to build a sustainable world." (UNESCO 2009: 7)
- "ESD calls for new kinds of learning that are not so much of a transmissive nature (i.e. learning as reproduction) but rather of a transformative nature (i.e. learning as change)." (UNESCO 2009: 65)

This is very quickly said and written, but it is necessary to spell out what reorientation means, because it has important implications for what is needed should ESD really be mainstreamed and implemented in any country.

The views and opinions expressed in this chapter are the full responsibility of the authors and do not reflect any official position.

R. Jucker (✉)
SILVIVA, Foundation for Experiential Environmental Education, Zurich, Switzerland
e-mail: rolf.jucker@bluewin.ch; http://www.silviva.ch; http://rolfjucker.net

F. Nuoffer
éducation21, Swiss Foundation for Education for Sustainable Development,
Berne, Switzerland
e-mail: Florence.Nuoffer@education21.ch; http://www.education21.ch

© Springer International Publishing Switzerland 2015
R. Jucker, R. Mathar (eds.), *Schooling for Sustainable Development in Europe*,
Schooling for Sustainable Development 6, DOI 10.1007/978-3-319-09549-3_9

1. *The current education systems (re-)produce unsustainability.* There can be no doubt that our current social and economic systems, the property-based institutional regime in which they are embedded (Steppacher 2008), our production and consumption patterns, and our lifestyles are unsustainable and therefore not fit for the future (see Living Planet Report 2012; Millennium Ecosystem Assessment 2005; IPCC 2007; Zukunftsfähiges Deutschland 2008; UNCTAD 2013; UNRISD n.D.; ECOSOC 2012; ILO 2013; OECD 2011). It is equally indisputable that current education systems the world over and on all levels are an integral element of this unsustainable system. By integral we mean that they are crucial and constituent elements in continuously reconstructing our unsustainable industrial societies, their fundamental ideologies and lifestyles. "(...) learning embedded in educational systems derived from worldviews that 'sustain unsustainability' is a significant part of the problem" (Davis and Cooke 2007: 348), not part of the solution. Or in the words of UNESCO, when reflecting on the ten lost years between Rio 1992 and Johannesburg 2002: "Society must be deeply concerned that much of current education falls far short of what is required." (UNESCO 2002: 9; see also Bowers 2001; Jucker 2002: 228–252, O'Sullivan 1999). Malcolm Plant writes in this context: "A curriculum committed to social critique is clearly at odds with dominant political and social practices and consumer-led economic systems. Moreover, (...) prevailing educational systems shut out the most effective teaching strategies, that of personal experience and dynamic learning, and appear to prohibit the attainment of the outcomes of informed and politically active citizenry." (1998: 96–97) It is crucial to be aware of this when designing implementation strategies for ESD which should be more than mere good intentions.

2. *Today's best education is mostly destructive for the future of the planet.* This becomes blatantly obvious when one looks carefully at those who are primarily responsible for the unsustainability of the current situation. It is first and foremost those with the 'best' education any country can provide, fully capable of using the key skills identified by the OECD (2005), i.e. the political and business leaders all over the world (see Jucker and Martin 2008). But secondly, it is the world's middle classes who shatter the planet with their aspirations and consumer decisions. There is enough evidence that those who know enough about the negative consequences of their lifestyles continue to leave a disproportionately large ecological footprint on the planet (Zukunftsfähiges Deutschland 2008: 152). And the link between high income through (in conventional terms) good education (i.e. well-paid jobs) and high ecological footprint is well established (see Mackenzie et al. 2008; Living Planet Report 2010: Figure 33). David Orr has made the point succinctly: "It is worth noting that this [unsustainability of the current world] is not the work of ignorant people. Rather, it is largely the result of work by people with BAs, BScs, LLBs, MBAs, and PhDs." (Orr 1994: 7)

3. *Paradigm change needed.* The consequences of the above are obvious. Wherever you might look, commentators always talk in terms of radical change

which is needed to achieve a transition to sustainability (radical in the original Greek sense of "beginning at the roots"). The reports from the IPCC are a good case in point (see above), but it doesn't seem coincidental that

- Obama is talking about change in the context of sustainability: "Nor can we consume the world's resources without regard to effect. For the world has changed, and we must change with it" (Obama 2009),
- the most authoritative study on a sustainable Germany has as its subtitle "step change" (Zukunftsfähiges Deutschland 2008: see especially 305–454),
- in the field of technology Hawken and the Lovins's are advocating "the Next Industrial Revolution" (Hawken et al. 2000),
- in the context of Peak Oil, Campbell states: "Petroleum Man will be virtually extinct this Century, and Homo sapiens faces a major challenge in adapting to his loss." (Campbell 2008), and
- similar terminology can be found with reference to the banking crisis of 2008: "A banking crisis mustn't take the heat off climate change action. (...) And we need urgent action to stop environmental catastrophe" (Cameron 2008),
- as far as the economy is concerned, the Stern Review simply states "Climate change is the greatest market failure the world has ever seen" and recommends an "urgent global response" in the form of "strong and early action" (Stern 2006: viii, vi).

There is the much quoted but still apt saying of Albert Einstein that "we can't solve problems by using the same kind of thinking we used when we created them". This means that we cannot use the current dominant ideologies, world views and knowledge systems (including the education systems) to solve the problem of unsustainability. Indeed, the shift needed has to be akin to the dimensions of the Copernican revolution. Nebel and Wright make clear what happens in times of such paradigm shifts:

> Steeped in the old world view, not only did people ignore the new theory, but anyone who suggested that it had merits was vigorously attacked by the existing power structure, which was dominated by the Catholic Church and which had a vested interest in maintaining the old beliefs. (Nebel and Wright 2000: 11)

4. *Fundamental redesign of education*: According to Capra, this is the challenge we face: "The main task in the years to come will be to apply our ecological knowledge and systemic thinking to the fundamental redesign of our technologies and social institutions" (Capra 2000: 19). If Capra is right, then this also applies to the education system. We will have to redesign our education systems so that they become capable of contributing to (re)producing a sustainable society. In systems thinking terminology this means that we need "second-order change", systemic change, not "first-order change", tinkering at the edges of existing systems (Steen 2008: 228). Sterling has underlined the necessity of such a paradigm change:

> Sustainability does not simply require an 'add-on' to existing structures and curricula, but implies a change of fundamental epistemology in our culture and hence also in our educational thinking and practice. Seen in this light, sustainability is not just another

issue to be added to an overcrowded curriculum, but a gateway to a different view of
curriculum, of pedagogy, of organisational change, of policy and particularly of ethos.
(Sterling 2004: 50; see also Sterling 2001)

Orr is blunter about it, echoing Einstein: "The crisis [of education] cannot be solved
by the same kind of education that helped create the problems" (Orr 1992: 83).

5. *Systemic approaches based on ecological insights.* The direction of the mentioned paradigm shift has been indicated by Capra above: we need to foster
systemic thinking and compatible action on the basis of profound and substan-
tiated understanding that "there can be no long-term economic or social development on a depleted planet", as UNESCO has stated in an early draft version of
the *Framework for a Draft International Implementation Scheme* (UNESCO
2003: 7) (the sentence has been removed in subsequent versions, see UNESCO
2005).[1] Or as a key text for the German speaking discussion has phrased it:
"Only those models of prosperity will be capable of producing justice which
don't overstretch the biosphere. There will be no justice in the twenty-first
century without ecology." (Zukunftsfähiges Deutschland 2008: 89)

As Bowers (2009), Sterling (2009) and Shiva (2005) have argued, this has
further consequences: a systemic approach reveals that we are relational beings.
Systems need to be understood as interactive, co-evolving, dynamic wholes
where everything is in constant relationship, thus moving and interdependent.
And Bateson has shown that intelligence is embedded in the system as a whole,
not in any individual (2000). In other words, a worldview based on the notion of
the autonomous individual, consumerism as the measure of happiness, and
progress on the back of exploiting the world's resources is incompatible with
sustainability. Currently, we are simply thinking in concepts, mental models and
metaphors which are not suitable for sustainability. There is a need to develop an
ecologically informed intelligence that relies upon cultural languages that are free
of the misconceptions of earlier times when there was no awareness of environmental limits (Bowers 2008). Basing reforms (educational or otherwise) on a
systemic view of the ecological commons (i.e. cultural and natural commons
combined) becomes vitally important. We will only be able to reduce our ecological footprint if we can revitalise non-monetised, non-commodified communities
and natural resources. Educational approaches that enable learners to recognise that
the health of the cultural commons requires a healthy environmental commons will
also provide them with the language necessary for exercising the communicative
competence necessary for resisting corporate and political enclosure of what
belongs to nobody but all species, past, present and future—an enclosure which
is literally destroying the planet.

In this context the distinction between property and possession is very helpful
since it allows us to understand that a property-based institutional regime

---

[1] The sentence can still be found on UNESCO's ESD site under "Environment" (http://www.
unesco.org/en/esd/themes/environment/) or as part of UNESCO's online learning tool "Teaching
and Learning for a Sustainable Future", under "Key Themes in ESD" (http://www.unesco.org/
education/tlsf/extras/desd.html?panel=3#themes) (both accessed 29 March 2013).

(i.e. capitalism) "imposes exponential growth" in a finite material system (Steppacher 2008: 340). The impacts of such growth are aggravated, firstly, since it can only be based on a strategic exploitation of mineral resources (ibid.: 342–344), and secondly, because social and ecological understanding is subordinated under the logic of the market and economic rationality (ibid.: 335–340). This means that exponential growth is by necessity ecologically and socio-economically unsustainable. Possession-based institutional systems or societies, on the other hand, base their economies and their energy needs on "living [i.e. renewable] resources [which] do *not* allow for exponential growth" (ibid.: 341). This 'weakness' is, of course, the ultimate strength of such systems since "people using living resources whose economies are determined by funds rather than mineral stocks know that 'their' resources are finite; they know they must continuously adapt ends and means according to the availability and accessibility of such resources" (ibid.: 348).

6. *Accepting the fact that we live on a planet with physical and material limits.* Any ESD which does not...

- lead to a profound understanding that "the Earth ecosystem is finite, non-growing, materially closed, and while open to the flow of solar energy, that flow is also non-growing and finite" (Daly 1996: 49), *and*
- does not make sure that this understanding informs all of what we think and *do*,

... cannot in earnest be called an education which contributes to a sustainable future. In other words, without a functioning and resilient biosphere there would be no human life on Earth, let alone any elaborate social, political or economic systems. From a systems perspective this means that the biosphere is the basic system on which all other sub-systems (such as the economy) depend, including the relevant rules and limits to those systems. Paul Hawken clarifies the point: "We have failed to recognize that, just as in the lives of cells, the conditions of ecological systems are not established by human laws but by Nature's rules, rules which are non-negotiable and fundamentally rooted in the laws of physics." (Paul Hawken, quoted in Porritt 2000: 103) These relationships can be best visualised with the model of strong sustainability shown in Fig. 9.1.

If we pull the above together we arrive at an uncomfortable conclusion: ESD and a sustainable society are possible only with truly radical change—"rethinking" in the words of UNESCO—, yet all our endeavours to implement ESD are within an existing education system and a society which show no intention of engaging seriously in this paradigm change.

So the basic question for Switzerland and many other countries is: how do we work for change, fully knowing that most people neither want the proposed change nor understand the necessity for it?

Before making suggestions for solutions to this problem it is necessary to review the current situation of ESD in Switzerland.

Fig. 9.1 Strong
sustainability model (phase[2]
2013; SANZ Inc. 2009)

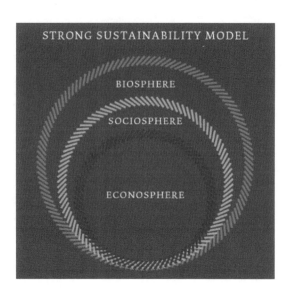

9.2 ESD in Switzerland: Where Are We?

The following is based on the Global ESD Indicators framework which has been developed for the UN Decade of ESD (GMEF Operational Plan UNESCO 2009: 19–23), as far as it concerns formal education.

ESD indicator	Fulfilled
Issue 1: policy, regulatory and operational measures that support ESD	
National coordinating body for the implementation of ESD	Yes, but...
Formal structures for interdepartmental government cooperation relevant to ESD at the national level government	Yes, but...

There is a coordinating body called the Swiss Coordinating Conference on ESD consisting of seven Government Departments and Agencies (environment, development, health, human rights, vocational training, research and development, spatial development) as well as the Swiss Conference of Cantonal Ministers of Education (EDK) (education is the responsibility of the federal states, called cantons, there is no national education ministry). It is tasked with coordinating the strategy and implementation of ESD on national and cantonal levels, in all areas of the education system. The problems with this body are manifold:

- There is no high level political mandate either from the Swiss Government or the Swiss Parliament for this body. In other words, there has never been a public or at least political discourse on the necessity of ESD in Switzerland, leading to a high level political legitimising of the process. This hampers implementation on all levels since there is no top-level political will, no firm legal ground, nor secure (i.e. legally guaranteed) financial commitment to ESD.
- As can be seen from the constitution of the body, the most important—in terms of where change is needed—Government Departments (finance and interior) are not even represented.

- This body is purely a civil servant affair. There is no representation on it from civil society, NGOs, the private sector and other important 'shadow educators' such as the media (as suggested by the UNECE ESD strategy for Europe (2005: §36, 7).
- In addition, the body is—despite its broad task—very much focussed on the compulsory school system (7–16 year olds and teacher training). Non-formal and informal learning are not represented at all.

The fact that there is a body—but not one in the spirit of the UN Decade or the UNECE-Strategy—is strategically clever for those who do not want to engage in serious second-order change: whenever somebody argues for the creation of a national coordinating body that deserves the name, they can block this with reference to the existing body. But the fact that the existing body is seriously limited in its legitimacy, mandate and ability to initiate real change means on balance that not much is happening which would move us into second gear. It is hard sometimes not to wonder whether this is intentional.

It is possible, though, that a development in 2012 might move us closer to a gear change: the three main government department which fund ESD activities (health, development, environment) and the above mentioned EDK (quasi education ministry) decided to found a new operational body called, in reference to Rio and the Agenda 21 process, éducation21. This independent foundation is tasked with implementing ESD in the school system. This move seems an important step forward, but since the new body started its work only in 2013, there is not much that can be said about its effectiveness before the end of the UN decade of ESD.

ESD indicator	Fulfilled
Issue 1: policy, regulatory and operational measures that support ESD	
ESD incorporated into national policy documents	(X)

Since 1998 sustainability is enshrined in articles 2 and 73 of the Swiss Constitution as one of the main objectives of the state. In 1997 Switzerland adopted its first strategy for sustainable development, which has been revised in 2002, 2008 and 2012. ESD is mentioned in each version:

2002 Version	2008 Version	2012 Version
(...) is it especially important for the Swiss Government to initiate decisive measures in order to embed sustainable development in compulsory schooling, at universities and in vocational training. (Strategie NE Schweiz 2002: 18)	This should also contribute to a comprehensive embedding and strengthening of the understanding of sustainable development in all areas and on all levels of education (in formal education including vocational education, as well as in non-formal and informal education, with a view to lifelong learning) and research. (Strategie NE Schweiz 2008: 34)	The [Swiss Government] has made education, research and innovation policy a priority, and has allocated a budget to this area that is increasing faster than the average. This measure should also help to root and strengthen an understanding of sustainability in all fields and at all levels of education, as well as in research. (Strategie NE Schweiz 2012–2015: 47)

Yet again, these fine words have to be read in context. The definition of sustainable development which underpins the strategy is based on the concept of weak sustainability, which starts from the assumption that all three areas of sustainability (economic, social and ecological dimensions) are of equal value and can be substituted one for the other. The Swiss Government admits that some environmental capital (*sic*) cannot be substituted (i.e. stable climate, biodiversity) and its loss would impact on the quality of life of future generations, but this does not stop it from advocating a growth strategy (Strategie NE Schweiz 2012–2015: 7)—even though it is now called "green growth" or "socio-economic growth" (ibid.: 31)—and the context of the educational proclamations above is the overall aim of (no, not sustainability), but to "strengthen the country's creativity and competitiveness" (Strategie NE Schweiz 2012–2015: 47). Given the fact that education is almost exclusively reduced to "knowledge" (ibid.: 47) and high status knowledge (e.g. the universities) is praised as the solution for sustainable development, it becomes apparent that no critical reflection has taken place to assess the underlying values and worldviews which inform such statements. They would have to be entirely rethought, were they to take into account what has been stated in points 1 and 2 above.

ESD indicator	Fulfilled
Issue 1: policy, regulatory and operational measures that support ESD	
Specific national ESD policy or strategy	Yes, but …

As indicated above there is no coherent national ESD policy which, on the one hand, covers all areas and levels of education, including informal or vocational education, or, on the other hand, brings together all necessary stakeholders.

But there is a document, prepared by the EDK (conference of cantonal education ministers), which was a reaction to the demand for a national action plan voiced at the beginning of the UN decade. It is called "Vorgehenspapier zur UNO-Dekade Bildung für Nachhaltige Entwicklung 2005–2014" (which can be roughly translated as "road map to the UN Decade"). The document is interesting in so far as it unconsciously reconstructs the tension between the need to reorient the education system (if the main documents of UNESCO and UNECE are taken seriously as claimed) and the explicit statement that reorientation is not on the cards for Switzerland and that only ESD that fits into existing educational structures and content will be implemented (Vorgehenspapier Plattform EDK-Bund 2005: 14, 16).

Two years later the above mentioned Government ministries and the EDK agreed a so-called action plan on ESD ("Massnahmenplan 2007–2014 Bildung für Nachhaltige Entwicklung") which came into effect in 2007 (Massnahmenplan 2007). On the one hand (apart from the limitations in reach beyond compulsory schooling and stakeholder engagement) it is quite good because it attempts a systemic approach at mainstreaming: integration into the new regional curricula, integration into teacher training, integration into quality development of schools on a whole school basis. On the other hand, the mentioned missing political and top-level support for ESD clearly shows: the integration into the regional curricula is limited by the fact that the existing disciplines would not have decreased weekly hours and ESD is not accepted as the overall new perspective with all the corollary measures of timetable allocation, compulsory curriculum elements and student testing. Instead it remains a 'nice to have' add-on, given over to those teachers who feel they want to do it or who happen to be teaching geography. The debate has not shifted from the level of "Yes, but there are so many other worthy things we could do" and "health education/vocational education/computer literacy is more important than environmental education/global education" [take your pick]. This clearly indicates two things:

- The necessary discussion has not happened on why we need ESD and why ESD makes no sense other than as the new guiding principle/perspective/paradigm which informs whatever we do in education.
- That this discussion is not happening has much to do with the missing equivalent discussion on sustainability in general, in terms of overarching guiding principle for all our societies. Since leading this discussion would pose many very unattractive questions (addressing our overdevelopment in the West, talking about limits and loss and giving up, about contraction not growth, about global justice and equity [see Selby 2007]), the evading strategy is to talk about structures, not content, about how health education relates to global or environmental education. This amounts to nothing but tinkering at the edges.

What we would need instead is a discussion about visions: where do we need to be in 2050 in order to be a sustainable society? What do people need to be able to do, think, feel? From this, you can gather the learning processes you need in order to achieve it. Accordingly you would reorient the existing systems. But for fear that ESD and SD really do mean (*paradigm*) *change* we remain entrenched in discussions about organisational structures or teaching materials.

ESD indicator	Fulfilled
Issue 1: policy, regulatory and operational measures that support ESD	
ESD incorporated into relevant legislation	No.

So far there is no legal basis for ESD, which—as indicated above—is a major stumbling block.

ESD indicator	Fulfilled
Issue 1: policy, regulatory and operational measures that support ESD	
public budgets and/or economic incentives available to specifically support ESD	Yes, but...

The above mentioned "Massnahmenplan" has been allocated up to 1.5 Million Swiss Francs in total for the period 2007–2014. Additionally, global education, environmental education and health education are funded on a national level to the tune of roughly 8.5 Million Swiss Francs a year (even though it has to be borne in mind that not all of this is ESD; much, if not most is spent in a sector specific way). This leaves us at a very optimistically calculated maximum of about 0.002 % of GDP.

ESD indicator	Fulfilled
Issue 2: measures taken to promote SD through formal education	
Include ESD in the curricula	Not really...

Up until now the situation has been very uneven in Switzerland. Since every canton can work to its own curriculum there has been a very large discrepancy between some where you could conceivably argue that ESD was taught under the guise of EE or global learning and others where there was no provision at all. In autumn 2012 the French speaking cantons introduced a new, common curriculum (called Plan d'Études Romand) which also claims to integrate ESD. The question whether ESD will figure in the new curricula in any meaningful way is still open, due to the following problems:

• ESD is integrated into something called "general studies" which is a transversal approach oriented essentially on the DeSeCo competencies (OECD 2005). On first sight this is wonderful, allowing a transdisciplinary approach. On second sight, it poses tremendous problems, because these general studies lack the same weight, time allocation and testing/evaluation regime as the traditional disciplines. In the eyes of the teachers, parents and pupils this means: not important.
• ESD is not conceptualised in any meaningful way as integrating the various sectorial approaches (e.g. EE, global education, citizenship education and health education). It is not visible as an overarching concept which reorientates the curricula under a new perspective, namely sustainability.
• Taking ESD seriously as an integrated learning process would mean, at least, allocation of time, space, money, teaching resources, and access to learning environments outside school. None of this is foreseen in a mandatory way for the new curriculum for French-speaking Switzerland.

Partly based on the experiences in the French-speaking part, there was a concerted effort not to replicate these problems in the German-speaking part. Since 2008 all 21 German-speaking cantons are in the process of elaborating for the very first time a common curriculum as well. In the context of the above mentioned Massnahmenplan (action plan) a project was launched with the intention of collaborating very closely with the curriculum project team right from the word go in order to integrate ESD in a systemic way already at the conceptual stage of the new curriculum. This did not quite work out because the person to implement this ESD project was only appointed 8 months after the start of the overall project, at which time the conceptual framework was nearly developed. But still: there was a project mandate to integrate ESD and to monitor and evaluate the integration. The mandate was supported by an expert group, providing ESD concepts, themes and learning approaches to the curriculum project team. In the consultation on the conceptual framework for the new curriculum which finished in Spring 2010 it became clear, though, that again ESD was not in any way the central concept around which the new curriculum would be organised. Similar problems as in the French-speaking part emerged:

- The lobbying for traditional disciplinary approaches is so strong, that it is not quite clear where in the timetable the necessary time and space for transdisciplinary ESD projects can be created.
- The approach to treat ESD as a cross-disciplinary theme but at the same time to demand that it is largely implemented within existing disciplines leads again to a marginalisation of ESD. The good intentions are declared but the structure of the timetable and the largely disciplinary orientation will make meaningful ESD either impossible or a voluntary fringe phenomenon, with marginally better implementation in certain disciplines such as Natur-Mensch-Mitwelt (nature-human-society).
- The necessity of a reorientation as stated above has completely vanished from the project. ESD is not the new lens through which we should view the entire curriculum, as suggested by Sterling above, but it is one of several cross-disciplinary themes which ideally dissolve into thin air through successful disciplinary 'treatment'.

This lack of revisioning is also apparent in vocational training. There has been some success in this area as well, in the sense that ESD is part of the general studies curriculum for all apprentices in Switzerland (about 65–70 % of the post-16 age group). But again, it is treated as an add-on, together with other things, not as a reframing perspective.

ESD indicator	Fulfilled
Issue 2: measures taken to promote SD through formal education	
Explicit learning outcomes which support SD in the curriculum	Not yet. . .
SD addressed explicitly in the curriculum	Not yet. . .

Again, this is planned for the new curricula in the German-speaking region, but it has not materialised yet and it is doubtful how specific, let alone effective such learning outcomes will be.

ESD indicator	Fulfilled
Issue 2: measures taken to promote SD through formal education	
Promote/address ESD through a whole-school or whole-institutional approach	Not yet. . .

With regard to this, the above mentioned action plan has also proposed a concrete project which was originally suggested by the Swiss Foundation for Environmental Education. The main aim is to embed ESD into a whole school approach through hijacking the school and quality development processes which are now mandatory in Switzerland, on the understanding that you cannot meaningfully talk about quality management of a school outside an SD and ESD framework.

This project could conceivably have the largest impact on the ground and there are very positive experiences in this direction within the healthy schools network in Switzerland (currently roughly 1,800 out of a total of 6,141 schools are participating) and within a pilot scheme for eco-schools in the canton of Zurich, originally introduced by the Swiss Foundation for Environmental Education. The national project has now a good chance of getting off the ground, as it is a main priority within the new ESD organisation éducation21.

This is potentially very good news, given the fact that effective ESD seems only possible in a communal learning process, based on real world experiences, which could (and elsewhere does) very easily happen in schools.

ESD indicator	Fulfilled
Issue 3: measures taken to equip educators with the competencies to include ESD in their teaching	
ESD as part of the initial educators training	Not yet. . .
ESD as part of educators' professional development	Not yet. . .

Measure 2 of the action plan is supposed to take care of this and the recommendations from this project are now available. How far these recommendations will go towards a systemic integration is not yet known. The project also showed that at present there is a wide spectrum of provision. Some teacher training institutions do very little, others offer compulsory modules on ESD. As far as CPD is concerned, the same situation applies: there are here and there very interesting courses on offer, but since ESD is not part of the curriculum nor demanded by schools, they have a very low priority for teachers and therefore more often than not have to be cancelled.

Viewed more systemically, it holds that there is not a single teacher training institution as yet which has remodelled its initial teacher training or its continued professional development programme within an ESD perspective.

ESD indicator	Fulfilled
Issue 3: measures taken to equip educators with the competencies to include ESD in their teaching	
ESD as part of the training of leaders and administrators of educational institutions	No.

This discussion hasn't reached Switzerland yet. Partly due to the very widespread notion that ESD is something we have to inflict on the next generation (so that they do a better job than us and mop up the mess we created) the understanding has not developed that ESD is first and foremost a necessity for the decision makers of today, the top and middle management positions which shape in large measure institutions like teacher training colleges. ESD leadership training and capacity building for these people is one of the most pressing needs in Switzerland (and elsewhere, as Jucker and Martin (2008) have argued repeatedly).

ESD indicator	Fulfilled
Issue 3: measures taken to equip educators with the competencies to include ESD in their teaching	
National networks/associations of educators involved in ESD	Yes. . .

As explained above, there is no high-level, formally established and mandated network or commission on ESD. Yet there is a considerable number of networks and fora which are involved in ESD:

- The Swiss Foundation for Environmental Education and the Swiss Foundation for Education and Development (merged in January 2013 to form éducation21) have, since 2003, organised a national ESD exchange forum three times a year. This has been open to anybody and has established itself as the main, informal meeting place of ESD stakeholders, mostly, but not exclusively working within the formal education sector.
- Up until the end of 2012, the Swiss Foundation for Environmental Education also facilitated various other networks, initially conceived as EE networks, but in reality mostly concerned with ESD: an initial teacher trainers' network, meeting twice a year in both the French- and German-speaking regions, a network for tutors concerned with teachers' continuous professional development, and a network of EE and ESD professionals, working in NGOs, cantonal education ministries etc.

Our experience shows that this networking part is one of the most important elements of developing a strong community of ESD practitioners because it relies on regular face-to-face contact, mutual support, exchange of up-to-date information and sharing of practical examples. Yet at the same time, there is an inherent limitation in all the networks mentioned above: none of their members are officially mandated (and paid!) as members of the networks. They come out of personal (or institutional) interest, because they are highly motivated to make ESD happen. These should not be belittled since organic, self-motivated networks like the ones mentioned will always be necessary because they are intrinsically motivated by the subject matter and therefore have a far higher ethical commitment. But they should be complemented by officially mandated networks if main-streaming is to occur because ESD cannot be thoroughly embedded within say teacher training institutions if those ESD champions are not visible in the organisational chart, on the pay lists, as members of decision-making bodies in the institution etc.

None of these indicators can tell us anything of necessary depth about the long-term success (or not) of the learning processes they hopefully at least capture. We will dwell a little more on this question in the conclusion, but we would like to shift the focus here and present two concrete projects which embody successful, effective ESD (yet in both cases the projects have never labelled themselves as such).

9.2.1 Solar Youth Project, Greenpeace

(http://www.greenpeace.org/switzerland/de/Themen/Jugendsolar/, accessed 5 Aug 2014)

The project has been running since 1998 and has tried to refocus ESD under the slogan "deeds instead of words". More than 10,000 young people have participated in the installation of more than 240 photovoltaic panels on public buildings, schools and farms, including the biggest one ever installed in Switzerland.

This active approach has positive consequences on three levels:

1. Education happens through experiential learning in the real world, through cooperation with farmers, electricians, photovoltaic specialists etc.
2. This learning produces concrete results in terms of installed and working PV systems and, through this, CO_2 savings. It also provides young people with direct experience that solar generation works.
3. Young people are actively involved in the planning and running of the project.

Solar energy is clearly key to a sustainable future. It can serve as an entry point to the whole question of energy provision in the future. The project approaches this topic in a way that:

- is positive, productive and future proof;
- focuses on real experiences;
- shows that experimenting with solar energy is fun;
- motivates young people because they experience how a group of people can actively change energy dependence into energy independence;
- has direct transfer relevance for their future daily life.

9.2.2   Young People Restoring Mountain Forests

(http://bergwald.ch/, accessed 5 Aug 2014)

This is a fascinating educational endeavour which works on various levels:

Economic
- Secondary school children/apprentices work for one week in the Swiss Alps
- They perform necessary maintenance work in forests which protect villages from avalanches. This is necessary work which Swiss society is less and less willing to fund. But it is real work, not invented for educational purposes. Pupils realise this instantly and feel valued accordingly.

Social
- Because the work is demanding and often dangerous pupils learn very quickly that it is of vital importance that everybody works together and is reliable in what s/he is doing.

Ecological
- Most pupils, having been brought up in an urban environment, arrive in the forest with preconceived ideas about foresters "killing trees". During the week, they experience first hand the interdependence between functioning ecosystems and humans.

Values
- During the week pupils learn and experience a number of values which are not easily accessible in contemporary consumer society any more. Most pupils start the week with reservations: often there is a daily walk of an hour or more into the forest where they work, there is a lot of sweat, a lot of hard physical work to which most of them are totally unaccustomed and which they usually detest. By the end of the week, their comments often include statements like: "This is the first time in my life that I have done something really useful"; "Hard work is worthwhile if it is of real value to others"; "The trees I selected will be here in 100 years' time, long after I'll have died"; "I much better understand the interdependence of forest and human beings"; "The class is totally different now: we have learned how dependent we are on each other and this has positively changed the class morale".

9.3 Conclusion: Strategies for Change

So what can we do in the face of the insight that reform of the current system will only prolong the agony, but not lead to the necessary change? To my mind, we need to be cunning, as the great German poet Bertolt Brecht has always said: we need to work inside the system to loosen its grip on power and mental models, so that change becomes possible. But we also need to work outside of it: A good strategy is to heed Gandhi's advice to be the change you want to see in the world (Gandhi 1999). Another is to provide oneself and others with the necessary learning opportunities, knowledge, skills, values and action competencies to put Gandhi's words into practice. This means to enable every one of us to engage in learning processes which *are* actually part of the required change.

To end, we would like to suggest some approaches which might help overcome the kind of deadlock we have experienced in Switzerland. At the 2009 World Conference on ESD in Bonn it became quite obvious that the two things we are still lacking are a decidedly holistic, systemic approach and the move of ESD from the margins to the centre, by which we don't mean—as should be clear by now—that ESD is embedded in the existing system, but that ESD *creates a new, different* system.

9.3.1 A Holistic, Systemic Approach to ESD

As far as the first issue is concerned we have come to the conclusion (which is reflected above) that enabling systemic thinking and matching practices is paramount. It is infuriating to follow the same old discussions over and over again, if ESD should be ESC (Education for Sustainable Consumption) or ECC (Education for Climate Change) or if we better stick with EE (Environmental Education) or EGC (Education for Global Citizenship). All of these are sectorial, subject-specific and reductionist approaches which miss the whole point of ESD. They are more often than not dictated by the relative (financial) powers of the relevant Government ministries supporting one or the other. But fact is that no approach which starts from the complexity of real world problems or any meaningful conceptualisation of what education should be about—as defined in the Implementation Scheme of the UN Decade of ESD: "ESD prepares people of all walks of life to plan for, cope with, and find solutions for issues that threaten the sustainability of our planet." (UNESCO 2005: Annex I, 3)—would ever end up with such a plethora of adjectival educations and a claim that schools should implement all of them individually.

So what we really need is a new push for an integrated, transdisciplinary, holistic concept of education (which is, in fact, ESD). We need to un-learn the way we thought about education for the last 250 years and we need a new social movement which pushes for a similar feat of social engineering as was the introduction of compulsory schooling achieved in the context of the Industrial Revolution. Only this time, this re-orienting of society should not serve the powerful political and economic elites as it did then, but its underlying values ought to serve the resilience of the biosphere and a just and equitable human society on Earth.

9.3.2 ESD as the New Mainstream

This leads us to the second point, and it is one where we are not sure we have many promising tools to bring about the desired outcomes. The 2009 gathering in Bonn to celebrate the first half of the UN Decade of ESD has underlined what ESD practitioners all knew if they dared to admit it. ESD is *not* at the centre of education, it and sustainability are *not* at the top of political agendas worldwide (as every G20 or post-Kyoto protocol Climate Change Summit goes to show). Bonn was a gathering of mostly like-minded people who lured—as at numerous conferences in the past—themselves into a false self-congratulatory mood. Education is and always was a tool of society to reproduce itself. So, if society as a whole and the economy as the main political driver in particular, are not performing the paradigm shift to sustainability, education and even more so ESD will only ever tinker at the edges. This insight again calls for a systemic view. Of course, we as ESD practitioners should continue to push for all those changes in education systems which we rightly deem necessary (see again World Conference on ESD 2009). But doing only this is like claiming that Education for Climate Change equals ESD. Unless all of us turn into competent change-agents on all levels of our interaction with society, economy and the biosphere, the required paradigm shift will not come about. In ESD we have to become far more switched-on to complex, multi-stakeholder, multi-level, multi-method approaches to tackle the issues at stake. As Ernst has aptly stated with regard to changes in behaviour relevant to the environment:

> Environmental behaviour takes place in a technological, social and institutional context, whose factual constraints and inertia largely determine our behaviour. This context is currently not sustainable. A shift in our behaviour therefore has to be accompanied by a change in the technological, social and institutional incentives and infrastructures; otherwise it is bound to fail. (Ernst 2008: 57–58, our translation)

Thinking outside the box should not just be on the list of competencies for students or pupils to acquire. All of us should practice it ourselves.

9.3.3   Spheres of Influence and Leverage Points

To do this effectively, we have to bear in mind both where our spheres of influence are and how successful change processes happen (see Jucker et al. 2008: 60–62). We can all contribute our share to the above mentioned new social movement, but within a sustainable society small is beautiful and regional (see Jucker 2002: 40–41). This means that the most effective and long-lasting changes will happen within our communities where both our sphere of influence and possibility for real change is larger than on the national or international level, and more effective than on the individual level. As the Transition Town movement has in places shown (http://www.transitiontowns.org) and as Bowers (2001) and (Shiva et al. 1997; Shiva 2005) have argued convincingly, we need to revitalise our communities, especially the cultural and natural commons, if we are to achieve the two things we need for a sustainable society: resilient, self-reliant communities and sustained communal learning.

9.3.4   Revitalising the Cultural Commons

Revitalising the cultural commons is important because in a sustainable society we won't be able to rely on fossil fuel energy slaves (as Hermann Scheer has aptly called it) and the natural and labour resources of Third World peoples. Also, as Steppacher has emphasised, "in most cases, commons are possession rather than property regimes" (2008: 349) which opens up the potential for ecological and socio-economic sustainability in the first place (see above). This means that we need to focus less on consumerism and more on non-monetised, non-commodified activities in face-to-face contact with others. Becoming aware of the intergenerational knowledge and skills that are at the core of the local cultural commons will be crucial to enable us to critically assess which traditions contribute to the development of skills, values and lifestyles that have a smaller ecological footprint, while also contributing to a morally coherent community. In other words, whatever we do and learn needs to be framed by sustainability principles (see Jucker 2002: 31–42). These are communal learning processes where all involved are learners and where all need to exercise ecological intelligence (Bowers 2009; Sterling 2009).

For those of us living in Euro-American societies, this is possibly the biggest and most difficult challenge, as we need to balance conserving our civil liberties while moving beyond the idea of the autonomous individual. The ecological crisis confronts us with a fundamental choice: we either engage in a cooperative learning journey which results in "whole system learning" or we will invariably fail the challenge of sustainability.

9.3.5 Initiating Communal Learning

In other words, the notion that learning could take place in separate entities called schools, removed from the hands-on, practical endeavour of creating our sustainable communities in a specific place in real-time, is quite absurd. Indigenous communities often realise this when they come into contact with the Euro-American reality of schooling and education. They find that it alienates their children from real-world networks of community support and intergenerational knowledge and skills (such as growing food), from knowing their place, from linking into systemic traditional wisdom and knowledge, and from respect towards the Earth. Instead the Western approach to schooling fills their minds with abstract knowledge, bearing no relation to their specific communities (see Bowers/Apffel-Marglin 2005; Norberg-Hodge 2000).

To avoid a similar form of alienation, we need to wean ourselves from our over-dependence and over-reliance on a virtual, print- and computer-based abstract world and become fully present in the real world with other people living and learning in mutually supportive ways. This, it seems to me, is the challenge: turning our communities into sustainable ones by initiating long-term communal learning processes in which all of us play an active part. As Scott has put it:

> The process that we call sustainable development makes no sense other than as a social learning process of improving the human condition that can be continued indefinitely without undermining itself. In this sense, sustainable development doesn't, instrumentally, depend on learning; rather it's inherently a learning process of making the emergent future ecologically sound and humanly habitable, as it emerges, through the continuous, responsive learning which is the human species' most characteristic endowment. (William Scott, in a contribution to the EAUC's SHED-SHARE email discussion list on 05.10.2009)

If we manage to grow these seeds all over our countries, we might get ESD and move towards sustainability in the end. Schools will then serve as spaces where students can ask questions about traditions that have contributed to overshooting the sustaining capacity of natural systems, about those aspects of the cultural commons that should be reformed or abandoned entirely, or learn about the characteristics of cultures that have learned to live within the limits of their bioregions.

Given the fact that on a very fundamental level we are faced with a contradiction—ESD is only possible with a radical paradigm change; according to all indicators such a radical change within society, economy and the education system seems currently impossible—we might need to heed the advice of an unknown graffiti artist in Zurich: "You have no chance; so use it!" Therefore all we can and should do is demonstrate to ourselves that our proposed change is indeed possible and desirable and to open up opportunities for others to live their sustainable change (not *any* change!) on as many levels of the education system and society as possible (see Jucker 2014).

References

Bateson, G. (2000). *Steps to an ecology of mind: Collected essays in anthropology, psychiatry, evolution, and epistemology to an ecology of mind*. Chicago: University of Chicago Press (originally published in 1972).

Bowers, C. A. (2001). *Educating for eco-justice and community*. Athens: The University of Georgia Press.

Bowers, C. A. (2008). *Toward a post-industrial consciousness: Understanding the linguistic basis of ecologically sustainable educational reforms*. https://scholarsbank.uoregon.edu/xmlui/bitstream/handle/1794/7411/Book%20on%20language.pdf?sequence=1. Accessed 29 Mar 2013.

Bowers, C. A. (2009). *Educating for ecological intelligence*. https://scholarsbank.uoregon.edu/xmlui/bitstream/handle/1794/9268/Book%20on%20E-Intell.pdf?sequence=1. Accessed 29 Mar 2013.

Bowers, C. A., & Apffel-Marglin, F. (Eds.). (2005). *Rethinking Freire. Globalization and the environmental crisis*. Mahwah: Lawrence Erlbaum Associates.

Cameron, J. (2008, October 16). A banking crisis mustn't take the heat off climate change action. *The Times*.

Campbell, C. J. (2008). *Understanding peak oil*. Uppsala: ASPO International. http://www.peakoil.net/about-peak-oil. Accessed 29 Mar 2013.

Capra, F. (2000). The challenge of our time. *Resurgence, 203*, 18–20.

Daly, H. E. (1996). Consumption: Value added, physical transformation, and welfare. In R. Costanza, O. Segura, & J. Martinez-Alier (Eds.), *Getting down to earth: Practical applications of ecological economics* (International Society for Ecological Economics), (pp. 49–59). Washington, DC: Island Press.

Davis, J. M., & Cooke, S. M. (2007). Educating for a healthy, sustainable world: An argument for integrating Health Promoting Schools and Sustainable Schools. *Health Promotion International, 22*(4), 346–353.

ECOSOC (United Nations Economic and Social Council). (2012). *Dialogues at the Economic and Social Council 2012*. http://www.un.org/en/ecosoc/docs/pdfs/dialogues_at_ecosoc_2012.pdf. Accessed 6 Feb 2014.

Ernst, A. (2008). Zwischen Risikowahrnehmung und Komplexität: Über die Schwierigkeiten und Möglichkeiten kompetenten Handelns im Umweltbereich. In I. Bormann & G. de Haan (Eds.), *Kompetenzen der Bildung für nachhaltige Entwicklung. Operationalisierung, Messung, Rahmenbedingungen, Befunde* (pp. 45–59). Wiesbaden: VS Verlag für Sozialwissenschaften.

Esteva, G. (2004). *Back from the future*. http://gustavoesteva.com/english_site/back_from_the_future.htm. Accessed 29 Mar 2013.

Gandhi, M. (1999). *The collected works of Mahatma Gandhi*. New Delhi: Publications Division, Ministry of Information & Broadcasting, Government of India (100 Vol. CD-Rom version).

GMEF Operational Plan. (2009). *UN DESD Global Monitoring and Evaluation Framework (GMEF): Operational plan*. Paris: UNESCO. http://portal.unesco.org/education/en/files/56743/12254714175GMEFoperationalfinal.pdf/GMEFoperationalfinal.pdf. Accessed 29 Mar 2013.

Hawken, P., Lovins, A. B., & Lovins, L. H. (2000). *Natural capitalism. The next industrial revolution*. London: Earthscan.

ILO (International Labour Organisation). (2013). *World of work report 2013*. http://www.ilo.org/wcmsp5/groups/public/—dgreports/—dcomm/documents/publication/wcms_214673.pdf. Accessed 6 Feb 2014.

International Panel on Climate Change (IPCC). (2007). *Fourth assessment report: Synthesis*. http://www.ipcc.ch/publications_and_data/publications_ipcc_fourth_assessment_report_synthesis_report.htm. Accessed 29 Mar 2013.

Jucker, R. (2002). *Our common illiteracy: Education as if the earth and people mattered*. Oxford/New York/Lang: Frankfurt/M. [=Environmental Education, Communication and Sustainability; Vol. 10]. Accessible at: http://books.google.de/books?id=whJ_AAAAMAAJ&source=gbs_navlinks_s. Accessed 29 Mar 2013.

Jucker, R. (2014). *Do we know what we are doing? Reflections on learning, knowledge, economics, community and sustainability*. Newcastle upon Tyne: Cambridge Scholars Publishing.

Jucker, R., & Martin, S. (2008). Educating earth-literate leaders. In B. Chalkley, M. Haigh, & D. Higgitt (Eds.), *Education for sustainable development. Papers in honour of the United Nations decade of education for sustainable development (2005–2014)* (pp. 13–23). London: Routledge.

Jucker, R., Martin, S., Martin, M., & Roberts, C. (2008). Education and sustainable development—Learning to last? In J. E. Larkley & V. B. Maynhard (Eds.), *Innovation in education* (pp. 51–92). Hauppauge: Nova.

Living Planet Report. (2010). http://assets.panda.org/downloads/lpr2010.pdf. Accessed 29 Mar 2013.

Living Planet Report. (2012). http://assets.wwf.ch/downloads/lpr_2012_as_printed.pdf. Accessed 29 Mar 2013.

Mackenzie, H., Messinger, H., & Smith, R. (2008). *Size matters*. Canada's ecological footprint, by income. http://www.policyalternatives.ca/sites/default/files/uploads/publications/National_Office_Pubs/2008/Size_Matters_Canadas_Ecological_Footprint_By_Income.pdf. Accessed 29 Mar 2013.

Massnahmenplan. (2007). *Massnahmenplan 2007–2014 Bildung für Nachhaltige Entwicklung*. Bern: EDK and SK BNE. http://edudoc.ch/getfile.py?docid=5727&name=massnahmenplan_BNE_d&format=pdf&version=1. Accessed 29 Mar 2013.

Millennium Ecosystem Assessment. (2005). *Overall synthesis report*. http://www.maweb.org/documents/document.356.aspx.pdf. Accessed 29 Mar 2013.

Nebel, B. J., & Wright, R. T. (2000). *Environmental science: The way the world works*. Upper Saddle River: Prentice Hall.

Norberg-Hodge, H. (2000). *Ancient futures: Learning from Ladakh*. London: Rider.

O'Sullivan, E. (1999). *Transformative learning—Educational vision for the 21st century*. London: Zed Books.

Obama, B. (2009). *Inaugural address*. http://www.whitehouse.gov/blog/inaugural-address/. Accessed 29 Mar 2013.

OECD (Organisation for Economic Co-operation and Development). (2005). *The definition and selection of key competencies.* Paris: OECD. http://www.oecd.org/pisa/35070367.pdf. Accessed 29 Mar 2013.

OECD. (2011). *An overview of growing income inequalities in OECD countries: Main findings.* http://www.oecd.org/social/soc/49499779.pdf. Accessed 6 Feb 2014.

Orr, D. W. (1992). *Ecological literacy: Education and the transition to a postmodern world.* Albany: State University of New York Press.

Orr, D. W. (1994). *Earth in mind: On education, environment, and the human prospect.* Washington, DC: Island Press.

phase². The strong sustainability think tank. (2013). *What is strong sustainability?* http://nz. phase2.org/what-is-strong-sustainability. Accessed 15 Feb 2013.

Plant, M. (1998). *Education for the environment. Stimulating practice.* Dereham: Peter Francis Publishers.

Porritt, J. (2000). *Playing safe: Science and the environment.* London: Thames & Hudson.

SANZ Inc. (2009). *Strong sustainability for New Zealand: Principles and scenarios.* Auckland: Nakedize.

Selby, D. (2007). As the heating happens: Education for sustainable development or education for sustainable contraction? *International Journal of Innovation and Sustainable Development, 2*(3/4), 249–267.

Shiva, V. (2005). *Earth democracy. Justice, sustainability, and peace.* Cambridge, MA: South End Press.

Shiva, V., Jafri, A. H., Bedi, G., & Holla-Bhar, R. (1997). *The enclosure and recovery of the commons: Biodiversity, indigenous knowledge and intellectual property rights.* New Delhi: Research Foundation for Science, Technology and Ecology.

Steen, S. (2008). Bastions of mechanism—Castles built on sand. A critique of schooling from an ecological perspective. In J. Gray-Donald & D. Selby (Eds.), *Green Frontiers: Environmental educators dancing away from mechanism* (pp. 228–240). Rotterdam: Sense Publishers.

Steppacher, R. (2008). Property, mineral resources and "sustainable development". In O. Steiger (Ed.), *Property economics—Property rights, creditor's money and the foundations of the economy* (pp. 323–354). Marburg: Metropolis.

Sterling, S. (2001). *Sustainable education—Re-visioning learning and change.* Darrington: Green Books [=Schumacher Briefing No. 6].

Sterling, S. (2004). Higher education, sustainability and the role of systemic learning. In P. Blaze Corcoran & A. E. J. Wals (Eds.), *Higher education and the challenge of sustainability: Problematics, promise, and practice* (pp. 49–70). New York: Springer.

Sterling, S. (2009). Ecological intelligence: Viewing the world relationally. In A. Stibbe (Ed.), *The handbook of sustainability literacy* (pp. 77–83). Dartington: Green Books. http://www.sustain ability-literacy.org/. Accessed 29 Mar 2013.

Stern, N. (2006). *Stern review: The economics of climate change. Summary of conclusions.* http://www.wwf.se/source.php/1169158/Stern%20Summary_of_Conclusions.pdf. Accessed 5 Aug 2014.

Strategie NE (Nachhaltige Entwicklung) Schweiz. (2002). *Bericht des Schweizerischen Bundesrates: Strategie Nachhaltige Entwicklung 2002.* Bern: Bundesamt für Raumentwicklung (ARE). http://www.are.admin.ch/dokumentation/publikationen/00014/ 00215/index.html?lang=de&download=NHzLpZeg7t,lnp6I0NTU04212Z6ln1acy4Zn4Z2q ZpnO2Yuq2Z6gpJCDen93fmym162epYbg2c_JjKbNoKSn6A–. Accessed 29 Mar 2013.

Strategie NE (Nachhaltige Entwicklung) Schweiz. (2008). *Schweizerischer Bundesrat: Strategie Nachhaltige Entwicklung: Leitlinien und Aktionsplan 2008–2011.* Bern: Bundesamt für Raumentwicklung (ARE). http://www.are.admin.ch/dokumentation/publikationen/00014/ 00271/index.html?lang=de&download=NHzLpZeg7t,lnp6I0NTU04212Z6ln1acy4Zn4Z2qZpn O2Yuq2Z6gpJCDfIR4gmym162epYbg2c_JjKbNoKSn6A–. Accessed 29 Mar 2013.

Strategie NE (Nachhaltige Entwicklung) Schweiz. (2012–2015). *Schweizerischer Bundesrat.* Bern: Bundesamt für Raumentwicklung (ARE). http://www.are.admin.ch/themen/nachhaltig/00262/

00528/index.html?lang=en&download=NHzLpZeg7t,lnp6I0NTU042l2Z6ln1ad1IZn4Z2qZpn O2Yuq2Z6gpJCEd319gGym162epYbg2c_JjKbNoKSn6A. Accessed 29 March 2013.

UNCTAD. (2013). *Trade and Development Report of the United Nations Conference on Trade and Development (UNCTAD)*. http://unctad.org/en/PublicationsLibrary/tdr2013_en.pdf. Accessed 5 Aug 2014.

UNECE. (2005). *UNECE strategy for education for sustainable development. Adopted at Vilnius, 17–18 March 2005*. Geneva: United Nations Economic Commission for Europe. http://www.unece.org/env/documents/2005/cep/ac.13/cep.ac.13.2005.3.rev.1.e.pdf. Accessed 29 Mar 2013.

UNESCO. (2002). *Education for sustainability. From Rio to Johannesburg: Lessons learnt from a decade of commitment*. Paris: UNESCO. http://unesdoc.unesco.org/images/0012/001271/127100e.pdf. Accessed 29 Mar 2013.

UNESCO. (2003). *United Nations Decade of Education for Sustainable Development (January 2005–December 2014): Framework for a Draft International Implementation Scheme*. Paris: UNESCO. http://portal.unesco.org/education/en/file_download.php/9a1f87e671e925e0df28d8d5bc71b85fJF+DESD+Framework3.doc. Accessed 29 Mar 2013.

UNESCO. (2005). *Draft International Implementation Scheme for the United Nations Decade of Education for Sustainable Development (2005–2014)*. Paris: UNESCO. Download: http://unesdoc.unesco.org/images/0014/001403/140372e.pdf. Accessed 29 Mar 2013.

UNESCO. (2009). *United Nations Decade of Education for Sustainable Development (DESD, 2005–2014) Review of contexts and structures for education for sustainable development 2009*. Paris: UNESCO (Author Arjen Wals).

UNRISD (United Nations Research Institute for Social Development). (no Date). *Post-2015 Development Agenda*. http://www.unrisd.org/80256B3C005BB128/%28httpProjects%29/38DF80F450689724C1257A7D004BD04B?OpenDocument. Accessed 6 Feb 2014.

Vorgehenspapier Plattform EDK-Bund. (2005). *UNO-Dekade 'Bildung für Nachhaltige Entwicklung' 2005–2014 : Vorgehenspapier der Plattform EDK-Bund/Schweizerische Konferenz der kantonalen Erziehungsdirektoren (EDK)*. Bern: EDK.

World Conference on ESD (31.03.–02.04.2009, Bonn). (2009). *Bonn declaration*. http://www.esd-world-conference-2009.org/fileadmin/download/ESD2009_BonnDeclaration080409.pdf. Accessed 29 Mar 2013.

Zukunftsfähiges Deutschland in einer globalisierten Welt. Ein Anstoss zur gesellschaftlichen Debatte. (2008). Frankfurt/M.: Fischer Taschenbuch Verlag.

Chapter 10
Education for Sustainable Development in Austria: Networking for Innovation

Franz Rauch and Günther Pfaffenwimmer

10.1 Introduction: Education and Sustainable Development

The lines of reasoning currently being pursued in Austria focus on the notion of sustainable development, on environmental education, on development education (or global learning or global citizenship education) and international peace as well as civic education which have sparked a debate on the nature of education in general (Rauch and Steiner 2006). As with human rights, sustainable development may be regarded as a regulatory idea (Kant 1787/1956). Such ideas do not determine how an object is made up but serve as heuristic structures for reflection. They give direction to research and learning processes. In terms of sustainability this implies that the contradictions, dilemmas and conflicting goals inherent in this vision need to be constantly re-negotiated in a process of discourse between participants in each and every concrete situation (Minsch 2004). This implies a great challenge but also has considerable potential to enhance innovative developments in education.

F. Rauch (✉)
Institute of Instructional and School Development, Alpen-Adria-University,
Klagenfurt, Austria
e-mail: Franz.Rauch@aau.at

G. Pfaffenwimmer
Sub-Department for Environmental Education, Austrian Federal Ministry of Education and
Women's Affairs, Vienna, Austria
e-mail: Guenther.pfaffenwimmer@bmbf.gv.at

© Springer International Publishing Switzerland 2015
R. Jucker, R. Mathar (eds.), *Schooling for Sustainable Development in Europe*,
Schooling for Sustainable Development 6, DOI 10.1007/978-3-319-09549-3_10

10.2   Theoretical Concept of Networks in Education

According to Castells' (2000) notion, networks constitute a new social morphology in society, where dominant functions and processes are increasingly organized around networks. New information technologies provide the material basis for its pervasive expansion throughout the entire social structure. Castells (2000) conceptualizes his notion of network as a highly dynamic, open system consisting of nodes and flows.

In the wake of these general social trends and structural transformations, networks have also become increasingly attractive in educational systems. In the 1990s, systemic school modernization processes were launched by policymakers, prompted by the need for reformatory change in the light of the results of international assessments (like the Trends in International Mathematics and Science Study (TIMSS) and the Programme for International Student Assessment (PISA) studies). Having proclaimed 'school autonomy', the central administration in Austria has been focusing more and more on strategic steering whilst delegating responsibilities to decentralised units (Posch and Altrichter 1993; Fullan 2007; Rauch and Scherz 2009). Less bureaucratic guidance generates a need for different coordination (Altrichter 2010). Intermediate structures (Czerwanski et al. 2002) such as networks are expected and conceived to fill a structural gap and take over functions traditionally assigned to hierarchy. Ideally, networks are conceived as an interface and an effective means of pooling competencies and resources (Posch 1995; OECD 2003). As intermediate structures they manage autonomy and interdependent structures and processes and try to explore new paths in learning and cooperation between individuals and institutions.

In order to understand the development of ESD in Austria as networks, social network theories might help. In this respect the authors consider the following aspects crucial:

- Mutual Intention and Goals (Liebermann and Wood 2003)
- Trust Orientation (McDonald and Klein 2003; McLaughlin et al. 2008)
- Voluntary Participation (Boos et al. 2000; McLaughlin et al. 2008)
- Principle of Exchange (Win-Win Relationship) (OECD 2003; McCormick et al. 2011)
- Steering Platform (Dobischat et al. 2006)
- Synergy (Schäffter 2006)
- Learning (Czerwanski et al. 2002; O'Hair and Veugelers 2005)

Per Dalin's (1999) description of how networks function in education is an important theoretical basis underlying the formation of regional ECOLOG networks.

According to Dalin, networks have an informative function, which becomes visible in a direct exchange of practice and knowledge for teaching and schools and as a bridge between practice and knowledge. Through networking members are encouraged to engage in further opportunities for learning and competence development (professionalization). Trust is a prerequisite for cooperation within a network. It is the basis for the psychological function of a network which encourages and strengthens individuals. In a fourth function of networks, the political function, enforceability of educational concerns increases, following the motto "together we achieve more" (Rauch 2013).

Table 10.1 gives an overview of the ESD-developments in Austria. In the following sections we describe these initiatives accompanied by critical appraisals.

10.3 The Austrian National Strategy for Education for Sustainable Development

The Austrian Strategy for Education for Sustainable Development was proposed by three Federal Ministries[1] and passed by the Austrian Council of Ministers on November 12, 2008. The strategy was developed in an open participative process in the years 2005 and 2006. It involved some 350–400 stakeholders from the formal, non-formal and informal educational sector in ten events. The whole process involved a number of methods:

- The first step was an expert paper which also included the results of the participative process (Heinrich et al. 2007).
- The participative process involved two major tools: On the one hand expert round tables on formal, non-formal and informal education were held. On the other hand five regional workshops with mixed stakeholder participation took place.
- As a fourth step a number of workshops were held with the working group on education of the Austrian committee for sustainable development. This committee was installed by the Federal Government and consisted of nominated representatives of all ministries and special interest groups, such as the chamber of commerce and chamber of labour.

The participation process provided information and contributions to the expert paper and thus to the draft strategy. The strategy comprises the following relevant elements:

- Mainstreaming within the education system
- Partnerships and networks
- Competence development among teachers
- Research and innovation
- Scenario development
- Monitoring and evaluation (Austrian Federal Ministry of Agriculture, Forestry, Environment and Water Management; Austrian Federal Ministry for Education, Arts and Culture and Federal Ministry of Science and Research 2008: 4).

---

[1] The Federal Ministry of Agriculture, Forestry, Environment and Water Management, Federal Ministry of Education and Women's Affairs as well as the Federal Ministry of Science, Research and Economy.

Table 10.1 Overview of the ESD-developments in Austria

	ESD-implementation and relevant legal developments	ECO school network (ECOLOG)	Teacher education	Higher education	International initiatives
1995		Start of ECO-school concept			ENSI decision on focus topics: ECO-schools, teacher education, (IT)-networking and quality assurance
1996		Start of ECO-school pilot phase (1996–1998)			
1997			UMILE-research project (Environmental Education in Teacher Education)		
1998					
1999		Concept for ECO-school network			
2000					
2001		Start of ECO-school network	UMILE-network		
2002	ESD platform in Ministry of Education (2002–2008)	National Environmental Performance Award for Schools and Teacher Training Universities			EU-SEED-network project (2002–2005)
2003					
2004			First National Teacher Training University Course "Innovation in Teacher Education—Education for Sustainable Development" (BINE)		EU CSCT-project (2004–2007)

Year					
2005	Signing of Vilnius declaration ESD Strategy process (2005–2007)			International conference "Committing Universities to Sustainable Development"	UNECE Vilnius declaration
2006	EU-ESD-Conference		Research project "Competences for Education of Sustainable Development" (KOM-BiNE) (2006–2008)		UNECE evaluation
2007	UNESCO Award			Sustainability award established	EU-SUPPORT network project (2007–2011)
2008	ESD strategy decision Austrian agency for education for sustainable development ("Dekadenbüro")		Second National Teacher Training University Course BINE		
2009					UNECE evaluation
2010		300 ECO-schools	ECO-school-network with Teacher training universities		
2011	Legislation on quality management in schools				EU-CoDeS-network project (2011–2014)
2012		400 ECO-schools	Third National Teacher Train-ing University Course BINE	Alliance of Sustainable Universities	
2013					

Appraisal: *The ESD strategy is based on a rather broad and systematic involvement of stakeholders. Nevertheless some of the participants felt under informed at some points due to time constraints. Another fact was that the participative and the political process developed somewhat independently from each other. The participative process came up with issues and aspects which could not be fully included in the final policy paper. Finally, it took until autumn 2008 to decide on the strategy due to two national elections in this period. Another problematic aspect was that the parliament was not involved formally in the process.*

A number of initiatives which started in the first half of the decade have continued, but rather on a constant or slowly growing level. This is also due to a lack of focused political interest in the ESD strategy and its implementation, a fact which could also be related to external developments, such as the financial crisis.

The Award of official Decade Projects by UNESCO Austria (Österreichische UNESCO-Kommission 2010, 2012) contributes to making the UN-Decade's aims and concerns more visible. It also supports local and individual ESD stakeholders and helps to better establish projects and safeguard their financing.

10.4 International Connections

Two international examples of Austrian activities deserve to be mentioned specifically:

UNECE-ESD (United Nations Economic Commission for Europe): Austria has been an active partner in the UNECE ESD process since the drafting of the UNECE ESD Declaration in 2004. It has signed the Vilnius declaration and organised its translation into German (as the first non-UN language version; www.unece.org/env/esd.html). According to the UNECE implementation scheme and work plan the Ministry of Education together with an expert group has organised the monitoring of ESD in Austria based on the UNECE ESD indicators in the years 2006 and 2008/2009 (and up-coming in 2014) and reported back nationally as well as also to UNECE (Rammel 2010).

Environment and School Initiatives (ENSI http://www.ensi.org) is an international network which, through research and international exchange of experiences, has supported educational and pedagogical developments, environmental understanding, insight into learning for sustainable development, active approaches to teaching and learning, as well as education for citizenship. Based on an initiative by the Austrian Federal Ministry of Education and Women's Affairs ENSI was implemented in 1985 to introduce environmental education (EE) into the educational programme of OECD/CERI (Organisation for Economic Co-operation and Development/Centre for Educational Research and Innovation) and was launched in 1986. It is still the only existing network on government level (since 2008 organised as a Not-for-Profit-Organisation (NPO) in the field of Environmental Education, ESD and school development.

In June 1995 the ENSI country coordinators decided to focus on: ECO-schools, teacher education, (IT)-networking and quality assurance. This decision has provided a guiding orientation on the development of ESD initiatives in Austria which will be described in detail below.

Another main international development based on ENSI has been a series of successful EU networking projects to which Austria has contributed and also has gained interesting impulses: SEED: School Development through Environmental Education (2002–2005) (http://www.ensi.org/Projects/Former_Projects/SEED/); CSCT: Curriculum, Sustainable development, Competences, Teacher Training (2004–2007) (http://www.ensi.org/Projects/Former_Projects/CSCT/); SUPPORT: Partnership and participation for a sustainable tomorrow (2007–2013) (http://support-edu.org/); CoDeS: Collaboration of schools and communities for sustainable development (2011–2014) (www.comenius-codes.eu).

Appraisal: *The international co-operation, esp., within ENSI, has proved an important resource for impulses, points of reference, a platform for mutual exchange and influence on policy decisions based on international expertise for ESD approaches in Austria, mainly in the Ministry of Education, but also with regard to research initiatives at the university level.*

10.5 The Austrian ECO-Schools Programme—Education for Sustainability (ECOLOG)

ECOLOG, a key action programme and network for the greening of schools and education for sustainability, was developed in 1996 by an Austrian team of teachers working on the international project Environment and School Initiatives (Posch 1999). It is a national support system with the aim of promoting and integrating an ecological approach into the development of individual schools and attempts are being made to embed the programme in Austria's federal states through regional networks. Overall coordination is ensured by the FORUM Umweltbildung (FORUM Environmental Education) which operates as a contractor with the Federal Ministry of Education and Women's Affairs and the Austrian Federal Ministry of Agriculture, Forestry, Environment and Water Management. In this setting the ECOLOG programme may itself have become sustainable and can be seen as an interface between environmental education and school development.

ECOLOG is based upon an ENSI approach: Schools—so called ESD-Schools—analyse the ecological, technical and social conditions of their environment and, on the basis of these results, define objectives, targets and/or concrete activities and quality criteria, to be implemented and evaluated. Students as well as all the other stakeholders of a school should be involved in a participatory way, and collaboration with authorities, businesses and other interested parties is encouraged. The measures concern, among others, areas like saving resources (e.g. energy and water), reduction of emissions (i.e. waste, traffic), spatial arrangement (from the classroom to the campus), the culture of learning (communication, organisational structure) and health promotion as well as opening of the school to the community. All in all over 400 schools with about 90.000 students are currently part of the network. Many others are reached through the web site, teacher in-service-training seminars and newsletters.

A special focus is placed on the (re)orientation of technical and vocational education and training in support of sustainable development and the transition to a green economy. Since 1992 a whole range of training activities on environmental, health and social aspects of sustainability have been offered in the Austrian vocational education and training system and curricula have been developed accordingly. Environmental education, health promotion, civic education and gender equality are integrated in all curricula as cross-curricular principles as well as in specific subject areas where appropriate. Special developments include: 104 (out of 690) vocational schools have joined the ECO-school network, 34 have reached the National Environmental Performance Award for Schools and University Colleges of Teacher Education, two upper secondary vocational schools have compiled sustainability reports where projects and activities are documented: Sustainable Development at the Upper Secondary Technical School (htl) Donaustadt, Vienna, Austria since 2005[2] and International Business College Hetzendorf since 2009/10[3].

10.5.1 How Are ECO-Schools Supported and What Are the Incentives?

In order to provide support the Federal Ministry of Education and Women's Affairs has organised networking structures:

- ECOLOG regional teams, having a coordinating function in the regions;
- ENSI-Teacher team, having an advisory and development function for the Ministry as well as for ECOLOG regional teams;
- Scientific advisory board, having an advisory function for the Ministry;
- Network of representatives of University Colleges of Teacher Education, having an advisory and development function, especially within their institutions.

---

[2] http://www.htl-donaustadt.at/info/umweltmanagement/nachhaltigkeitsbericht/. Accessed 7 Feb 2014; and www.unece.org/fileadmin/DAM/env/esd/GoodPractices/Submissions/Countries/Austria/SCPT/2009AusSchoolSustinability.pdf. Accessed 7 Feb 2014.

[3] www.ibc.ac.at: http://unternehmen.oekobusinessplan.wien.at/unternehmen/5217. Accessed 7 Feb 2014; and http://www.ibc.ac.at/website/index.php?id=206&tx_ttnews%5Btt_news%5D=606&cHash=51528cc4ad72a9dab21798bb6a350f1d. Accessed 7 Feb 2014

These groups have met twice a year centrally but operate within their regions. Most of the people involved know (about) each other and have collaborated on various occasions.

Central support is provided by the Ministry of Education and by the FORUM Umweltbildung. This comprises the central co-ordination of the regional support teams, including two meetings per year for exchange purposes, the maintenance of the web site www.oekolog.at, the publication of a monthly electronic newsletter and the provision of a manual on didactics and teaching methods, a file full of information and checklists, the organisation of events, a scheme for extra-curricular certification of student achievements and the financing of regional in-service training workshops (two per year and region). In total, the costs amount to Euro 93,000 per year.

For reporting a standard framework is provided and additionally some writing workshops are offered to assist teachers in writing readable and informative reports.

On a regional level support is provided by the ECOLOG regional teams. Their major task is to organise further education and training and—closely connected to that—to promote the exchange of experiences between schools in order to derive maximum benefit from the pool of competencies which accumulate at the schools. These teams are constituted by nominees from the regional school boards, the regional University Colleges of Teacher Education and a member of the ENSI-teacher team. In some provinces, the ECOLOG regional teams managed to establish co-operation with the Environment Departments of the provincial governments and with NGOs and were able to get some financial support for the ECOLOG network schools.

There are three additional support measures:

- Seminars for Heads and Coordinators to Enhance Quality of ESD/ECO Schools
    The intention of these seminars was to enhance innovative potentials at the schools. It was the aspect of ESD being "part of everyday school life" or part of the mission statement for the schools that led the Austrian ENSI Team (http://ensi. bmbf.gv.at) to think of innovative approaches. Additionally, the aim is the integration of the quality criteria as a framework for ESD within regular teacher education (Fritz et al. 2009; Lechner 2011; Lechner and Rauch in print).
- Education Support Fund for Health Education and Education for Sustainable Development
    Since 1992 the Fund for Health Education and Education for Sustainable Development has financed and promoted environmental education projects in schools, and since 1996 also health education projects. Schools and NGOs cooperating with schools are invited to submit projects, which are then evaluated. The fund is dedicated to smaller low-cost projects (€ 500 per school) which promote direct participation of students and hands-on learning opportunities.
- National Environmental Performance Award for Schools and University Colleges of Teacher Education
    This is a national, government-based award to acknowledge top level performance. Its criteria were put in force in January 2002. About half of the

120 criteria relate to Environmental Education (EE) and ESD, the school curriculum and school development. The other half refers to technical aspects, such as energy saving. The award is valid for 4 years, after that the compulsory external evaluation has to be renewed. By June 2013 98 schools received the award, 47 schools had renewed once, 22 schools had renewed twice. In 2014 up to 10 schools will have their third evaluation[4].

10.5.2 Evaluation of the ECOLOG Programme

Throughout the past 14 years of the ECOLOG Programme, a series of evaluations, inquiries and studies have been written: Thonhauser et al. (1998), Ehgartner (1999), Rauch and Schrittesser (2003) and Rauch and Dulle (2011) which are based upon interviews with teachers, head teachers, facilitators of schools as well as on observational data gathered at the schools and on the analysis of material produced by the schools. Payer et al. (2000), Schober-Schlatter (2002) and Knoll and Szalai (2009) used questionnaires, Heinrich and Mayr (2005) did a cross-case analysis of the reports of the regional networks.

The results can be summarised as follows:

- At the school level communication has proved to be a central element of ecologically oriented steering in order to produce a common understanding of ESD and the precondition for learning of all members of the school community.
- The head teachers play an important role with their 'official' support of the project, e.g. by putting it on the agenda of teachers' conferences, and by repeated statements of support in public. Additionally, heads enhance motivation by recognizing small steps (e.g. with photos or an information wall); maintaining contacts outside the school (e.g. public relations and use of media) and by providing incentives (e.g. by co-ordinating and negotiating financial support with the body responsible for maintaining and financing schools).
- The backing of the initiative by the Ministry is seen as a motivating factor. The homepage of the network (http://www.oekolog.at) is an important source of information. Regular in-service training workshops for teachers provide time and space for meetings, bringing people together face-to-face and giving them a sense of identity. It seems crucial to keep up funding available, to develop local advisory support further and to find structural links and cooperation to quality assessment and educational standards. Support should also include a revised political mandate, strengthened quality assurance (agreements on goals that are clear and achievable), early feedback on feasibility, quality monitoring, associated scientific support, and safeguarding the resource base (material and non-material incentives).

---

[4] See http://www.umweltzeichen.at/cms/home/bildung/schulen/content.html. Accessed 7 Feb 2014.

- The openness of the ECOLOG concept allows a wide range of issues and fosters creativity. Its impacts are seen in different areas, such as changes in teaching methods (e.g. more project work and social learning), the increased integration of health, ecological and social issues in lessons, the design of buildings (e.g. school yard, energy optimization of the school), or how school life is organised (e.g. healthy snacks).
- The participation in ECOLOG results in an enhancement of the school image and a further development of the external relations (e.g. with the community).
- ECOLOG schools, who are living a sustainable everyday culture, can influence the environmental consciousness and competencies of pupils positively (e.g. for a sustainable use of resources).
- Addressing values and the interconnection of knowledge are especially important. Anyway, the importance of the school as one influencing factor among many should not be overestimated. Some schools succeeded in involving the parents (more primary schools than secondary schools), which is seen as another supporting factor.
- On the one hand, ECOLOG lives due to the personal commitment and efforts of individual teachers. On the other hand, it is necessary to establish a culture of teamwork to enable the development of a sustainable school culture. This is a challenge for schools.
- ECOLOG schools are committed to quality development and assurance. The production of annual reports, correlating with the concept of the school development plan, caused some difficulties in the beginning. After 10 years of experience, schools accomplish this task increasingly well. Meanwhile, the ECOLOG annual report is seen as a helpful tool for reflection and planning. ECOLOG supports quality development through the definition of visions and aims, e.g. the shaping a liveable world and the perception of nature as a whole. Furthermore, ECOLOG offers a broad range of evaluation and reflection methods. Therefore, ECOLOG contributes to the implementation of legal provisions like the quality management of educational standards (especially in science).

Appraisal: *The ECOLOG programme has been growing continually for many years, being the oldest network supported by the Ministry of Education. One reason for this is that ESD/EE is always connected with current developments in the Austrian education system, such as quality evaluation and quality assurance. Other factors of success are the support system of the network which keeps the projects going as well as an active evaluation culture. External, formative evaluation provides feedback and confirmation for the Ministry as well as for the participants in the network. ECOLOG network schools have been ready to take part in the evaluations as they made sense to them and as schools were guaranteed to receive the results of the respective studies.*

10.6 Environmental Education and ESD in Teacher Education—The ENITE Network

As a reaction to the OECD country reports in 1991 and 1992 teacher education became one of the focal points within ENSI. The Austrian contribution was the start of the ENITE (Environmental Education in Teacher Education) network. ENITE is a research and development network which supports the development and study of initiatives in teacher education.

In its first phase (1997–2000), teams of lecturers, teachers and students at several teacher-training institutions worked in environment-related teacher education as part of a research project. In accordance with the action research approach, the initiatives were analytically studied by those who actually implemented them, the aim being the continuous development of these projects. The initiatives studied within ENITE had to include at least some of the following components:

- students learning relates strongly to environmental initiatives in schools (cooperation between teacher training institutions and schools);
- learning experiences build on the previous experiences of students and are influenced by them. This implies an active participation of the students in developing contents and methodology of a project (from problem definition to quality evaluation);
- learning is designed as an interdisciplinary process and not fragmented into disciplines;
- learning includes a research component based on systematic reflection on actual teacher practice (action research);
- the impact on and changes in work cultures and organisational structures are taken into account in the action and reflection processes (Rauch and Kreis 2003).

The outcomes have been published in two books (Posch et al. 2000; Kyburz-Graber et al. 2003).

In its second phase (2000–2004) this research project has triggered an ENITE-network initiative which has provided a platform for mutual exchange of experience and ideas in order to support the stabilisation of existing initiatives and their expansion to additional institutions of teacher training. From 2000 to 2004 the FORUM Umweltbildung provided a home base for the ENITE network. Scientific supervision is provided by the Institute for Instructional and School Development of the Alpen-Adria-University Klagenfurt. Since then the network has crystallised into the development of a university course which involves all network partners (see below).

10.6.1 Teacher Training University Course "Innovation in Teacher Education—Education for Sustainable Development" (BINE)

The main outcome of the ENITE-network so far is the National Teacher Training Course "Innovation in Teacher Education—Education for Sustainable Development"

(BINE). The first 2-year course started in 2004 and invited teacher-trainers from university colleages of teacher education and universities to work on sustainable development issues and their educational challenges. The course offers three 1-week seminars plus regional mentoring meetings. The aim is to improve pedagogical research competences (mainly action research), to research and reflect on educational practice in teacher education in diverse educational subjects, and to implement sustainable development issues in the teacher education curriculum. The ENITE principles (see above) form the basis of the BINE course of studies. The course is evaluated by a formative and summative self-evaluation with internal (question-naires, feedback by participants) and external (interviews with participants at the beginning and the end of the course) components. In the second course (2007–2009) all unviersity colleges of teacher education in Austria were involved as well as the Austrian Educational Competence Centre for Biology at the University of Vienna. In 2012 and 2013 the BINE course was successfully offered a third time.

The results of the evaluation have shown that the BINE course offers an adequate instructional and learning strategy for the participants to construct the meaning of the complex issues of Sustainable Development (SD) and ESD by researching, reflecting and exchanging in the learning group focused on concrete examples. The course has enabled the creation of a learning community. It is a challenge not to simplify ESD and lose its potential to identify the inter-connections between the ecological, social, economic and cultural-political spheres more clearly and adequately. The action research process provides a basis for learning in order to further develop the participants' concepts of ESD as well as research and implementation competencies (Rauch and Steiner 2005; Erlacher 2006; Steiner 2006). It is planned and discussed to support research projects at university colleges of teacher education, colleges and universities within the ENITE Network in the next phase (Rauch et al. 2010).

Networking Between the ECO-School Programme and University Colleges of Teacher Education

In order to support and strengthen this networking systematically, a separate network structure for university colleges was established within the ECO-school network in 2009/2010. Since March 2010 nominated representatives of all 14 Austrian university colleges meet twice a year for dialogue and exchange of information about developments within the ECO-school network and to get encour-aged to start ECO-initiatives in their institutions.

Appraisal: *The ENITE-project and especially the BINE-courses are successful examples of long-term collaborations between the Alpen-Adria University Klagen-furt and a majority of the university colleges of teacher education and with FORUM Umweltbildung, an initiative of the Austrian Federal Ministry of Agriculture, Forestry, Environment and Water Management and the Austrian Federal Ministry of Education and Women's Affairs.*

Since 2006–2007 teacher education is involved in a dynamic reform process based on new legislation for teacher training. A positive result of the ENITE-network and the BINE courses is, that, despite the constraints due to the structural reform, communication and collaboration and even participation between university colleges and the ECO-school network could be stabilised and enhanced.

A main challenge for university colleges is to integrate contents of the ECO-school approach into pre-service as well as into in-service curricula/study programmes. Another challenge is the 'whole school approach', since its tasks, structures and management are more complex than normal school management.

Another aspect of continuous and successful networking is the collaboration of Austrian stakeholders in Environmental Education and ESD in Teacher Education within the International Network Environment and School Initiatives ENSI (see above). Based on OECD/CERI country reports in 1991 and 1992 teacher education became one of the focal points within ENSI. This was the stimulus for the ENITE/UMILE (Umweltbildung-Innovation-LehrerInnenbildung-Environmental Education-Innovation-Teacher Training) research project which resulted also in an international publication (Kyburz-Graber et al. 2003) and which was also interlinked with the EU-SEED-network project (2002–2005). Out of this interaction the EU-CSCT-project (2004–2007) developed, then focusing on competencies for ESD (Sleurs 2008). The Austrian research project "Competencies for Education of Sustainable Development" (KOM-BiNE) was stimulated by and also strongly contributed to this international CSCT endeavour (Rauch et al. 2008; Steiner 2011). Finally the combined results were integrated into the development process of the UNECE document Learning for the future: Competences for Education for Sustainable Development (UNECE 2011).

10.7 ESD in Higher Education

On the occasion of launching the United Nations Decade of Education for Sustainable Development, an international conference on "Committing Universities to Sustainable Development" was held from 20 to 23 April 2005 in Graz, Austria. The conference was jointly organised by COPERNICUS CAMPUS, the Karl-Franzens-University Graz, the Technical University Graz, oikos International, and was sponsored by UNESCO. The objective of the conference was to discuss the role of universities and other higher education institutions in an overall societal transition towards sustainable development as well as strategies for the necessary opening of universities to society. This conference and the *Graz Declaration— Committing Universities to Sustainable Development* (BMBWK 2005) gave an important stimulus for ESD in higher education in Austria.

10.7.1 Sustainability Award for Higher Education

Austrian universities and higher education institutions are committed by law to develop concepts for sustainable development of society and also for their immediate environment. In order to support and encourage this task the Austrian Federal Ministry of Agriculture, Forestry, Environment and Water Management together with the Austrian Federal Ministry of Science, Research and Economy established the Sustainability Award for Higher Education. The award is an integral part of the Austrian Strategy for ESD and puts sustainability on the agenda of all Austrian universities.

The target group includes all universities and higher education institutions in Austria. The main objective is to strengthen and to integrate the issues of sustainable development in the daily life of higher education institutions in Austria. After a starting phase of networking and communication with the relevant stakeholders, the project aims at: (1) raising the motivation for sustainable higher education; and (2) supporting the pioneers in this field by the development of a nationwide 'sustainability award' for higher education institutions in the public sector.

The general approach is to develop a bench marking and nationwide learning process among public higher education institutions. This is done by the 'sustainability award contest' which is held every second year. Its main focus is on continuous processes of 'sustainable higher education' and not on temporary projects, single persons or singular events. Additionally, participation, open learning and innovation should be at the very heart of such processes. As the award is divided into eight different action fields (such as curricula, operations or student initiatives), universities can submit their contributions to this contest according to their individual strengths to win the award in one particular action field related to their own opportunities. Subsequently, Universities are encouraged to use this experience to improve their performance in other fields and to strive for more awards in other action fields in the long run. The first awards were announced in 2008, in March 2010 and April 2012, the announcement for 2014 has been published. One hundred and sixty-five initiatives and projects were submitted and each time eight projects/initiatives were awarded. The award helps to make sustainability not only part of the mission statement but, more importantly, of everyday university life. It adds visibility to already existing projects led by ambitious and committed pioneers. It fosters internal networking and coordination of these projects in the university as a whole, thus embedding sustainability more strongly in the overall university culture. Furthermore, it encourages systematic exchange of good practice between higher education institutions and makes them aware that sustainability is a core dimension of university development in many parts of the world[5].

---

[5] http://www.bmwf.gv.at/startseite/forschung/national/nachhaltigkeit/. Accessed 7 Feb 2014.

There are four additional initiatives in higher education:

- Alliance of Sustainable Universities: The Alliance of Sustainable Universities is a national association of currently nine universities committed to sustainability (www.openscience4sustainability.at).
- SUSTAINICUM: This project develops a pool of materials for teaching SD and ESD in Higher Education Institutions (http://www.boku.ac.at/sustainicum/index.htm).
- Regional Centre of Expertise on Education for Sustainable Development (RCE): the aim of RCEs is to establish networks on a regional level as well vertically, i.e. among educational institutions of all levels as well as horizontally among educational, public (administration, politics, etc.) and also private institutions (enterprises, associations, etc.). At the moment RCEs exist in Graz (http://rce.uni-graz.at) and Vienna (http://www.rce-vienna.at/).
- Sustainability as an inter-faculty focus at Alpen-Adria-University Klagenfurt: The status quo of sustainability in research, teaching, and administration at the Alpen-Adria University Klagenfurt was studied in 2010 (Hübner et al. 2010). As a result a research focus on SD and an interdisciplinary elective open for all Masters and PhD students seemed to be an appropriate step to implement sustainability at the University across disciplinary and structural borders (Hübner et al. 2014).

Appraisal: *The strengths of this award are: (1) the development of a national network of university stakeholders; (2) the focus on processes; (3) the possibility of starting with sustainability in 'easy' areas (according to the individual strengths of the institution); (4) bridging a wide spectrum of different stakeholders and experts in the award commission. Hindering factors are that no monetary prize is awarded, as well as the high complexity and heterogeneity among higher education institutions.*

At the beginning of the Sustainability Award, SD and ESD were hardly known. Meanwhile several universities declared sustainability an important concern and integrated it into their agreement of work with the Federal Ministry of Science, Research and Economy. Additionally several universities have compiled a sustainability report annually.

Internationally the Sustainability Award is acknowledged as an innovative initiative by being quoted in the inventory of innovative practices in education for sustainable development in the European Union (Danish Technology Institute 2008).

Networking on all levels—from central, Ministry supported, to regional and to local, i.e. intra-institutional—proved a successful approach to SD and ESD at the university level. Social contacts are indispensable for the creation of structures and the transmission of information. Such developments, however, can only happen in small steps. Support is quintessential for the continued development of regional identities in networks. The duties of the steering committee and its coordinator (s) are diverse and can only be accomplished by teamwork.

10.8   Conclusions and Outlook

Austrian networks in ESD have carried out creative projects and thereby tried to raise the attractiveness of ESD. Based on the examples presented, the following general findings can be drawn:

- Good practice cannot be cloned, but exchanging experience on a personal level promotes learning and innovation.
- Networks in education offer goal-oriented exchange processes among teachers (information function) which support the professional development of teachers (i.e. fresh ideas for classroom teaching, interdisciplinary cooperation at schools) (learning function).
- Networks have the potential to create a culture of trust, with the effect of raising self-esteem and risk-taking of teachers (psychological function) and upgrading science at school (political function).
- It is necessary to maintain a balance of action and reflection (goal-directed planning and evaluation) as well as autonomy and networking (analysis of one's own situation, but also support by 'critical friends' in order to set up a sustainable support system for schools).
- Evaluation and research need to be driven by an interactive link between an interest to gain new knowledge and an interest to develop. A culture of self-critical and collective reflection might flourish, but reflection should not hamper a project from being taken forward (see previous aspect).
- There are a number of risks, e.g. that

  – a network moves away from the interests of its stakeholders,
  – common visions and goals disappear,
  – the network fails due to weak coordination and steering,
  – the network fails due to a lack of resources (money and time),
  – the network mutates into a bureaucratic structure.

The overall challenge might be described as keeping a balance between structures and processes or, in other words, between stability and flow to enable sustainable development and learning.

Sustainable Development is a theme transcending social and structural realities. It cannot be implemented through education alone. The fact that the European Commission did not support the Decade hampered the dynamic the Decade could have acquired in Europe.

Experts developed a national educational strategy for the Decade. In Austria, the Decade was primarily supported by the Federal Ministry of Education and Women's Affairs, Federal Ministry of Science, Research and Economy and the Federal Ministry for Agriculture, Forestry, Environment and Water Management. It did not succeed in including relevant departments adequately, such as the economy or social areas.

Education for Sustainable Development requires a redefinition of education. It is based on pedagogical principles such as interdisciplinarity, value orientation, cultural awareness, problem-solving, methodological diversity, participation and local relevance. The aim remains to empower the individual to shape society in a reflected, responsible manner.

References

Austrian Federal Ministry of Agriculture, Forestry, Environment and Water Management, Austrian Federal Ministry for Education, Arts and Culture & Austrian Federal Ministry of Science and Research (Eds.). (2008). *Austrian strategy for education for sustainable development. Short version.* Vienna.

Altrichter, H. (2010). Netzwerke und die Handlungskoordination im Schulsystem. In N. Berkemeyer, W. Bos, & H. Kuper (Eds.), *Schulreform durch Vernetzung. Interdisziplinäre Betrachtungen* (pp. 95–116). Münster: Waxmann.

BMBWK. (2005). *Graz declaration on committing universities to sustainable development.* http://www-classic.uni-graz.at/geo2www/Graz_Declaration.pdf. Accessed 29 Sept 2013.

Boos, F., Exner, A., & Heitger, B. (2000). Soziale Netzwerke sind anders. *Journal für Schulentwicklung, 3*, 14–19.

Castells, M. (2000). *The rise of the network society* (The information age: economy, society and culture 2nd ed., Vol. 1). Oxford: Blackwell Publishers Ltd.

Czerwanski, A., Hameyer, U., & Rolff, H. G. (2002). Schulentwicklung im Netzwerk—Ergebnisse einer empirischen Nutzenanalyse von zwei Schulnetzwerken. In H. G. Rolff, K. O. Bauer, K. Klemm, & H. Pfeiffer (Eds.), *Jahrbuch der Schulentwicklung* (pp. 99–130). München: Juventa.

Dalin, P. (1999). *Theorie und Praxis der Schulenwicklung.* Neuwied: Luchterhand.

Danish Technology Institute. (2008). *Inventory of innovative practices in education for sustainable development.* http://ec.europa.eu/education/more-information/doc/sustdev_en.pdf. Accessed 29 Sept 2013.

Dobischat, R., Düsseldorf, C., Nuissl, E., & Stuhldreier, J. (2006). Lernende Regionen—begriffliche Grundlagen. In E. Nuissl, R. Dobischat, K. Hagen, & R. Tippelt (Eds.), *Regionale Bildungsnetze* (pp. 23–33). Bielefeld: Bertelsmann.

Ehgartner, M. (1999). *Ökologisierung von Schulen* (Ein Projekt des BMUK— Diplomarbeit). Wien: Universität Wien.

Erlacher, W. (2006). *Evaluationsbericht Universitätslehrgang BINE 2004–2006.* Klagenfurt: University of Klagenfurt.

Fritz, S., Lackner, E., Lechner, C., & Zimmerhackl, K. (2009). *Quality criteria for ESD schools: An innovative approach for teacher education in Austria.* Vienna: Austrian Federal Ministry for Education, Arts and Culture.

Fullan, M. (2007). *The new meaning of educational change.* London: Routledge.

Heinrich, M., & Mayr, P. (2005). *ÖKOLOG-Oekologisierung von Schulen—Bildung für Nachhaltigkeit. Analyse und Ausblick. Zusammenfassender Bericht über die systematischen Reflexionen von Erfahrungen in den ÖKOLOG-Schulen.* Linz: University of Linz.

Heinrich, M., Minsch, J., Rauch, F., Schmidt, E., & Vielhaber, C. (2007). *Bildung und Nachhaltige Entwicklung: eine lernende Strategie für Oesterreich.* Muenster: Monsenstein & Vannerdat.

Hübner, R., Hadatsch, S., & Rauch, F. (2010). *Nachhaltige Entwicklung an der Universität Klagenfurt—Ist Stand und Profilierungsmöglichkeiten.* Gefördert vom Forschungsrat der University of Klagenfurt, Forschungsbericht. Klagenfurt: University of Klagenfurt.

Hübner, R., Rauch F., & Dulle, M. (2014). Implementing an Interfaculty Elective "Sustainable Development": An intervention into a University's culture between organized scientific rationality and normative claim. In K. D. Thomas, & H. E. Muga (Eds.), *Cases on pedagogical innovations for sustainable development* (pp. 510–522). IGI Global: Hershey.

Kant, I. (1787/1956). *Kritik der reinen Vernunft.* Hamburg: Felix Meiner Verlag.

Knoll, B., & Szalai, E. (2009). *ÖKOLOG und Gender—ÖKOLOG-Schulen aus dem Blickpunkt Gender betrachtet* (Studie im Auftrag des Bundesministeriums für Unterricht, Kunst und Kultur). Wien: BMUKK.

Kyburz-Graber, R., Posch, P., & Peter, U. (2003). *Challenges in teacher education—Interdisciplinarity and environmental education.* Innsbruck: StudienVerlag.

Lechner, C., & Rauch, F. (in print). Quality Criteria for Schools focussing on Education for Sustainable Development (ESD). A case study on seminars for teacher in Austria. In F. Rauch, A. Schuster, T. Stern, M. Pribila, & A. Townsend (Eds.), *Promoting change through action research: International case studies in education, social work, health care and community development.* Rotterdam: Sense.

Lieberman, A., & Wood, D. R. (2003). *Inside the national writing project. Connecting network learning and classroom teaching.* New York: Teacher College Press.

McCormick, R., Fox, A., Carmichael, P., & Procter, R. (2011). *Researching and understanding educational networks.* London: Routledge.

McDonald, J., & Klein, E. (2003). Networking for teacher learning: Toward a theory of effective design. *Teacher College Record, 8,* 1606–1621.

McLaughlin, C., Black-Hawkins, K., McIntyre, D., & Townsend, A. (2008). *Networking practitioner research.* London: Routledge.

Minsch, J. (2004). Gedanken zu einer politischen Kultur der Nachhaltigkeit. Aufbruch in vielen Dimensionen. In F. Radits, M. Braunsteiner, & K. Klement (Eds.), *Bildung für eine Nachhaltige Entwicklung in der LehrerInnenbildung* (pp. 10–18). Baden: Teacher Education College Baden.

OECD (Ed.). (2003). *Schooling for tomorrow. Networks of innovation.* Paris: OECD.

O'Hair, M. J., & Veugelers, W. (2005). The case for network learning. In W. Veugelers & M. J. O'Hair (Eds.), *Network learning for educational change* (pp. 1–16). Maidenhead: Open University Press.

Österreichische UNESCO-Kommission. (2010). *Sustainability in action, Band 1.* Wien. http://www.unesco.at/bildung/dekadenbroschuere.pdf. Accessed 29 Sept 2013.

Österreichische UNESCO-Kommission. (2012). *Sustainability in action.* Band 2. Wien. http://www.unesco.at/bildung/dekadenbroschuere2.pdf. Accessed 29 Sept 2013.

Payer, H., Winkler-Rieder, W., & Landsteiner, G. (2000). *Ökologisierung von Schulen. Umwelteffekte und Wirtschaftsimpulse.* Wien: ÖAR-Regionalberatungs GmbH.

Posch, P. (1995). Professional development in environmental education: Networking and infrastructure. In OECD (Ed.), *Environmental learning for the 21st century* (pp. 47–64). Paris: OECD.

Posch, P. (1999). The ecologisation of schools and its implications for educational policy. *Cambridge Journal of Education, 29*(3), 341–348.

Posch, P., & Altrichter, H. (1993). *Schulautonomie in Österreich.* Innsbruck: StudienVerlag.

Posch, P., Rauch, F., & Kreis, I. (Eds.). (2000). *Bildung für Nachhaltigkeit. Studien zur Vernetzung von Lehrerbildung, Schule und Umwelt.* Innsbruck: StudienVerlag.

Rammel, C. (2010). *UNECE-report 2010, summary and final report.* Vienna: Austrian Federal Ministry of Education, Science and Culture.

Rauch, F. (2013). Regional networks in education: A case study of an Austrian project. *Cambridge Journal of Education, 43*(3), 313–324.

Rauch, F., & Dulle, M. (2011). *Auf dem Weg zu einer nachhaltigen Schulkultur—15. Jahre ÖKOLOG-Programm, 10 Jahre Netzwerk ÖKOLOG.* Wien: BMUKK.

Rauch, F., & Kreis, I. (2003). The project "Environmental Education in Teacher Education" (ENITE): An Austrian initiative. In R. Kyburz-Graber, P. Posch, & U. Peter (Eds.), *Challenges in teacher education—Interdisciplinarity and environmental education* (pp. 127–139). Innsbruck: StudienVerlag.

Rauch, F., & Scherz, H. (2009). Regionale Netzwerke im Projekt IMST: Theoretisches Konzept und bisherige Erfahrungen am Beispiel des Netzwerks in der Steiermark. In K. Krainer, B. Hanfstingl, & S. Zehetmeier (Eds.), *Fragen des Bildungswesens—Antworten aus Theorie und Praxis* (pp. 273–286). Innsbruck: StudienVerlag.

Rauch, F., & Schrittesser, I. (2003). *Networks as support structure for quality development in education*. Enschede: CIDREE.

Rauch, F., & Steiner, R. (2005). University course "Education for Sustainable Development—Innovations in Teacher Education" (BINE): Reasons, concept and first experiences. In Karl Franzens University Graz, Graz University of Technology, Oikos International & Copernikus Campus (Eds.), *Proceedings of the Conference "Committing Universities to Sustainable Development"* (pp. 359–368). Graz: Technical University Graz.

Rauch, F., & Steiner, R. (2006). School development through education for sustainable development in Austria. *Environmental Education Research, 12*(1), 115–127.

Rauch, F., Steiner, R., & Radits, F. (2010). Der Universitätslehrgang Bildung für Nachhaltige Entwicklung—Innovationen in der Lehrer/innenbildung (BINE): Ein Instrument zum Aufbau von Forschungskompetenz an Paedagogischen Hochschulen. *Erziehung und Unterricht, 1*(6), 92–96.

Rauch, F., Steiner, R., & Streissler, A. (2008). Kompetenzen für Bildung für nachhaltige Entwicklung von Lehrpersonen: Entwurf für ein Rahmenkonzept. In B. Bormann & G. de Haan (Eds.), *Kompetenzen der Bildung für nachhaltige Entwicklung. Operationalisierung, Messung, Rahmenbedingungen, Befunde* (pp. 141–158). Wiesbaden: VS Verlag.

Schäffter, O. (2006). Auf dem Weg zum Lernen in Netzwerken—Institutionelle Voraussetzungen für lebensbegleitendes Lernen. In R. Brödel (Ed.), *Weiterbildung als Netzwerk des Lernens* (pp. 29–48). Bielefeld: Bertelsmann.

Schober-Schlatter, P. (2002). *Schule auf dem Weg zur Nachhaltigkeit. Bedingungen und Hemmnisse eines ökologie-orientierten Wandels von Schulen*. Dissertation. Linz: University Linz.

Sleurs, W. (Ed.). (2011). *Competencies for ESD (Education for Sustainable Development) teachers*. A framework to integrate ESD in the curriculum of teacher training institutes. Comenius 2.1 project 118277-CP-1-2004-BE-Comenius-C2.1: Brussels. http://www.ensi.org/media-global/downloads/Publications/303/CSCT%20Handbook_11_01_08.pdf. Accessed 29 Sept 2013.

Steiner, R. (2006). *Universitätslehrgang "Bildung für nachhaltige Entwicklung—Innovationen in der LehrerInnenbildung" (BINE) SS 04 – WS 05/06. Report*. Salzburg: FORUM Environmental Education.

Steiner, R. (2011). *Kompetenzoriente Lehrer/innenbildung für Bildung für Nachhaltige Entwicklung*. Münster: Monsenstein & Vannerdat.

UNECE (Ed.) (2011). *The competences in education for sustainable development "Learning for the future: Competences in education for sustainable development"*. United Nations Economic Commission for Europe (UNECE), (ECE/CEP/AC.13/2011/6): Geneva. www.unece.org/fileadmin/DAM/env/esd/ESD_Publications/Competences_Publication.pdf. Accessed 29 Sept 2013.

Thonhauser, J., Ehgartner, M., & Sams, J. (1998). *Ökologisierung von Schulen. Evaluation eines OECD-Projekts*. Salzburg: University of Salzburg.

Chapter 11
Moving Schools Towards ESD in Catalonia, Spain: The Tensions of a Change

Mariona Espinet, Mercè Junyent, Arnau Amat, and Alba Castelltort

11.1 Contextual Background of EE and ESD in Catalonia, Spain: A Brief Sketch of a Trajectory

As elsewhere in Spain, EE (Environmental Education) in Catalonia was born in the 1970s and early 1980s, when the country emerged from a long dictatorship that had lasted over 40 years.

From 1975 to 1990, EE in Catalonia passed through two different phases (Franquesa 1997; Junyent 2001; López 2003). First, there was a boom, coinciding with the return of democracy to Spain, which was embodied in the construction of trails and facilities with a naturalistic approach. These aimed above all to bring the environmentalist and conservationist ideas that had emerged in Europe some years previously to the world of education. Secondly, there was a formalisation process, as various professionals quickly organised conferences and established associations

M. Espinet (✉)
Science Education, Gresc@ and LIEC Research Groups, Universitat Autònoma de Barcelona (UAB), Barcelona, Spain
e-mail: Mariona.Espinet@uab.cat

M. Junyent
Science Education, Complex Research Group, Universitat Autònoma de Barcelona (UAB), Barcelona, Spain
e-mail: merce.junyent@uab.cat

A. Amat
Science Education, Universitat de Vic,
Gresc@ and LIEC Research Groups (UVic), Barcelona, Spain
e-mail: arnau.amat@uvic.cat

A. Castelltort
Environmental Education, LIEC Research Group, Universitat Autònoma de Barcelona, Barcelona, Spain
e-mail: alba.castelltort@uab.cat

© Springer International Publishing Switzerland 2015
R. Jucker, R. Mathar (eds.), *Schooling for Sustainable Development in Europe*,
Schooling for Sustainable Development 6, DOI 10.1007/978-3-319-09549-3_11

in order to share their ideas and begin new projects and new initiatives throughout the country.

In the 1990s, several national and international factors emerged which led to a change of approach in EE in Catalonia. A key factor was the Rio Summit of 1992, which created a new perspective on the world based on the two major concepts of globalisation and sustainability. At a national level, the Government of Catalonia created a Ministry of the Environment for the first time, which fostered the institutionalisation of EE in Catalonia. In addition, local authorities made waste management a priority, which led to significant investment in municipal campaigns and materials for schools (Generalitat de Catalunya 2003). Finally, the change in formal education following educational reform meant that EE became part of the school curriculum (López 2003).

11.2   Large Scale EE and ESD Programmes in Catalonia

The current large scale EE and ESD (Education for Sustainable Development) programmes in Catalonia have been developed over the last 40 years. Despite their different origins, all of them are currently active and contribute in different ways to the development of the ESD Decade in relation to schools. The financial crisis experienced as a consequence of today's neoliberal world economy has affected considerably all EE and ESD large scale programmes in Catalonia, dramatically challenging their survival. The programmes include: EE Centres; Green Schools; School Agenda 21; and the Catalan Schools towards Sustainability Network (XESC). What follows is a narrative supported with quantitative data and qualitative interpretations of the changes and tensions experienced for each of these programmes.

11.2.1   Catalan Environmental Education Centres:
An Explosion After the Spanish Dictatorship

In Catalonia, there is an extensive range of facilities with various names (such as Nature Classrooms, Nature Schools, EE Centres, Environmental Interpretation Centres), which organise visits to schools to work on various topics related to EE. The birth of EE in Catalonia occurred at the same time as the creation of the first facilities. It is therefore usually dated to 1975 with the creation of the First Spanish nature trail, named Santiga Forest Trail, which was inspired by various English Nature Trails (Terradas 2010).

From the very beginning, these facilities were the result of both private and public initiatives, and facilities of both types still exist today. The private facilities were created mostly by professionals working in the natural sciences, such as biologists and geologists, but also by educational professionals, such as teachers and educationalists, who saw an opportunity to find employment at a time when this was difficult. The public facilities, mainly created by government authorities responsible for the management of natural parks, were regarded as a tool to build new social, environmental and educational initiatives after 40 years of dictatorial repression. It is not surprising that the pedagogical approach of EE centres in Catalonia was, and in many cases still is, focused on nature themes such as wildlife, vegetation and geology (Generalitat de Catalunya 2003). It can, therefore, be said that initially environmental activities almost always took place out of school and were very similar to informal education.

The number of facilities and professionals in EE greatly increased over the next 20 years. For example, there were a dozen EE facilities in Catalonia in the 1980s, a figure which rose to over 200 by early 2000. The capacity in 2003 provided activities for around 10,000 students every day, and more than 650 professionals were employed. This rapid proliferation of facilities led to a debate on the quality of these facilities which is as yet unresolved. Some professionals call for official standardisation, in order to distinguish quality education centres from those of a purely commercial nature, the latter having neither clearly defined educational objectives nor an educational team (Generalitat de Catalunya 2003). The Government of Catalonia's register of EE organisations currently lists 114 institutions located all over Catalonia, ranging from associations to service companies.

This boom in facilities, institutions and professionals since the mid-1980s has led to a search for some kind of association to establish a common working framework and to share resources, experiences and information. Various conferences and exchange forums were created, eventually leading to the establishment of the Catalan Environmental Education Society (SCEA) which was an important step towards the institutionalisation of EE in Catalonia. The SCEA includes a large proportion of the professionals in very diverse fields, such as environmental educators who work for service companies, freelance educators, and NGO workers, teachers, researchers and workers in the public sector. The SCEA had 100 members in 1990. It currently has more than 200 members (SCEA 2012).

A long-standing tension in this area has been how to adapt the more nature-focused vision of EE to the vision of ESD. During the DESD, some of these centres have adapted their educational projects in an attempt to highlight the environmental, social and economic dimensions of protected natural areas. Other facilities have also appeared which are often close to metropolitan areas, where students can work on different topics such as energy, waste and consumption more in line with ESD. There are both facilities with a nature-focused approach and other facilities with objectives that are more focused on the challenges of sustainability.

Another tension, in this area is related to the economic crisis, which is causing major upheavals in EE facilities and organisations. First, many schools have reduced their visits and stays at EE centres. This means that many of the country's facilities have seen one of their major revenue streams reduced. Secondly, many companies have relied mainly on public funds for the management of environmental programmes and campaigns. Both local and regional governments are making drastic cuts in campaigns and programmes of this type. Studies to assess the real impact of the crisis are currently in progress, but quite dramatic results are expected in the sector.

As a result, the economic crisis has increased uncertainty in this sector. An as yet unresolved debate on the professionalisation of environmental educators still persists today, and has intensified in recent years. The professional sector has called for the work of environmental educators to be regulated on several occasions, and above all has called for qualifications providing accreditation of a minimal level of training and a collective working agreement that regulates the rights and obligations of workers.

11.2.2 Green School Programme: The First Attempt in Catalonia to Develop an ESD Programme Which Is School-Centred

The Catalan Green School Programme was born in 1998 and it constituted the first programme aimed at greening schools in a systematic and coordinated manner. It was inspired by various similar initiatives all over Europe, which were mainly based on the idea of creating environmental audits in schools, in order to foster various initiatives based on a diagnosis (Sanglas 2003). Until that point, the focus of interest in EE lay mainly in activities outside schools, while, as a result of this programme, schools themselves become leading actors in EE in Catalonia.

The Green School Programme was the result of a joint initiative by the private SCEA, and the public Ministry of the Environment of the Government of Catalonia. The Ministry of Education subsequently also joined the management of the programme. The initial pilot project involved 12 schools across Catalonia. Currently, more than 450 schools are affiliated to the programme (Generalitat de Catalunya 2012).

The Programme has three main objectives, first, to promote the exchange of experiences between schools; secondly, to ensure the educational community's active participation and involvement in improving the environment; and, above all, thirdly, to help schools to incorporate the values of ESD in all aspects of life at the institution. The programme separates these areas into four different fields. The first field is curricular, in order to integrate ESD in all areas of the curriculum on a multidisciplinary basis. Another field is organisation and participation, in order to promote more democratic participation by the entire educational community. Another field is management, in order to use resources in the most sustainable way possible with a focus on the educational project. The final field is commitment to the environment, so that the school undertakes actions to improve its environment, but at the same time uses the environment as an educational context.

A school that wants to become a Green School is asked to establish an Environmental Committee, with representatives from the entire educational community, who will act as facilitators of the programme within the school. Through the committee, the school is initially asked to consider the school's current state and its proposed commitments. This initial consideration takes place by means of an audit, which will be used to draft a strategic plan for each of the four contexts of action (the environmental cohesion plan) to meet the commitments proposed by the school.

Through the Environmental Committee, each school then has to draft an action plan, which sets out the actions that the school will take over the next 3 years. The actions must be consistent with the school's strategic plans, and must include actions in the four dimensions within the programme. These actions also include evaluation strategies used for monitoring the action plan once it has been completed. The programme provides schools with support from environmental educators acting as facilitators of the schools' action plan and providing consultancy services.

Finally, the Green School Programme has, since its inception, provided various forums to encourage networking. For example, its monitoring committees act as forums for training and the exchange of information between Green Schools in a small geographic area. However, the forums and symposiums, which have now been adopted by the entire Catalan Schools towards Sustainability Network, are used for the exchange of experiences and training in larger geographic areas.

During the first 15 years of this programme some difficulties have been identified which have resulted in changes in its management. The first is the relationship between the Education and the Environment Departments. The lacking of a strong and shared strategic vision has been considered a factor sometimes preventing the development of quality collaborative work necessary for programme coordination.

A second difficulty arises from the establishment of a stable technical support group constituted by environmental educators acting as school facilitators. Initially, such technical support was awarded directly to the Catalan Society of Environmental Education (SCEA). However, from 2005 changes in the Catalan funding regulations required the programme to use an external competitive award system. This condition changed the school technical support model and acted as a factor preventing the building of a stable technical support team of environmental educators. Currently, the technical support award system is territorialized and has the aim of incorporating local environmental education associations which are close to the schools. These local associations act as the authentic disseminators of the programme since they are the ones with a direct contact with the schools and a deep knowledge of the territory.

One of the most important achievements of the Green School Programme is that the programme's approach is sufficiently open and flexible and can incorporate many of the activities schools carry out, or include new initiatives in the action plans without difficulties. This means that many schools are part of a programme that is recognised nationally by the Government of Catalonia. It has therefore been possible to institutionalise and recognise the work in which many schools are engaged to make progress in applying the ESD programme (Guilera et al. 2005).

11.2.3 School Agenda 21 Programmes: Municipalities' Contribution to the Development of ESD in Schools

School Agendas 21 in Catalonia are local ESD programmes with a municipal scope. Although not all city halls in Catalonia have developed School Agenda 21 programmes, some of the big cities have done so or are in the process of doing so. In other Spanish autonomous regions, such as Galicia and the Basque Country, where Green Schools or Eco-Schools programmes are not widely established, School Agenda 21 initiatives have acquired a regional scope.

The first School Agenda 21 Programme (A21E) was created in Barcelona in 2001 as a result of a political commitment to disseminate the idea of sustainability and sustainable best practices in the city through the involvement of schools at all stages (Weissmann et al. 2013). A few years later, other municipalities which had also started their Local Agenda 21, began to develop their own School A21 programmes inspired by the model promoted by Barcelona City Council. Each municipality designed its School A21 model according to such features as the characteristics of its territory, budget, technical-teaching personnel, and educational community.

We briefly present here the educational model created by the Barcelona A21E in terms of its history, results and its impact on the community (Franquesa et al. 2011), which had an international recognition (UN Habitat 2012) and its proposals for future innovation that may be useful for other experiences. We also present the A21E model of Sant Cugat del Vallès in terms of its focus on research and the creation of a new field of work in ESD such as school agroecology (Espinet et al. 2010; Agroecologia Escolar 2014).

School Agenda 21 of Barcelona: The First Catalan Programme

The first guide of Barcelona School Agenda21 book was published by Barcelona City Council in 2001 (Weissmann and Llabrés 2001). It was published in Spanish a few years later by the Ministry of the Environment in 2004 (Weissmann and Llabrés 2004), and this helped to raise the profile for the experience of other municipalities in Spain. This first publication was a working proposal, which had to be implemented over the subsequent 10 years by schools with the support of the Programme Coordination. The experience during this period was included in a new guide (Weissmann and Franquesa 2011).

The Barcelona School A21 is a voluntary process which invites schools to prepare a project to involve the school's educational community in the diagnosis, contribution of solutions and adoption of commitments to a more sustainable city, starting with the immediate surroundings, i.e. the school itself. It is defined as an ESD programme, which encourages participation and emphasises the importance of encouraging students to take part in a real experience of transformation of the environment, which they have designed and implemented themselves with the help and guidance of their teachers. The School A21 emphasises the promotion of training for action (Sanmartí and Pujol 2002).

The proposed working plan for the Barcelona School A21 has a simple and flexible methodology which aims to guide each participating school in constructing its own way forward. At the beginning of the academic year, each school produces its own annual or triennial project depending on its starting point, objectives and the actions that it is planning. The project's working outline includes five phases (motivation, diagnosis, examination of alternatives/decision making, action and evaluation) which make up a model for analysis and intervention.

In order to help the schools to develop their ESD project, the programme's coordinating body provides specific support for teachers by means of various services (information and communication, training and advice, exchanges as well as teaching resources and materials). Sharing experiences and promoting innovative projects in this field is also highly valued. At the end of each academic year, the schools present a report on the implementation of their project to the A21 School General Coordinators, which includes indicators and evidence of the improvements achieved.

The first 12 years of the School A21 saw a significant increase in the number of schools participating. Sixty-nine schools began the project in 2001, and 320 nursery, primary, secondary and adult education schools in Barcelona carried out their projects in the 2012–2013 academic year. In this context of growth, the programme's coordinators conducted a joint strategic assessment in 2009 (Weissmann et al. 2013) in order to measure the level of efficiency and effectiveness of the programme's management and to identify future lines of action. We identify two of the main tensions highlighted in the research below.

The first arises from the challenge of how to continue adding new schools to the programme, while maintaining and stimulating the schools that have been involved since the first year, without any impairment to the quality of services and the personalised attention offered. New initiatives have been proposed, such as facilitating activities that can be decentralised from the coordination of the programme and which can be managed directly by teachers interested in sharing experiences in the same area (neighbourhoods, districts) or in a particular field (such as biodiversity, waste, water, energy, participation). In addition to fostering autonomy from the programme's coordination, this proposal also takes advantage of and recognises the experiences of schools and teachers that have been involved in the programme for many years.

The second tension derives from the voluntary nature of participation in the programme, and therefore the level of commitment of the people involved. This tension has been exacerbated by the current socio-economic situation, with the increased workload for teachers and mobility of teaching staff. Despite the programme's history, institutional recognition for the dedication and work of coordinating teachers of the ESD programmes has not yet been achieved. Although the participation figures show that the voluntary nature of the programme has not been a hindrance to participation by schools, teachers continue to call for adequate recognition for their work.

School Agenda 21 of Sant Cugat del Vallès: Building School Agroecology

The municipality of Sant Cugat del Vallès started its own School Agenda 21 programme in 2003 using an eco-audit approach to ESD. After an evaluation using an action research approach conducted by the environment department of the municipality and the Universitat Autònoma de Barcelona (Llerena and Espinet 2010) the programme moved towards a new ESD model focused on community development. It was decided to develop a local network including schools, local educational administrators, policy makers, researchers, environmental educators, and local farmers with the aim of developing urban agriculture using an agroecological approach (Llerena et al. 2010; Espinet et al. 2010). This move facilitated the creation of a community perspective which led to the development of a new field: School Agroecology (Espinet 2011).

All public schools from nursery to high school in the municipality of Sant Cugat del Vallès have participated in the School A21 programme. The network representatives have met once a month to develop and monitor agroecological changes in the school and the community. The programme has been led jointly by the Environment Department of the municipality and the Science and Mathematics Education Department of the Universitat Autònoma de Barcelona. The results of a case study of the programme show that the School Agenda 21 has become the motor of local agroecological development as the following impacts indicate Espinet and Llerena 2013; Llerena & Espinet in press):

- the number of ecological agricultural fields both in schools and the community has strongly increased;
- join projects among the schools have been developed;
- all schools have developed individual educational projects which affect the school's food system and they have started new collaborative projects;
- a new community actor has been established which is the agro-environmental educator;
- and the number of research projects on ESD and agroecology has increased.

The tensions of the change in the implementation of the School A21 in Sant Cugat del Vallès are twofold. The first deals with the municipality regulations which very often prevent schools and local administrators developing ESD projects adjusted to their needs. Local administrators very often need to juggle with regulations so that new rules, mandates and contracts are adjusted to the local culture of the schools and the community. The second tension deals with the introduction of a research dimension in the programme which pushes all actors to participate in both research and action towards ESD. The fact that the municipality behaves as a social laboratory where new community and educational projects can be developed creates an empowering as well as stressful environment for all participants.

To conclude, one of the key factors in the School A21 programmes has been their ability to adapt to each school, to evolve with them and to promote innovative approaches so that they can design their own School A21 'model' according to the characteristics of each town and city. After several years of working 'behind closed doors', the School A21 programmes are expanding their perspectives beyond the local arena through participating in other supra-municipal networks (XESC 2014; ESenRed 2014).

11.2.4 The Catalan Schools Towards Sustainability Network (XESC): The Coordination of Programme Initiatives at the End of the Decade

The Catalan Schools towards Sustainability Network (XESC) is a "network of school networks" that in 2012–2013 consisted of more than 1,100 schools in Catalonia (from nursery schools to adult education centres), 16 town councils (7 municipalities with School A21 and 9 with other EE programmes), the Government of Catalonia (the Ministry of Planning and Sustainability and the Department of Education) and the Barcelona Education Consortium. XESC was established in 2010, a decade after the launch of the Green School programme (Government of Catalonia) and the School Agenda 21 programme (Town Councils). The need arose to avoid overlaps in the municipalities where both the programmes were in place, and to begin a process of networking in order to share resources, experiences and training, to join forces and for mutual enrichment. XESC is an example of collaboration between regional and local governments promoted and encouraged by the various ESD programmes (Pérez 2013).

Some of the main objectives of the network are (XESC 2014):

- establish coordination and exchange mechanisms to facilitate collaboration between networks;
- organise spaces for exchanging experiences between schools and to collaborate in teacher training and the provision of educational resources;
- encourage internal debate to foster progress with the conceptualisation of ESD in the field of schools as well as research and evaluation.

In order to achieve these objectives, XESC provides the impetus for continuous training for teachers, the technical staff and consultants who manage and run the programmes, promotes the exchange of experiences between teachers and students, and provides resources via the Internet. Some of the initiatives that have been carried out during its short history are:

- EE teaching resources have been digitised to make them available to schools;
- exchange meetings (symposia) have been organised by teachers and students;
- participation in the international project "Let's Take Care of the Planet" has been enhanced by the organisation of two conferences at Catalan level and participation in the Spanish and European conference; and
- training initiatives on reflective practice for teaching/technical staff (in collaboration with the Ministry of Education) and on curricular greening aimed at teaching staff and technical and educational staff (in collaboration with the Universitat Autònoma de Barcelona).

XESC aims to continue its expansion in both quantity as well as quality through establishing evaluation and recognition mechanisms, promoting research and innovation in ESD, disseminating best practices in ESD, and improving the training of teachers, experts and consultants. In these early years, XESC has already identified some of the potential and the tensions involved in networking. One of the first sources of stress resulted from the management of the heterogeneous nature of the group itself, as it included both long-standing programmes with a large number of schools, and others that belonged to smaller municipalities with lower budgets. A working model was sought in which everybody could give and receive, and where the benefit of belonging to a network was mutual despite this heterogeneity of backgrounds and experience. This tension became a new opportunity to undertake exchange projects and shared work which directly benefited the schools concerned.

11.3 The Quality of ESD in Catalan Schools: Open Doors for Continuous Improvement

The programmes reviewed above point towards a common trend which is the wide acceptance by schools to participate in ESD programmes, the growing commitment of communities and administrations to contribute to these programmes, and the strengthening of partnerships developed among schools participating in ESD programmes. However, introducing ESD into schools and communities in Catalonia is still a difficult endeavour which needs to confront important quality challenges in relation to the school curriculum, the thematic choices, and the participatory culture of schools.

11.3.1 ESD and the School Curriculum in Catalonia

The school curriculum in Catalonia is dependent on the frames of the general education laws approved by the central Spanish parliament. The Spanish school curriculum has the particularity of being strongly dependent on the political focus of the government in power which promotes educational changes. This characteristic sets a context for ESD in schools which is quite unstable and thus prevents the educational communities from developing long and stable educational processes.

The Characteristics of the Catalan Educational System and School Curriculum

The Catalan Educational System is compulsory up to the age of 16, and it is organised by cycles: Infant Education (0–6 years), Primary Education (6–12 years), Secondary

Education (12–16 years) (which includes Compulsory Secondary Education, Baccalaureate and Middle Grade Vocational Training), Upper Grade Vocational Training and University Education. The school curriculum is organised by knowledge areas which in Primary Education include: languages; knowledge of the natural, social and cultural environment; arts education; physical education; and mathematics (Generalitat de Catalunya 2013).

The curriculum developed in each school and in each classroom has some degree of autonomy, that is, the teacher team responsible for its implementation has to adapt the curriculum to the characteristics of the group of children and the socio-economic and linguistic situation of the school and its social environment. In the educational system every child has the right to education, no matter his or her place of origin or residence. This right is independent of students' families' administrative status, even if they have no residence permit. The current school curriculum integrates the concept of competence as a crucial component and stipulates that the acquisition of competences by students is the basic reference of the educational action carried out by teacher teams.

A Competence-Based Curriculum as a Context for ESD

The Catalan school curriculum does not define an explicit area for ESD. However, the recent educational law which applies to Catalonia (Departament Educació 2009) adopts a competences based approach to education which constitutes an opportunity for the development of ESD:

> The purpose of education is to ensure that boys and girls acquire the necessary tools to understand the world and to guide them in their actions; to lay the foundations for them to become individuals able to become actively and critically involved in a plural, diverse and constantly changing society. In addition to developing the necessary knowledge, ability, skills and attitudes (knowledge, expertise, 'how to be' and 'how to do'), boys and girls must learn to mobilise all these personal resources ('how to act') to achieve personal fulfilment and thereby become responsible, independent and socially integrated individuals, in order to exercise active citizenship, join adult life satisfactorily and be able to adapt to new situations and undertake continuous lifelong learning. (Departament d'Educació 2009: 14)

The competences defined within the school curriculum in Catalonia are compatible with the eight Key Competences for Lifelong Learning defined by the European Parliament and Council (2006). They follow the recommendations of the influential Delors Report (UNESCO 1996) which established the pillars of a competence based approach to education. This publication highlighted the need to frame education through an integrative approach which stood in opposition to the tendency of formal education systems towards knowledge acquisition. This integrative approach was based on four pillars: learning to know; learning to do; learning to live together; and learning to be.

This approach is clearly compatible with ESD as conceptualised by different authors and the various fields associated with the concept: such as critical thinking and reflection, action- and value-based education, participation in decision making, collaboration, systemic thinking, local and global relevance, interdisciplinary and holistic approaches (Tilbury 1995; Sterling 1996; IES 1999; Junyent et al. 2003; UNESCO 2004; Tilbury and Wortman 2004). A competence-based school curriculum has the potential to provide an optimal context for the development of a coherent ESD in schools.

Greening the School Curriculum in Catalonia

The acceptance that the school curriculum in Catalonia offers a context for the development of ESD does not mean that schools are actually using this opportunity fully. The process of introducing ESD into the school curriculum is a difficult one and has been framed in Catalonia as a process of Curriculum Greening (Geli et al. 2007).

> Curriculum Greening is a process of reflection and action aimed at achieving education for sustainable development throughout the curriculum development, closely connected with the management of the teaching centre and intended to promote a more just, equity and participatory society. Curriculum Greening must allow the analysis of the social-environmental reality and the search for alternatives consistent with the values of sustainability. Curriculum Greening includes all areas of knowledge and promotes collaboration with different institutions. Curriculum greening involves acquiring competences of systems thinking and promote the responsibility, the commitment and the action of the educational community towards the development of its own environmental identity. (Geli et al. 2007: 73–74)

The greening of the school curriculum in Catalonia is not a compulsory process but an optional one which depends on the school and community initiatives. Thanks to the resources available to schools from the Catalan government, the municipalities, NGOs and other local associations, schools can develop ESD projects within the framework of the official school curriculum. The challenges deal with making these projects more sustainable, strengthening the relationship between schools and communities, and finally engaging schools in future visioning of all the essential components of ESD.

However, this desirable scenario is currently affected by new laws that can be considered regressive. *The Organic Law for the Improvement of Educational Quality* (BOE 2013) can easily lead to regressive changes in education. In the coming years, it will be necessary to consider how it will be possible to begin processes that offer resistance to this regression, while making progress in a critical and transformative education towards a more just, equitable and sustainable society.

11.3.2   The ESD Thematic Choices of Catalan Schools

Catalan schools involved in large scale ESD programmes have selected different themes which have oriented their ESD action projects within the school and the

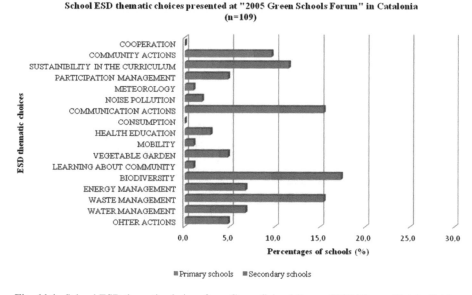

Fig. 11.1 School ESD thematic choices from Green School Forum 2005 (Generalitat de Catalunya 2005)

community. The thematic choices made by schools when developing ESD projects have reflected the social conditions, business or administration interests and local cultures surrounding schools at specific times and places. In order to understand the changes related to these choices we have compared data from the Green School Forum (Generalitat de Catalunya 2005) and the Forum of the Catalan Schools towards Sustainability Network (XESC 2014). Most ESD schools participate in these forums and they bring their most successful ESD experience to be shared with other schools of the network. Although these thematic choices do not represent a complete picture of what is being done in schools, it certainly provides an overview of the best ESD experience from the point of view of schools.

Figure 11.1 indicates that the most important ESD thematic choice of Catalan schools from the Green Schools Programme who participated in the 2005 Forum was Biodiversity, followed in second and third place by Waste Management and Communication Actions. The graph in Fig. 11.2 shows the same ESD thematic choices made by schools belonging to the XESC Network 6 years later. From the comparison between the two graphs it can be seen that there was an increasing percentage of schools selecting Waste Management and Food Gardening as ESD thematic choices. On the other hand, percentages of schools choosing Curriculum, Community Actions, and Communication decreased, while some ESD thematic choices such as Energy and Participation Management have been stable along the years monitored.

Waste management represents a very strong ESD thematic choice by Catalan ESD schools. More than 25 % of the experiences presented in the 2011 XESC

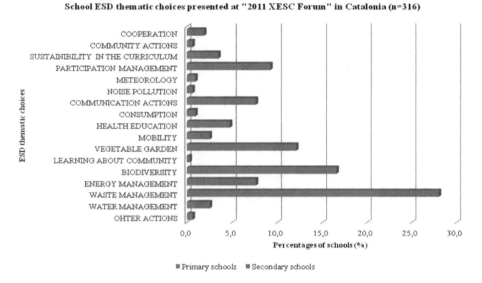

Fig. 11.2 School ESD thematic choices from Forum XESC 2011 (XESC 2014)

Forum were focused on Waste Management and represented a considerable increase compared to 2005. This could be the result of including local ESD School Networks run by municipalities into the ESD large scale programme of XESC. In the mid-1990s, many local governments and municipalities became strongly committed to improve local waste management. This led to major financial investments in educational campaigns and materials for schools which began to work extensively on waste management. The fact that municipalities are the responsible institutions for waste management in Catalonia and that they also support ESD projects on this issue could explain this change in the schools ESD thematic choice.

Energy represents an ESD theme chosen by only a minority of Catalan schools. Despite the efforts made by the local, national and international governments to promote ESD programmes on climate change and energy management Catalan schools have other preferences when engaging in ESD projects as shown in Figs. 11.1 and 11.2. A few years ago, some schools enrolled in programmes sponsored by various government authorities—sometimes with the help of non-governmental organisations—which helped them to install solar panels on schools. Although it was widely accepted in schools for some time, it soon became apparent that the solar panel installations were expensive and were often not used in the teaching and learning processes at the school. In addition, the level of action that the school could undertake was clearly limited by the financial resources allocated to it. The economic crisis currently affecting Spain has led to a swift decline in many of these major initiatives to promote the use of cleaner energies in Catalan schools, due to a lack of financial support from the government. Proposals for energy savings at schools are currently more successful, such as the European program Euronet 50/50 that encourages reduced energy expenditure and investment of the savings in school activities (Euronet 2014).

Food Gardening has become an increasing popular ESD thematic choice by Catalan schools. According to the Fig. 11.2, Food Gardening is the third most commonly chosen ESD theme by schools in 2011. Unlike energy or water management, the school garden is seen as an area that allows students and teachers at all educational levels to act directly and is therefore not limited by financial resources to the same extent. In addition, many teachers take advantage of the school garden as a context for learning in different areas, and especially those related to knowledge of the natural and social environment. Finally, the school garden is an inclusive learning environment that encourages the members of the community to participate both inside and outside the school (Amat and Espinet 2013).

11.3.3   The Participation Culture of Schools Working Towards ESD

There is a broad consensus around the idea that it is impossible to move towards sustainability without fostering high quality participation by citizens. However, there is still a great deal of progress to be made in order to enhance the participatory culture in schools. Recent research—focused on a case study of secondary schools in Barcelona participating in the School Agenda 21 programme—highlighted the idea that "people learn to participate by participating" and sets out the main characteristics and mechanisms of participation (Forestello 2013: 70). During the last 10–15 years, various initiatives and civic movements have fostered new forms of organisation and promotion of a participatory culture in different areas. This social context has, quite possibly, also encouraged schools to promote a new participatory culture in order to move towards a better society through the implementation of ESD programmes.

The Management of Participation in ESD School Programmes

Schools' participation in ESD programmes is managed by creating organisational structures that may have different names depending on the Programme concerned (an Environmental Committee in the case of Green Schools, and a Coordinating Committee for Agenda 21 Schools) and depending on the name given to them at each school (such as green committee, green group, eco group, environmental club, A21 School committee). The various programmes emphasise that student committees (in both primary and secondary schools) participate in and are actively involved in the decision-making process. The workings of each organisational structure to promote a culture of participation through the ESD programme is unique to each school, but some common issues for consideration have been identified (Forestello 2013):

- The frequency of meetings
- The time and scheduling of meetings
- The meeting place
- Coordination
- Committee members
- The tasks of the committee members
- Documentation.

Tensions Related to Participatory Processes in ESD

The main tension generated by the participatory processes and their organisation is that they initially appear to be slow and ineffective processes, but the experiences of the various programmes carried out in Catalonia present very positive evidence to the contrary. Recent years have confirmed the well-known African proverb "If you want to travel fast, travel alone. If you want to travel far, travel together." Schools that have committed to promoting a participatory culture have established more robust ESD programmes because they have made the transition from the initiative by a small group of people (or one person) to a shared and collective line of work. Moreover, these positive experiences are also feeding back into the process. The following examples are evidence of this: students from primary schools are interested in participating in the environmental committee when they reach secondary schools; new teachers, who have started teaching, have established committees because they were also committee members as students; and families who are looking for schools that run ESD programmes for their children. Although the path begun is certainly a positive one, there is still a long way to go in disseminating this culture of participation to all of the schools involved and their communities.

11.4 Conclusions: Challenges Ahead

Moving schools towards ESD in Catalonia can be seen as a dynamic process. The changes related to the integration of ESD in Catalan schools during the last 30 years have been important and innovative. Despite the tensions described above, they have been identified as the motor of those changes facilitating the confrontation of the necessary challenges to improve ESD. The most important challenges for schooling in ESD envisioned after the Decade of Education for Sustainable Development in Catalonia fall into the following areas: research, teacher education, and networking.

11.4.1 Undertaking Systematic Research on Schooling for ESD

Catalan universities have been slowly involved within the development of ESD large scale programmes. Research has also been slowly considered as a new professional dimension which is not only useful for the university community but also for the development of local and regional ESD programmes. Educational researchers have become partners and members of School Agenda 21 and Green School Programme thus developing new models for university, school and regional/local administration collaboration in ESD. It is necessary, however, to develop stable research structures within universities which focus on ESD research to ensure the continuity that high quality research needs. The Catalan research groups that focus their research on schooling for ESD are the following:

(a) Complex Research Group at Universitat Autònoma de Barcelona, which is working on ESD and complexity, teacher education and ESD, professional competences on ESD and curriculum greening at all levels of education. Some of the research undertaken by this group deals with the relationship between EE and complexity (Bonil et al. 2010) and the development and evaluation of a teacher training model aimed at building teachers' ESD competences (Bonil et al. 2012).

(b) Research Group GRESC@ (Research Group in Education for Sustainability, School and Community) at Universitat Autònoma de Barcelona, whose purpose is to create strategies that promote interdisciplinary research linking school education and community development oriented towards sustainability and democratic citizenship. Some of the research undertaken by this group has focused on the development of ESD evaluation approaches and tools that promote change in schools and communities (Grau and Espinet 2011), and the use of case study approaches in the building of school community networks for ESD (Llerena and Espinet 2010; Espinet and Llerena 2013).

(c) Research Group GRECA (Research Group on Scientific and Environmental Education) at Universitat de Girona has a research focus on professional training in ESD, the greening of university studies and compulsory education, the evaluation of educational programmes, especially in the field of natural and social sciences, EE and ESD. Some of the research undertaken by this group deals with the development of a curriculum framework for schooling on ESD (Geli et al. 2007) and the guiding documents for EE centres (Medir et al. 2013)

The research of these groups moves around interpretive and socio-critical paradigms, with a dominance of qualitative approaches. These chosen research approaches promote collaborative research strategies which include different agents involved in the investigation, thus being consistent with the principles of sustainability.

These research teams are part of the Edusost Network (Catalan Research Network on Education for Sustainability) funded by the government of Catalonia that aims to bring together the long tradition of the past and the present in Catalonia on EE and ESD. The Edusost network is based on the concept of sustainability as a new positive social value, mostly as an answer to the awareness of the limitations of the actual economic growth model. The network wants to help maintain continuity and develop joint projects enhancing network research as an added value and generating synergies between the different stakeholders that constitute the network (18 university research teams, three groups from the Public Administration, and 21 social entities) (Edusost 2014).

The challenge faced by Catalan research on schooling for ESD is the development of reliable research structures and models that facilitate the building of research teams including a diversity of agents such as school teachers and headteachers, environmental educators, teacher trainers and researchers. In these collaborative research models it is assumed that the knowledge built by all community agents on ESD is valuable and needs to be taken into account within the research process. This assumption is in line with Hart's (2003) reflection on the value of teachers' knowledge to promote ESD innovation and research. The purposes of such collaborative research projects should be: to evaluate ESD processes in the school and the community; to design innovative and sustainable school and community projects to enhance ESD; and to create new ideas, learning strategies, educational resources, monitoring and evaluation tools and processes which can be used to promote ESD in schools and communities (Junyent 2012).

11.4.2   Introducing ESD into the Catalan Teacher Education Curriculum

Pre-service and in-service teacher education is considered a crucial factor for the improvement of education in Catalonia as well as in other parts of the world. This belief is supported by the idea that the development of a sustainable society is a continuous process of learning and change, involving a variety of actors providing guidance and leadership in formal, non-formal and informal learning. According to UNECE (2012), this requires a corresponding enhancement in the competences of educators, leaders and decision makers at all levels of education. Teacher education institutions fulfil vital roles in the global education community (UNESCO 2005); they have the potential to bring changes within educational systems that will shape the knowledge and skills of future generations. Often, education is described as the great hope for creating a more sustainable future; teacher education institutions serve as key change agents in transforming education and society, so that such a future is possible. It is important to take into account that not only do teacher education institutions educate new teachers, they update the knowledge and skills of in-service teachers, create teacher-education curriculum, provide professional development for practising teachers, contribute to the writing of textbooks, consult with local schools, and often provide expert opinions to regional and national ministries of education. Teacher education institutions should function according to values of sustainability and set good examples of sound ESD practices with active participation of students and teachers.

Catalan teacher education institutions have made important efforts during the last decade to reorient teacher education curricula towards ESD. From 2000 to 2006, Curriculum Greening Seminars of Faculties of Education of the Catalan Universities have been held aimed at exchanging experiences in the field of curriculum greening and establishing strategies for interuniversity cooperation to incorporate ESD into the teacher education curricula. In addition and as a consequence of the creation of the European Higher Education Area, efforts have been made to redefine the professional competences in Sustainability, the greening of university curricula and more specifically the teacher competences in ESD (Cebrian and Junyent 2014). The new teacher education graduate programmes offered by all universities have created new and interesting opportunities for the development of ESD through practical experiences and the final projects for graduation. However, the highly competitive culture dominating universities at present and the strong departmentalisation of knowledge areas act as constraints towards the implementation of changes which would promote a quality ESD in the teacher education curriculum.

11.4.3   Building Bridges Between Regional, National and International ESD School Networks

The trajectory drawn by the evolution of Catalan large scale EE and ESD programmes has been characterised by a progressive connectivity with other local, regional, national and even international EE and ESD school networks. Important steps have been taken by different Catalan professionals in the field of EE and ESD to create new partnerships which have facilitated the birth of three wider networks where Catalan schools are participating:

(a) the Catalan Network XESC (Catalan Schools towards Sustainability Network) of a regional scope (XESC 2014);
(b) the Spanish Network ESenRED (Spanish Schools towards Sustainability Network) of a national scope (ESenRED 2014)
(c) the European Networks SEED, SUPPORT and CoDeS of an international scope, supported by ENSI (Environment and School Initiatives) (ENSI 2014).

The underlying belief of these networking initiatives is supported by the idea that ESD requires collaboration and partnerships, both at local and global levels. Partnerships help participants to bring together people of different contexts and perspectives, create synergy in their work, combine resources, build shared visions and knowledge, add value to local initiatives while maintaining relevance and motivate action for the future. Partnerships which share learning experiences provide a forum for mutual support and encouragement and can accelerate the process of change towards sustainability (Tilbury and Wortman 2004).

The creation of XESC and ESenRed has been an important step in the field of ESD in Catalonia and Spain. Drawing on the *Educació Ambiental en xarxa* [Networked Environmental Education] study carried out in Catalonia (Torras 2004), these examples are the first step towards the implementation of what used to be known as a "Utopian model of network of networks that enables full interaction between the various sub-networks and actors based on a common goal: environmental education in our country" (Torras 2004: 55). XESC, ESenRed and CoDeS (CoDeS 2014) are great opportunities to enhance the collective construction of knowledge and experiences, and to circulate them among the various nodes in the network (through consultants, training initiatives, digital resources, publications, etc.), thereby making networking much more effective than individual work. In the coming years we will certainly be able to identify and explain the benefits arising from these ESD networks.

The challenges sketched in these conclusions constitute key areas of action for integrating ESD into Catalan schools which are coherent with and can contribute to the *Global Action Programme on Education for Sustainable Development as follow-up to the United Nations Decade of Education for Sustainable Development (DESD) after 2014* (UNESCO 2014). However, the present economic, political, environmental and social situation in which Catalonia as well as Europe find themselves construes a difficult scenario for ESD and will probably demand more radical changes as a consequence of the strongest tensions.

Acknowledgments Supported by Spanish MEYC EDU2012-38022-C02-02 and the grant from the Catalan Government 2014SGR1492.

References

Agroecologia Escolar. (2014). http://agroecologiaescolar.wordpress.com. Accessed 21 Mar 2014.

Amat, A., & Espinet, M. (2013). *Collaboration between school and community through agroecology: A vegetable school garden as a boundary object.* In 10th ESERA conference, Cyprus, July 2013.

BOE. (2013). *Ley Orgánica 8/2013 para la mejora de la calidad educativa (LOMCE).* Boletín Oficial del Estado, 10 de diciembre de 2013, núm 295. http://www.boe.es/boe/dias/2013/12/10/pdfs/BOE-A-2013-12886.pdf. Accessed 20 Mar 2014.

Bonil, J., Junyent, M., & Pujol, R. M. (2010). *Educación para la Sostenibilidad desde la perspectiva de la complejidad.* EUREKA, 7, núm. Extra, 198–215.

Bonil, J., Calafell, G., Granados, J., Junyent, M., & Tarin, R. M. (2012). Un modelo formativo para avanzar en la ambientalización curricular. Profesorado. *Revista de Currículum y Formación del Profesorado, 16*(2), 145–163.

Cebrian, G., & Junyent, M. (2014). Competencias profesionales en Educación para la Sostenibilidad: un estudio exploratorio de la visión de futuros maestros. *Enseñanza de las Ciencias, 32*(1), 29–49.

CoDeS. (2014). *School and community collaboration for sustainable development.* http://www. comenius-codes.eu. Accessed 21 Mar 2014.

Departament d'Educació. (2009). *Currículum Educació Primària.* Barcelona: Servei d'ordenació curricular, Departament d'Educació, Generalitat de Catalunya.

Edusost. (2014). *Xarxa de Recerca en Educació per a la Sostenibilitat (Catalan research network on education for sustainability).* http://www.edusost.cat. Accessed 21 Mar 2014.

ENSI. (2014). *Environment and school initiatives.* http://www.ensi.org. Accessed 21 Mar 2014.

ESenRed. (2014). *Escuelas Sostenibles en Red (Schools towards networked sustainability).* http://confint-esp.blogspot.ch/p/esenred.html. Accessed 21 Mar 2014.

Espinet, M. (2011). *The making of a new STEM discipline: A sociocultural view of school agroecology. Ponencia en cultural studies of science education forum.* Orlando: University of Central Florida.

Espinet, M., & Llerena, G. (2013). *School Agroecology as a motor for community and land transformations: A case study on the collaboration among community actors to promote Education for Sustainability (ES) school networks.* In 10th ESERA conference, Cyprus, July 2013.

Espinet, M., Llerena, G., Grau, P., & Amat, A. (2010). *Trobant el propi camí cap a l'escola sostenible... (Canviant de model d'agenda 21 escolar). IV Jornada sobre L'Estat de la Recerca en Educació per a la Sostenibilitat.* Barcelona: Edusost. http://www.edusost.cat/ca/documents/ documents-propis-de-la-xarxa/doc_download/283-qtrobant-el-propi-cami-cap-a-lescola-sosteni blecanviant-el-model-dagenda-21-escolarq-. Accessed 5 May 2014.

Euronet. (2014). *EURONET 50/50 project.* http://www.euronet50-50.eu. Accessed 11 Apr 2014.

European Parliament and Council. (2006). *Recommendation of the European Parliament and of the Council of 18 December 2006 on transnational mobility within the community for education and training purposes: European quality charter for mobility (2006/961/EC).* http://eur-lex.europa.eu/LexUriServ/LexUriServ.do?uri=OJ:L:2006:394:0005:0009:EN:PDF. Accessed 21 Mar 2014.

Forestello, A. M. (2013). *La cultura de la participación en los centros de secundaria. Un estudio de casos en la Agenda 21 Escolar* (Monografías de Educación Ambiental, Vol. 13). Barcelona: Editorial Graó.

Franquesa, T. (1997). *Situació i reptes de l'educació ambiental. Conferència en el lliurament del I Premi Albert Pérez-Bastardas.* Barcelona: SCEA/Fundació Roca i Galés.

Franquesa, T., Pujol, T., & Muccio, G. (2011). *Education for sustainable development: The Barcelona School Agenda 21 Program.* http://sustainabledevelopment.un.org/index.php? page=view&type=1006&menu=1348&nr=2165. Accessed 21 Mar 2014.

Geli, A. M., Junyent, M., Medir, R. M., & Padilla, F. (2007). *Ambientalització Curricular de l'Ensenyament Obligatori. Una proposta de definició, caracterització i d'estratègies* (Monografies Universitàries, Vol. 7). Barcelona: Departament de Medi Ambient i Habitatge, Generalitat de Catalunya.

Generalitat de Catalunya. (2003). *ECEA: Estratègia Catalana d'Educació Ambiental: una eina per a la comunicació i la participació.* Document Marc. Barcelona: Generalitat de Catalunya.

Generalitat de Catalunya. (2005). *Escoles Verdes.* http://www.gencat.cat/mediamb/escolesverdes. Accessed 11 Apr 2014.

Generalitat de Catalunya. (2012). *Programa Escoles Verdes.* Document Marc. Barcelona: Generalitat de Catalunya.

Generalitat de Catalunya. (2013). *The educational system in Catalonia: Basic aspects.* Education Department. http://www.xtec.cat/web/projectes/alumnatnou/acollida/informacio2. Accessed 21 Mar 2014.

Grau, P., & Espinet, M. (2011). Estrategias interactivas para la evaluación de actividades de educación ambiental para la sostenibilidad en el ámbito municipal. *Ambientalia: Revista Interdisciplinar de las Ciencias Ambientales, 30*–45.

Guilera, M., Tarín, R. M., Pujol, R. M., & Espinet, M. (2005). Country report Spain—Catalonia. In F. Mogensen & M. Mayer (Eds.), *Eco-schools: Trends and divergences. A comparative study on ECO—Schools development process in 13 countries* (pp. 310–327). Vienna: Austrian Federal Ministry on Education, Science and Culture.

Hart, P. (2003). *Teachers' thinking in environmental education: Consciousness and responsibility*. New York: Peter Lang Publishing.

IES (Institute of Environmental Studies). (1999). *Education for sustainability. Integrating environmental responsibility into curricula: A guide for UNSW faculty.* Sydney: The University of New South Wales.

Junyent, M. (2001). *Educació Ambiental: un enfocament metodològic en formació del professorat.* Unpublished doctoral thesis. Girona: Universitat de Girona.

Junyent, M. (2012). Engaging schools into reflection on the quality of ESD. Building an ESD curriculum framework on compulsory educations: A collaborative project. In M. Réti & J. Tschapka (Eds.), *Creating learning environments for the future. Research and practice on sharing knowledge on ESD* (pp. 105–109). Kessel-Lo: Environment and School Initiatives (ENSI). http://www.ensi.org/media-global/downloads/Publications/341/Creating%20learning %20environments.pdf. Accessed 21 Mar 2014.

Junyent, M., Geli, A. M., & Arbat, E. (2003). *Proceso de Caracterización de la Ambientalización Curricular de los Estudios Superiores.* Girona: Universitat de Girona/Red ACES.

Llerena, G., & Espinet, M. (2010). Estudio de caso para la evaluación de una Agenda 21 Escolar: el caso de Sant Cugat del Vallès. In M. Junyent & L. Cano (Eds.), *Investigar para avanzar en Educación Ambiental* (pp. 159–179). Madrid: OAPN, Ministerio MAMRM.

Llerena, G., & Espinet, M. (in press). The collaboration between local administration and university to promote education for sustainability school networks on agroecology. In M. Espinet (Ed.), *CoDeS selected cases on school community collaboration for sustainable development.* Viena: CoDeS.

Llerena, G., Espinet, M., Martín-Aragón, A., & Fisher, K. (2010). La recerca sobre el parc rural de la Torre Negra: Oportunitats per a la reforma ambiental de Sant Cugat del Vallès. *Biblio 3W Revista Bibliográfica de Geografia y Ciencias Sociales, 15*(887), 1–10.

López, R. (2003). Panorámica de la evolución de la educación ambiental en España. *Revista de Educación, 331,* 241–264.

Medir, R. M., Heras, R., & Geli, A. M. (2013). Guiding documents for environmental education centres; an analysis in the Spanish context. *Environmental Education Research.* doi:10.1080/ 13504622.2013.833590.

Pérez, P. (2013). *La Red de Escuelas para la Sostenibilitat de Cataluña (XESC).Trabajar en red, una herramienta clave en los programas de educación ambiental dirigidos a centros educativos.* https://docs.google.com/file/d/0B7ODzRcrmKWPb253dWNOUk1fUlE/edit? usp=sharing&pli=1. Accessed 21 Mar 2014.

Sanglas, E. (2003). *Programa Escoles Verdes. Evolució i Estat Actual.* Unpublished work. Bellaterra: Universitat Autònoma de Barcelona.

Sanmartí, N., & Pujol, R. M. (2002). ¿Qué comporta capacitar para la acción en el marco de la escuela? *Investigación en la Escuela, 46,* 49–54.

SCEA (Societat Catalana d'Educació Ambiental). (2012). http://www.scea.cat. Accessed 21 Mar 2014.

Sterling, S. (1996). Education in change. In J. Huckle & S. Sterling (Eds.), *Education for sustainability* (pp. 18–39). London: Earthscan.

Terradas, J. (2010). *Ecologia viscuda.* València: Publicacions Universitat de València.

Tilbury, D. (1995). Environmental education for sustainability: Defining the new focus of environmental education in the 1990s. *Environmental Education Research, 1*(2), 195–212.

Tilbury, D., & Wortman, D. (2004). *Engaging people in sustainability.* Gland: IUCN (Commission on Education and Communication).

Torras, A. (Ed.). (2004). *L'Educació ambiental en xarxa. Barcelona: Institut de Ciències de l'Educació de la Universitat Autònoma de Barcelona i Societat Catalana d'Educació Ambiental.* http://www.scea.cat/forum2000_4.htm. Accessed 10 Apr 2014.

UN Habitat. (2012). *Best practice—Educating for sustainability: The Barcelona School Agenda 21 Programme: Barcelona.* http://www.unhabitat.org/bp/bp.list.details.aspx?bp_id=1116. Accessed 21 Mar 2014.

UNECE. (2012). *Learning for the future. Competences in education for sustainable development.* ECE/CEP/AC.13/20116. United Nations. Economic Commission for Europe. http://www.unece.org/fileadmin/DAM/env/esd/ESD_Publications/Competences_Publication.pdf. Accessed 21 Mar 2014.

UNESCO. (1996). *Learning: the treasure within. Education for the 21st century.* Delors Report. Paris: UNESCO.

UNESCO. (2004). *Decade for education for sustainable development (2005–2014).* Paris: UNESCO.

UNESCO. (2005). *Guidelines and recommendations for reorienting teacher education to address sustainability. Technical paper No. 2.* Paris: UNESCO Education Sector, Education for Sustainable Development Action.

UNESCO. (2014). *Global Action Programme (GAP) on ESD.* http://www.unesco.org/new/en/unesco-world-conference-on-esd-2014/esd-after-2014/global-action-programme/. Accessed 10 Apr 2014.

Weissmann, H., & Franquesa, T. (2011). *En el camí de l'escola sostenible. Una nova guia per fer l'Agenda 21 Escolar.* Barcelona: Ajuntament de Barcelona. http://www.bcn.cat/agenda21/a21escolar. Accessed 21 Mar 2014.

Weissmann, H., & Llabrés, A. (2001). *Guia per fer l'Agenda 21 Escolar.* Barcelona: Ajuntament de Barcelona.

Weissmann, H., & Llabrés, A. (2004). *Guia para hacer la agenda 21 escolar.* Ministerio de Medio Ambiente. Organismo Autónomo Parques Nacionales. http://www.magrama.gob.es/es/ceneam/recursos/documentos/guia-hacer-agenda-21escolar_tcm7-12827.pdf. Accessed 10 Apr 2014.

Weissmann, H., Rodríguez, C. L., Jurado, A., & Pomeroy, M. (2013). La Agenda 21 Escolar de Barcelona: una experiencia innovadora. In R. Canal (Ed.), *Ciudades y pueblos que puedan durar. Políticas locales para una nueva época* (pp. 59–72). Barcelona: Icaria Editorial.

XESC. (2014). *Xarxa d'Escoles per a la Sostenibilitat de Catalunya (Catalan Schools towards Sustainability Network).* http://www.xesc.cat. Accessed 21 Mar 2014.

Chapter 12
The Challenge of Mainstreaming ESD in Hungary

Mónika Réti, Dániel Horváth, Katalin Czippán, and Attila Varga

12.1 Transition Towards ESD

The history of education for sustainable development (ESD) is inseparable from the history of environmental education (EE) in Hungary, as there is a very tight correlation between the international recognition of the theme and its acceptance in Hungary. International processes in educational policy as well as pedagogy had made their impact in as much as they were adapted and adjusted into the Hungarian system of education in the past decades. Thus the current situation of ESD in Hungary, on the one hand, mirrors some global trends and, on the other hand, lessons learnt from the Hungarian case could be useful worldwide.

Several models make attempts to describe the relationship between EE and ESD. All of them agree that there is a learning journey guiding a transition. Education emphasising harmony with nature or gaining direct experience in outdoor learning have century-long traditions in Hungary, with future-leading initiatives in such aspects as competency development, a complex and problem-based approach and collaborative learning (Havas 1996). The holistic EE roots had a remarkable impact on the transition from a systemic emphasis on EE towards that of ESD.

According to Huckle and Sterling (1996), EE includes one, two Havas (2001) or all three elements of education in the environment, education about the

M. Réti (✉) • A. Varga
Hungarian Institute for Educational Research and Development, Budapest, Hungary
e-mail: Reti.Monika@ofi.hu; Varga.Attila@ofi.hu

D. Horváth
Library and Information Centre of the Hungarian Academy of Sciences, Budapest, Hungary
e-mail: horvathorszag@gmail.com

K. Czippán
UNECE Expert Group on Competences in ESD, Commission on Education and Communication of IUCN, Budapest, Hungary
e-mail: czippank@gmail.com

© Springer International Publishing Switzerland 2015   201
R. Jucker, R. Mathar (eds.), *Schooling for Sustainable Development in Europe*,
Schooling for Sustainable Development 6, DOI 10.1007/978-3-319-09549-3_12

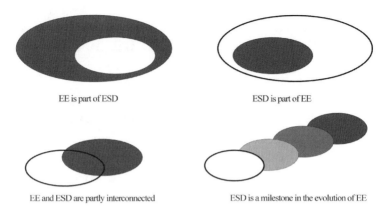

EE is part of ESD   ESD is part of EE

EE and ESD are partly interconnected   ESD is a milestone in the evolution of EE

Fig. 12.1 Four views about the connections between ESD and EE (After Hesselink et al. 2000)

environment and education for the environment. They envision the latter as a firm basis for education for sustainability; this also describes the way ESD evolved in Hungary.

In 2000, IUCN undertook an online survey that yielded responses from 70 participants from 25 countries. The findings produced the models illustrated in Fig. 12.1 (Hesselink et al. 2000).

András Victor (2002) claims that in Hungary, ESD originates from the evolution of EE, as the already holistic approach to EE gradually embraced human-made environmental and social issues in content and didactics, integrating ethical, moral and economic topics as well.

In the early 1980s, environmental issues had a special relevance in public life in Hungary: as the political system regarded these as "relatively harmless" issues, representing less "danger" to the oppressive system led by the communist government of the era. Initiatives supporting ecological consciousness were "tolerated" (instead of the other two categories being "forbidden" or "supported"). Thus, people with ambitions for civil activities gathered around the environmental movement, which—by the end of the 1980s—resulted in a relatively large number of civil gatherings connected to ecological movements. Environmental issues therefore became mainstream topics in early attempts to break through the restrictions of the communist regime. Conceived by scholars of pedagogy and practitioners in the field, the idea made its way into education. In the 1990s, EE became a field of experimentation for teachers and also for pioneers engaged in the green movement (Könczey 2006).

It also meant that the environmental movement (including EE) had a strong link to raising an active citizenship and participation, as well as to civil society involved in fostering democratic transition. It is only a small surprise therefore that, of the very first Hungarian NGOs, several are involved in environmental protection and nature conservation. The role of civil society organisations is still strong in Hungary (including forest schools, outdoor activities, teacher training events, summer camps and school campaigns).

Although the main guidelines for EE were created and published in the 1970s–1980s, legal frameworks clarifying and legitimising the position of EE were only adopted in the mid-1990s (Kiss and Zsiros 2006), including:

- the Act No. LIII of 1995 on the General Environmental Protection Rules declaring that "all citizens have right to acquire and develop knowledge about the environment" (LIII 1995: §54, Paragraph 1) and that EE is a responsibility of the state and the local government both to be introduced into the school system and in other opportunities for learning;
- the Act No. LIII of 1996 on Natural Protection that asserts that all institutions in public education should engage in EE (LIII 1996b);
- the Act No. 79 in 1993 (modified 62/1996) on Public Education in Hungary, which prescribes that "local curricula in all educational institutions should include EE" (Act 79 1993: §44–45);
- the Decision of Parliament No. 83 of 1997 that calls for creating a National EE Strategy;
- the first National Core Curriculum in 1996 (LIII 1996a, b), which highlights EE as a common cross-cutting developmental aim for all schools and the Government Decree No. 17 of 2004 on Frame Curricula defining EE as a horizontal content and approach relevant to all cultural domains in Hungarian public education, thus making all teachers responsible for introducing the ideas and applying relevant didactics at all levels and in all subjects;
- the Hungarian Act on Higher Education aiming to ensure sustainable development is implemented in HE, to disseminate health- and environment-conscious values and raise awareness (this act regulates the operation of teacher training universities).

These regulatory documents gave high priority to EE, making it a compulsory task for public education. But already in the late 1990s, a transition towards ESD began, which is well illustrated by the following statement by Havas, one of the most prominent national experts in the field:

> Many think that ESD is an important point in the agenda of educational development. Well, in the 21st century, it seems to be the agenda itself. (Havas 2002: 16–27)

The on-going revision and renewal of these documents introduced the idea of sustainability and ESD. While in the National EE Strategy the holistic approach and an opening towards social responsibility and global learning were gradually emphasised more and more (from 1998 to 2003) (Vásárhelyi and Victor 1998, 2003), in the last revised strategy the ideas of sustainability and learning for sustainability are implicitly and explicitly present: one chapter deals with EE for sustainability, another describes parallels between EE and ESD, and even states that this Strategy is meant as an ESD document while not undermining the importance of EE (Vásárhelyi 2010). One reason for this was that, due to difficulties with language use, publishers decided not to change the title of the document: however, it is a strategy supporting ESD in content and approach.

Since 2005 the spread of the concept of ESD in Hungary has been significantly supported by the UN Decade on ESD. Hungary has officially adopted the UNECE Strategy for ESD 2005 in Vilnius at a High-Level Meeting of Environment and Education Ministries, when the Decade was launched in the UNECE region (UNECE 2005) Although the Decade has not resulted in significant new ESD initiatives in Hungary, it has strengthened the position of those programmes which were running before, and made it easier for them to continue. The most significant effect of the Decade in Hungary has been the growing public under-standing of ESD. The continuous transition from the narrowly defined concept of EE to the broadly defined concept of ESD, and the growing integration of the social and economic aspects of sustainability into the existing educational programmes were crucial as well.

An influential document which demonstrates these tendencies was a White Paper written at the request of the President of Hungary, László Sólyom, which proclaims that ESD is one of the guiding principles for the renewal of the educational system (Csermely et al. 2009). Additionally, the studies described the beneficial effects of giving more support for ESD (Réti 2009). As the White Paper inspired educational policy-makers, the revised new National Core Curriculum (Goverment Decree 2012) and the Frame Curricula (Ministry of Human Reseources 2012) define sustainability and ESD as a cross-cutting principle, while keeping issues of EE also as priorities, but mainly with relevance to science education.

12.2 State of the Art—Islands of Innovation

However, the effect of the regulatory documents on educational practice is limited and in Hungarian public education the position of ESD is uneven. At present, ESD is dealt with in all public education institutions, but in this implementation there is a need for improvement. The regulatory documents (however forward-looking they may be) do not really influence everyday pedagogical practice. The most striking deficiency is that not all students receive ESD and there are substantial differences between schools in content and quality. There are several outstanding, interna-tionally networked schools that have incorporated EE into their whole school-life and combined it with citizenship, global, and cooperative education as well as other progressive educational fields, effectively doing ESD (Vajgerné Bőhm 2010). Unfortunately, these initiatives remain isolated examples. At best, they keep contact with each other, but do not deeply influence the mainstream.

Most schools do not know how to fulfil their legal obligations. Moreover, many teachers who are supposed to be responsible for ESD do not recognise its relevance or do not understand what it means. In practice, teachers often work with general aims, content and forms of EE and very rarely link it to local environmental problems and values or attempt to connect the global issues to the local possibilities and responsibilities. Therefore, their students are not really touched by ESD and will not be prepared for responsible decision-making concerning sustainability issues which affect their everyday life and their future (Réti and Varga 2008).

Fig. 12.2 Percentage of teachers in the primary schools collaborating for implementing ESD into daily practice (Majer and Réti 2013)

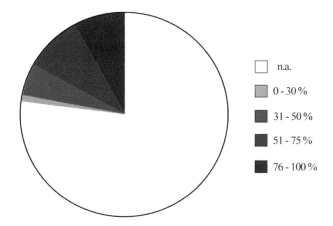

☐ n.a.

■ 0 - 30 %

■ 31 - 50 %

■ 51 - 75 %

■ 76 - 100 %

In a recent survey in primary schools in Hungary, only 21 % of the schools stated that environmental consciousness was an important aim to achieve in education, and only 7 out of 641 responding schools embedded more than 6 aspects of ESD in their daily work (Majer and Réti 2013).

An additional problem is that, despite being a cross-cutting priority, ESD is most often integrated only into science subjects, with the result that only science teachers deal with sustainability issues. Students are mainly acquainted with topics focused on the environmental crisis and environmental values (Havas 2006). ESD embedded in science lessons remains incapable of showing the social causes of the crisis or of preparing students for solving environmental problems or for preserving values, as both the causes and the possible solutions can be primarily found in the humanities or social sciences (Varga 2004). Responses from the Eco-schools in the earlier mentioned survey mentioned produce the data summarised in Fig. 12.2.

In Fig. 12.1 it can be seen that only 8 % of the schools engage regularly in collaborative practice among the teachers related to ESD, which (compared to the number of respondents) means only 11 schools. There is evidence to show that teachers express the need, but show little willingness, to participate in professional networks or to establish collaboration even with their colleagues in the same school or same city (Réti and Iker 2011).

In the light of the evolution of ESD in Hungary (namely the strong interconnectedness with civil movements) it might seem somewhat contradictory that participation as well as strengthening social skills and student autonomy are neglected parts of EE, despite the fact that educating for active democracy and entrepreneurship have also been key development tasks and therefore have also formed a crosscutting aspect in the National Core Curriculum since 1996. A possible reason behind this is that teachers' competencies need first be strengthened.

On the other hand, there have been several exemplary initiatives in the international context, for example:

• the Eco-school quality standards and the living network of qualified Eco-schools;
• mainstreaming the forest schools movement into public education;

- the NGO-initiated and regularly renewed National EE Strategy;
- the cooperation agreement between the relevant ministries jointly supporting EE and ESD as a whole in Hungary.

Despite the growing general public concerns for sustainability problems and the environmental crisis in recent years, the general, institutionalised strategic support from government towards ESD has declined. This slackening effort for catalysing, coordinating, communicating and networking has also contributed to deepening the gap between the elite (or pioneers) and the mainstream.

The increasing but not harmonised financial possibilities invited several actors to engage in ESD. Therefore there are well developed, sound educational programmes from experienced organisations together with the adventurers and inexperienced, naïve newcomers without any understanding for the needs of schools and their operation. But even the exemplary programmes are not reaching the majority of schools.

12.3   Causes and Conditions for Mainstreaming

Finding a remedy for the above problems is made difficult by several factors. Most teachers carry out ESD activities without professional commitment and enthusiasm. Professional control and output regulation do not assess adequately aspects of ESD and there is no easily accessible professional assistance for teachers. Teachers are left alone with the task of finding partners in society around the schools like NGOs, national parks, museums, cultural institutions and other organisations, and there is a lack of cooperation skills on both sides. Teacher training institutions place little emphasis on fostering inquiry-based approaches and in-service training events seldom focus on developing teacher competencies in fields related to inviting students to engage in co-operation, active participation and critical inquiry. There is also a lack of providing teachers with methods and empowering them in dealing with emotional commitment and shaping environmental attitudes (Radnóti 2012; Kiss 2012).

12.3.1   Financing and Methodological Issues

Asking a Hungarian teacher about the main obstacle for ESD in Hungary, one of the most likely answers is lack of resources, in spite of the availability of more and more opportunities to develop schools. An innovative way of eliminating financial obstacles could be the introduction of an ESD per capita grant as an earmarked part of the public education per capita grant provided by the state. This earmarking process, and an associated supporting and controlling mechanism, could guarantee that those schools that fulfil their obligations would automatically get the grant for serving the aims of ESD. In Hungary, dozens of different per capita grants are supporting different educational aims, from ICT (information and communication technology) development to integration of disadvantaged students, so per capita grants are an accepted way of expressing educational values and priorities. This solution would not mean looking for extra resources for ESD, but using the general cash-flow for schools in a more environmentally conscious way. Another currently less utilised opportunity for financing ESD developments in schools is the inclusion of strict environmental and sustainability cross-cutting conditions into all of the school development tender processes. Nevertheless, the first step should be to harmonise the growing financial resources for ESD and to set up national goals and priorities for ESD.

A comprehensive survey of the state of EE in Hungary (Albert & Varga 2004) showed that, with regard to ecological literacy, environmental attitude, environment protection and nature conservation activities, positive signs for EE can be observed.

Nearly inexhaustible opportunities for becoming familiar with and developing student groups and pedagogical practice are provided by the huge repository of quality assurance and action research (Bereczky & Varga 2004). The use of instruments facilitating pedagogical self-reflection (questionnaires, interviews, notes, video recordings, observations, photos, reports, problem analyses) is common in both approaches but quality assurance pays more attention to institution-level phenomena while action research deals with classroom-level phenomena. Initiatives compounding the advantages of these two methods for ESD, based on reflection on educational work could be beneficial.

12.3.2 Creating a Professional Background for ESD

An urgent task to cope with the challenges of ESD is to create an appropriate professional information network, which would enhance the improvement and deepening of the relationship between local civic and public professional organisations and with the educational and local authorities. With the help of this network, schools could get information about tenders, receive help for preparing tender documents, view sample curricula, learn about ideas and best practices, and get access to teaching aids.

All these developments are conceivable only if based on appropriate research. There is a need for research related to ESD, which would describe everyday practice, on the one hand, while, on the other hand, formulating evidence-based suggestions that can easily be introduced into everyday practice.

New developments can only reach schools if there is an appropriate system of in-service training. All accredited in-service teacher training should contain a compulsory element of ESD, which would guarantee that the in-service training affects the practical work of the teachers. Supporting shorter, more flexible forms of in-service training would also be necessary.

12.3.3 Personal Conditions for ESD

Teachers should not need to cope alone with the tasks of ESD. However, currently there is little partnership-like co-operation between teachers, students and parents. Students and parents are usually passive receivers. The teacher is the actor who plans, controls and evaluates the educational process. The pedagogical approaches, which serve EE and ESD, should pay special attention to involving students and parents as partners from planning through realisation to evaluation.

Teachers take up tasks of EE and ESD motivated by professional commitment, moral and voluntary motives. Seldom do school principals call upon teachers and financial motivation is also scarce. In the majority of schools teachers receive only moral recognition. In addition to the financial solutions discussed separately, a system of professional recognition (awards, prizes, certificates of merit) would also contribute to honouring experts in ESD and to the development of the profession. Such a system would make it possible that excellent and effective educators receive recognition increasing their prestige. Thus such outstanding ESD activities of certain teachers would become an instrument for developing teachers' individual careers as well.

12.3.4 Development of ESD at School Level

The efficiency of ESD is greatly enhanced if students experience it as a well-thought-out and well-designed process. Therefore, it is necessary to enhance school-level development of ESD.

A basic condition for school-level developments is that the school has a programme for EE and that the local curriculum requires the school to include ESD in all subject areas. The pedagogical programmes of schools do not regulate the EE and ESD learning activities in sufficient detail and with due professional support.

In order that local curricula be efficiently realised and continually renewed, it is inevitable that they are tightly connected to the quality assurance plan of the school. However, the connection between local educational programme and quality assurance is usually very loose, in most cases it does not exist at all. It would be advisable to integrate best practice examples of connecting ESD and quality assurance into pedagogical systems.

An organic part of quality assurance, which is dealt with by very few schools indeed, is the detailed documentation, dissemination, and presentation of the system and achievements of their own EE. The successful practice of ESD of individual institutions is isolated; unknown to the wider professional community and therefore it cannot be multiplied/re-used. This situation could be improved if the biggest possible number of schools joins regional or national networks providing opportunities for them to share experiences and to find common solutions for their problems.

The majority of public education institutions have no access to such support and resources that could guarantee school-level development of ESD. The situation could be improved by launching training courses about exploring and obtaining resources and support, where schools could get assistance for utilising the opportunities available from civil society. Operating charity funds helping the school, intense grassroots fundraising, finding sponsors and supporters, obtaining 1 % of personal income tax are all opportunities that use resources of the society while enhancing contacts between the school and society. At the same time these resources are much more flexible than the support of a given framework granted by successful competitions. Hungarian schools usually find it very difficult to move towards school-level development of ESD and environment-conscious, sustainable operation and seldom build partnerships with local authorities and firms.

All forms of local networking within and between groups of teachers of different subjects should also be empowered—including learning more about good practices in different fields (Réti and Iker 2011). There are schools with ongoing traditions of theme days, interdisciplinary projects related to raising environmental consciousness: these good examples should receive more attention (Kézy and Varga 2007). Moreover these projects could provide teachers themselves with experiences about co-operation and in this way contribute to developing their social competencies and shaping their attitudes to appreciate entrepreneurship and social responsibility.

12.3.5   Teacher Training

Newly trained (often young) Hungarian teachers emerge from science and information technology-centred higher education (Kiss 2012) and they will work in ESD in the same way as they carry out their pedagogical work in general, unless they get special coaching. Related to this phenomenon there has been a nearly decade-long debate among Hungarian experts of education about the following issues: is it enough if pedagogy tries to affect knowledge or should it directly influence attitudes and actions as well? (Réti 2012) The first view is easier to accept from both scientific and practical perspectives. Information constituting the basis of environmental awareness is much easier to define and study scientifically than attitudes or even actions. What is more, there always should *per definitionem* be some kind of consciousness and knowledge underlying environmentally conscious attitudes and actions. The relationship between knowledge and emotions is not unambiguous. If we want to develop environmentally conscious attitudes and actions—even if we accept the most radical cognitive view—it is not only lexical knowledge that is required, but also the interpretation of basic information in context, the ability to take in new information and the entire dimension of procedural and meta-cognitive abilities. Without dealing with these procedural or meta-cognitive elements ESD will hardly be successful. International surveys have drawn attention to the excessive knowledge-centeredness of the Hungarian school system (OECD 2007), which seems to be a misleading way of raising awareness

(OECD 2009). Current educational policy initiatives represent great steps from knowledge-centred schools towards action- and competence-centred schools (Horváth et al. 2006). Even the early versions of the National Core Curriculum radically broke away from the idea of knowledge-transferring pedagogy. Section 45.§ (5) of the amended Act on Public Education opens up the way to various methods of teaching different from traditional lessons (project work, thematic week), which are more advantageous from the perspective of acquiring competencies (Czippán 2013).

The desirable transformation of public education from the perspective of sustainability is inconceivable without a similar reform of teacher training. It is important that each teacher trainee has the opportunity to study subjects connected to ESD and to acquire knowledge, methods and competencies that can be used in their pedagogical practice. As part of educational professional competences, teacher trainees should acquire methods of self-evaluation and self-development as well. ESD can only renewing itself if it is continuously reflecting on its own assumptions and mindsets.

Both pre-service and in-service teacher training should foster teacher competencies for networking and forming partnerships with stakeholders, supporting trainees in how to communicate and co-operate with other sectors in society. Such programmes could provide trainees with tools and support for their own school practice and therefore could contribute to the delivery of local projects.

12.4   The Possible Future—Good Practice Exemplars

As the previous section tried to show, there are possibilities of high quality ESD in the Hungarian educational system. Fortunately more and more school start to use these possibilities. This section describes the most important initiatives helping schools to improve their ESD work, and then, on the basis of the experiences of these initiatives, summarises the main lessons and challenges of mainstreaming ESD in Hungary.

12.4.1   The Hungarian Eco-school Network

The Hungarian Eco-school Network has been operating as a locally adapted realisation of the Eco-school project of ENSI (Environment and School Initiatives) since 2000. The coordinator of the Network is the Hungarian Institute for Educational Research and Development (HIER), an institute of the Ministry of Education and Culture. The Ministry of Environment and Water also supports and strengthens the governmental support of the Network, thus the network is a joint programme of the two ministries.

The Ministry of Education and Culture and the Ministry of Environment and Water have, since 2005, awarded the "Eco-school Title" as the highest level of acknowledgement of Eco-school activity. Since then schools receiving this title have automatically become members of the Hungarian Eco-school Network. The Network is open to every Hungarian public educational institution.

Schools can join the Network every school year via a tender for the Eco-school title, and they have to renew their title every 3 years by reporting their developments and update their plans.

All fields of school life are concerned with aspects of sustainability. Civic education and participatory democracy are considered as vital aspects of the philosophy of eco-schools.

In order to become an eco-school, a school has to apply for recognition to the Ministries. Until 2009 they had to develop their work plans following the eco-school criteria system which consisted of 52 criteria with 90 sub-criteria regarding all areas of school-life. The main points of these criteria are: environmentally friendly behaviour, saving of resources and care for health. Schools do not have to meet all the criteria at the same time. To obtain the title only 25 criteria have to be met within 3 years. The compulsory criteria were the easiest to achieve. Most of the schools have met the criteria.

Small-scale changes were introduced in 2007 when criteria developed by the schools themselves were integrated into the system.

The eco-school work plan can be easily integrated into the compulsory work plan and quality assurance system of the schools.

In 2010 the application system changed, mainly due to the growing number of Eco-schools, with a focus on quality control of Eco-schools, introducing a self-evaluation approach. Applicants are no longer expected merely to formulate and fix their commitments and plans for the development of ESD, but also to describe already achieved results. Compared with the past, most of the criteria have been included after a reformulation on the basis of 5 years' experience with the Eco-School title. The criteria are grouped into nine larger thematic units. The minimum expected score is also defined by thematic units. Documentation and a responsible person should be defined for each fulfilled criteria.

Applicant schools identify at least one specialty themselves, which characterises the school in addition to the nationwide common criteria.

The main aim of the title is to institutionalise ESD activities on the school level and to ensure continuity of ESD in a school. It does not require difficult and expensive innovations or investments, which only few schools could afford. Eco-schools are not supposed to be an elite group of schools, but a group of schools committed to a learning journey towards ESD. The Network helps them to work in any area of school education and enforce environmental and sustainability principles of pedagogy and to serve as an example for all. Thus the title becomes both a compass for educational journeys inwards and outwards. The title is also a compass in the sense that it does not involve direct financial support, but is more a way of generating resources to help school development. One example of the resource generator effect is the case of school tenders supported by the EU Structural Funds: during the evaluation of tenders (aiming to develop school infrastructure, for instance) the Eco-school title means extra points for applicants.

At the moment the network has 711 member schools, so the network covers a little more than 15 % of the entire Hungarian public school system. Eco-schools can be found in the entire country, every region and every county, and are coordinated through HIER and its website www.okoiskola.hu (accessed 26 March 2014). The site is constantly updated with news, publications, teaching aids, and databases available to all interested and a weekly electronic newsletter is produced, which currently reaches more than 2,300 addresses.

The Network continuously helps its members to participate in national and international in-service teacher training programmes. The members' work is also supported via action research and pedagogical-psychological research, providing a theoretical basis of the pedagogy of sustainability and by supporting connections between green NGOs and schools to deepen the idea of sustainability in the everyday life of schools.

The main strength of the Hungarian Eco-school Network is that this initiative links research, governance and pedagogical development. This wide range of professional input is combined with a holistic approach to sustainability, so the Network helps its member schools to develop their work in any field of sustainability as well as in educational and technical issues.

The Eco-school Network system provides a professional framework for schools to develop and to implement their own sustainability projects. In this way the system is very cost-effective because the central investment generates many local projects based on local resources.

HIER is still responsible for the coordination of the Network, but gradually the initiatives are more and more shifted from top-down to horizontal ones, starting from the participating schools.

12.4.2   The Green Kindergarten Network

The green kindergarten system is very similar to Eco-Schools and more than 250 kindergartens have acquired a Green Kindergarten Title since the launch of the programme in 2006.

The main aim of the Green Kindergarten Programme is to ensure high quality EE for children before the school age. There is close co-operation between the Eco-school and Green Kindergarten Networks in order to develop a continuous transition for children from kindergarten life to school life in terms of EE as well. The term EE is here (as in some policy documents) used in a sense that it refers to ESD as well.

12.4.3   Forest Schools

In Hungary a forest school is:

> a specific learning and educational organizational unit that builds on the features of a particular environment. It refers to a learning arrangement comprising an uninterrupted period of several consecutive days during school time, outside the boundaries of the

headquarters of the organizing educational institution that aims at exploiting the interactive, grass-roots initiative, two-ways co-operation of the learners in the educational process. In terms of content and curricula, the teaching process is closely and integrally linked to the natural, artificial (built-by-humans) and socio-cultural environment of the chosen location. Its prime educational task is to develop healthy, unified behavioural capabilities that are in harmony with nature and the environment, and promote socialization rooted in community activities. (Erdei Iskola Program 2003: 1)

Forest schools are organised outside educational institutions, but despite their name, are not necessarily located in a forest, as a forest school may equally well operate on a lakeshore, in a grassland habitat or even in a municipal environment. A school can organise its forest school activities by itself or teachers can cooperatively develop and buy some programme elements and other services from a forest school service provider.

After many years of development and growing interest of different stakeholders, a governmental Forest School Programme was officially launched in 2003 with the intention of a 6-year long continuous developmental process financed by the ministries. The two most important target groups of the programme were the institutions of public education—in particular primary schools—and the organisations providing forest schools services.

It was a fundamental goal of the programme to promote contact building between the different players of forest schooling with respect to the following professional domains:

- schools;
- forest school providers;
- organisers of training linked to forest schooling;
- providers of expert services related to forest schooling.

The creation and operation of forest school networks was facilitated by internet mailing lists, joint events, and workshops and consulting. It is important that in the longer run each network has regional and local partners so that local issues can be resolved and tasks can be performed on the spot without the involvement of national centres.

The main objectives of the Forest School Programme included establishing an appropriate professional background for forest schooling in public education, particularly in primary schools; creating a group of service providers ensuring a quality of forest school programmes and suitable to provide other services of EE and eco-tourism; creating appropriate information and communication background for forest schooling; and establishing funding conditions to allow every child of primary school age to visit a forest school away from home at least once during his/her studies.

In order to ensure the quality of forest school services offered by service providers to schools, a qualification system was developed within the framework of the programme with the participation of the different stakeholders, such as service providers, schools, government and NGOs. The qualification system was successfully launched and still operates with NGO coordination. Since 2009 the qualification system of Forest Kindergarten also operates on a very similar basis, but naturally with an adapted criteria list.

Forest schooling in Hungary is now an integral part of the pedagogical programme of educational institutions, which serves to implement the objectives of the local curricula during school time. The didactics, tools and content of teaching and learning in a forest school differ from those familiar in classrooms. The content of learning is focused around revealing and understanding the relationship between humans and the environment. Knowledge transfer is driven by the curiosity and absorbing attitude of the children. Forest schools provide a scene for social learning, and promote intellectual and community development. The premises of the forest school can be interpreted simultaneously as a real and a virtual symbolic environment, an observation which is in favour of the methods of the pedagogy of drama. The experience of forest schooling has an impact on the entire school community. If the school community succeeds in sharing these experiences with each other and incorporating them into the school programme and toolkits available to teachers, then it can be able to set a trend along which new initiatives can be built in the future. This may contribute to improving the working atmosphere and positively influence the attitude of the parents and the school funding agencies towards the institution.

The programme was based and built on a wide range of interest groups in society. Institutions of public education, civil organisations, plus both governmental and entrepreneurial sectors got involved. Gradually, the programme has become quite independent of governmental resources. Although the final form of the programme has been shaped strongly with the help of governmental support, it is not primarily a top-down initiative but is strongly built on the common and simultaneous effort of different sectors of society.

12.4.4 Eco-universities

Universities and colleges, as trainers of future professionals, are important institutions for raising environmental and sustainability awareness. In order to help achieve this aim, the Conference of European Rectors (CRE) initiated the Copernicus Charter, which was launched in Geneva in 1994 (CRE 1994). Several Hungarian higher education institutions have joined the Charter and there are more preparing to do so. They are dedicated to take steps according to the three main objectives of the Charter: to activate higher education institutions in the field of sustainability, to encourage co-operation among them in this area, and to help build co-operation between the economy and the higher education institutions concerning environmental issues.

Beside the voluntary participation in ESD initiatives, most Hungarian higher education institutions have developed their sustainable development strategies to meet the requirements of European funding for financing their developmental projects. Although these strategic documents so far have a limited effect on the daily work of the institutions they could be a starting point for a ESD-centred transition of Hungarian higher education (TÁMOP 2012).

12.4.5 *National Environmental Education Strategy*

In the last few years, several countries have produced EE strategies for short- or medium-term development. Such a governmental strategy does not exist in Hungary but the Hungarian Society for Environmental Education (HSEE) has since 1997 coordinated the revision and continuous development of a National Environmental Strategy on an NGO basis (Vásárhelyi and Victor 2003). The strategy is edited in a participatory way with the contribution of several dozen NGOs and other stakeholders, involving several hundred experts in the process of revision and amendment of the actual version of the strategy every 6 years.

The introduction of each new version of the strategy introduces the topic of EE (now in fact ESD) as a whole, and the reasons why the strategy was developed. Each chapter begins with a state-of-the-art assessment followed by proposals, identifying the largest gaps, errors, and the needs for action in relation to the topic, offering recommendations and solutions. Finally, the authors point out the connections to other chapters of the Strategy.

The dominant element of the Strategy is to try to reach a wide section of society, including political and economic actors as well as socially or culturally marginalised people, and those not currently interested in the issues at stake.

12.5 Main Challenges of Mainstreaming of ESD in Hungary

The previous sections provide an overview of the main nationwide ESD initiatives in Hungary. The overall picture is ambiguous. On the one hand, from an optimistic point of view, a rapid development can be detected regarding the number and also the quality of work in involved institutions in different areas devoted to the development of EE and ESD. On the other hand, from a pessimistic point of view, it is clear that the vast majority of Hungarian educational institutions and people are still not reached by ESD.

The real challenge is to involve those thousands of schools, kindergartens, and higher education institutions which still operate without real commitment to develop their usually weak ESD work. The main lessons from the development of existing ESD are as follows:

12.5.1 *Step By Step*

Although a vision of ideal ESD work in schools can be determined on the basis of theories and the literature, the real school development work has to begin in each school at their actual level of development. If a development project aims at an ideal ESD school which is far removed from the practice in a given school and if it is not showing a clear step by step way, it will probably create resistance and incomprehension in the school community.

12.5.2   Snowball Effect, Peer Tutoring, Networking

The most efficient way to attract new schools into the world of ESD is to establish contacts between schools so that successful schools can share their success stories in ESD with others. No central training, no supplementary material, no control system can be as effective as another committed and successful school or a teacher having had positive experiences in the field of ESD. This approach means that the majority of the schools are not in direct contact with the official centre (or the bureau) of the ESD development initiative. There exists an indirect contact though through the teachers of other schools who serve as trainers for the initiative. That is why training the trainers is a key element of successful networking.

12.5.3   Reward

Motivation of schools and teachers are key elements of ESD development work. Official appreciation does not mean extra financial resources. In many cases professional and moral appreciation is enough and even perhaps more impor-tant—and from the point of view of the donors, it is easier to manage and maintain in the long run.

12.6   Support

But not even rewards or expectations or controls can cause a significant develop-ment in practical ESD work in schools. The history of development of ESD in Hungary clearly shows that without professional support the majority of schools would not change their old EE practices. The most effective support is not adding additional resources to the system but turning the existing financial and professional support systems into support for the development of ESD. This does not require extra efforts from schools to take part in the development and shows official recognition of the ESD development efforts.

12.6.1   Empowerment

Teachers need personal experience of team-work, partnership, participation and social responsibility. This provides a basis for them to feel dedication for applying student-centred approaches and initiate projects which focus on inquiries about local sustainability problems or ecological values.

12.6.2   Building an Institutional Framework

The integration of support for ESD into existing support systems is also a crucial step towards systemic developments in schools. Long lasting ESD development requires an appropriate institutional framework. Even if a new teaching method, a new co-operation or any other element of the development of ESD is included in the official documents and processes of the school, it is almost certain that the development will disappear at the very moment when the initial project ends or when the teacher who plays a central role in the innovation leaves the school. That is why long term support systems such as the qualification systems in Hungary are needed. Two or 3 year development projects cannot give the schools enough of a chance to embed an innovation. There is a need for long term development projects which can serve as a brand (like Eco-schools or forest schools), so that schools and teacher can identify with them.

However, these conclusions are not the Philosopher's Stones. No one can be sure that heeding these lessons is enough to achieve the basic goal of ESD, namely to provide quality ESD for all pupils and students in order to enable all pupils, students and staff to act sustainably and live their lives accordingly. The history of the existing networks clearly shows that development efforts usually follow the easiest way where quick success can be achieved. None of these initiatives are aimed directly at educational institutions and educators that are still resisting ESD. The philosophy of these initiatives is based on voluntary participation. The question is: is it possible to reach all the institutions, in other words to really mainstream ESD on this voluntary base? Can this voluntary approach reach all the teachers? It would be ideal as voluntary participation is much more desirable than coercion, but there is a high probability that it will not work. There is no guarantee that all the schools and other educational institutions can be integrated in this way into systemic improvement. This is why the main challenge for the development of ESD in Hungary (and probably also worldwide) is to create conditions within the educational systems which do not allow actors to escape ESD. The still strong resistance towards ESD can be diminished only by a combination of voluntary opportunities and strong expectation for ESD from the stakeholders of the educational systems, from authorities through to parents, to local communities and children. ESD cannot be mainstreamed in any school system against the will of the society. In other words: what is needed is a strong commitment of society toward ESD.

References

Act No. 79 in 1993 (modified 62/1996) on Public Education in Hungary, §44–45, www.okm.gov. hu/letolt/kozokt/kozokt_tv_070823.pdf. Accessed 23 Mar 2014.

Á. Majer, A., & Réti, M. (2013). *A fenntarthatóság pedagógiájának megjelenése a termé szettudományos nevelésben*. Presentation at Kiss Árpád Emlékkonferencia, 6 Sept 2013 (in press).

Albert, J., & Varga, A. (Eds.). (2004). *Lépések az ökoiskola felé*. Budapest: Országos Közoktatási Intézet.

Bereczky, R., & Varga, A. (2004). Akciókutatás a Remetekertvárosi Általános Iskolában. In É. Csobod & A. Varga (Eds.), *Fenntartható közösségek és iskolafejlesztés. Innováció a tanárképzésben, Az akciókutatás és a környezeti nevelés lehetőségei* (pp. 59–64). Budapest: Országos Közoktatási Intézet. http://documents.rec.org/publications/Fenntarthato_kozossegek_ es_iskolafejlesztes_HU_2004.pdf. Accessed 1 Sept 2013.

CRE. (1994). *Copernicus—The University charter for sustainable development*. http://www.iisd. org/educate/declarat/coper.htm. Accessed 1 Sept 2013.

Csermely, P., Fodor, I., Joly, E., & Lámfalussy, S. (2009). *Wings and weights. Proposals for rebuilding the education system of Hungary and combating corruption*. Budapest: Wise Men Committee.

Czippán, K. (2013). Science, sustainability and the big issues. Keynote lecture at "exploring effective approaches to outdoor learning" real world learning network conference, 21–24 January 2013, Sluňákov, Czech Republic, http://www.rwlnetwork.org/events/exploring-effective-approaches-to-outdoor-learning.aspx. Accessed 1 Sept 2013. Also in exploring effective approaches to outdoor learning. conference report (p. 9). http://www.rwlnetwork.org/ media/10961/rwl_slunakov_report_v1.pdf. Accessed 1 Sept 2013.

Decision of Parliament No. 83 of 1997 on the National Environmental Protection Program in Hungary http://njt.hu/cgi_bin/njt_doc.cgi?docid=30389.48706. Accessed 23 Mar 2014.

Erdei Iskola Progam. (2003). *Fogalomtár*. http://www.nefmi.gov.hu/letolt/kozokt/Erdei_ Iskola.../fogalomtar.do. Accessed 9 Sept 2013.

Government Decree No. 17 of 2004 (V. 20) on Frame curricula. http://njt.hu/cgi_bin/njt_doc.cgi? docid=83590.257939. Accessed 23 Mar 2014.

Goverment Decree. (2012). Goverment Decree No. 110 of 2012 on national core curriculum. http://net.jogtar.hu/jr/gen/hjegy_doc.cgi?docid=A1200110.KOR. Accessed 5 Aug 2014.

Havas, P. (1996). *A környezeti nevelés gyökerei Magyarországon*. Budapest: Infogroup Rt.

Havas, P. (2001). A fenntarthatóság pedagógiája. In K. A. Wheeler & A. P. Bijur (Eds.), *A remény paradigmája a 21. század számára* (pp. 9–40). Budapest: Tan-Sort Bt.

Havas, P. (2002). A fenntarthatóság pedagógiája. Elméleti és gyakorlati kérdések. *Fejlesztő Pedagógia, 2–3*, 16–27.

Havas, P. (2006). A természettudományi kompetenciákról és a természettudományi oktatás kompetencia alapú fejlesztéséről. In K. Demeter (Ed.), *A kompetencia. Kihívások és é rtelmezések* (pp. 199–216). Budapest: Országos Közoktatási Intézet.

Hesselink, F., Kempen, P. P., & Wals, A. (2000). *ESDebate: International debate on education for sustainable development*. Gland: IUCN CEC.

Horváth, D., Száraz, P., & Varga, A. (2006). A környezeti kompetenciák fejlesztése Magyarországon. Eredmények és lehetőségek. In K. Demeter (Ed.), *A kompetencia. Kihívások és értelmezések* (pp. 229–244). Budapest: Országos Közoktatási Intézet.

Huckle, J., & Sterling, S. R. (1996). *Education for sustainability*. London: Earthscan.

Kézy, Á., & Varga, A. (2007). Az ökoiskolák szerepe a közoktatás reformjában. *Új Pedagógiai Szemle, 12*, 41–52.

Kiss, G. (2012). *Az új felsőoktatási törvény szerinti tanárképzési rendszer elemzése és a várható változások*. Szeged: Magyar Biológia Tanárok Országos Egyesülete.

Kiss, F., & Zsiros, A. (2006). *A környezeti neveléstől a globális nevelésig. Oktatási segédanyag*. Nyíregyháza: MPKKI. http://www.nyf.hu/ttik/sites/www.nyf.hu.ttik/files/doc/kornyezeti_ neveles.pdf. Accessed 1 Sept 2013.

Könczey, R. (2006). Az európai környezeti nevelési törekvések és a magyar környezeti nevelés. In A. Varga (Ed.), *Tanulás a fenntarthatóságért* (pp. 24–48). Budapest: Országos Közoktatási Intézet.

LIII. (1995). Act No. LIII. of 1995 in the General Environmental Protection Rules: §54, paragraph 1, http://www.njt.hu/cgi_bin/njt_doc.cgi?docid=23823.239026. Accessed 23 Mar 2014.

LIII. (1996a). *Act No. of 1996 on the National Core Curriculum in Hungary*. http://www.complex. hu/kzldat/t9600062.htm/t9600062.htm. Accessed 23 Mar 2014.

LIII. (1996b). *Act No. LIII. of 1996 on Nature Conservation in Hungary.* http://faolex.fao.org/cgi-bin/ faolex.exe?rec_id=007942&database=FAOLEX&search_type=link&table=result&lang=eng &format_name=@ERALL. Accessed 23 Mar 2014.

Ministry of Human Reseources. (2012). Ministry of Human Resources Decree No. 51 of 2012 on Frame Curricula. http://net.jogtar.hu/jr/gen/hjegy_doc.cgi?docid=A1200051.EMM. Accessed 5 Aug 2014.

OECD. (2007). *PISA 2006 Science competencies for tomorrow's world.* OECD PISA database 2006, Tables 2.7, 2.8, 2.9 and 2.10. http://www.oecd.org/pisa/pisaproducts/pisa2006/ pisa2006results.htm. Accessed 1 Sept 2013.

OECD. (2009). *Green at fifteen? How 15-year-olds perform in environmental science and geoscience in PISA 2006.* Paris: OECD. http://www.oecd.org/pisa/pisaproducts/pisa2006/ 42467312.pdf. Accessed 1 Sept 2013.

Radnóti, K. (2012). *A jelenlegi, úgynevezett Bologna típusú, vagy osztott és a korábbi, vagy osztatlan természettudományos tanárképzési rendszer bemutatása és értékelése SWOT analízissel.* Szeged: Magyar Biológia Tanárok Országos Egyesülete.

Réti, M. (2009). A fenntartható fejlődés támogatása intézményes neveléssel. In M. Réti (Ed.), *Szárny és teher. A magyar oktatás helyzetének elemzése* (Háttéranyag). Budapest: Bölcsek Tanácsa Alapítvány. http://mek.oszk.hu/07900/07999/pdf/szarny-teher-oktatas-hatteranyag.pdf. Accessed 1 Sept 2013.

Réti, M. (2012). *The state of science education with special respect to teacher training, the situation of novice science teachers and the teachers' carrier model* (Synthesis report). Szeged: Magyar Biológia Tanárok Országos Szövetsége.

Réti, M., & Iker, J. (2011). A SINUS programcsomag bevezetésének lehetőségei. In J. Iker & M. Szihay (Eds.), *Záró tanulmányok a Pedagógiai szolgáltató és kutató hálózat kialakítása a pedagógus-képzésben a nyugat-dunántúli régióban című project eredményeiből* (pp. 82–97). Szombathely: NYME PSZK.

Réti, M., & Varga, A. (2008). Új tendenciák a fenntarthatóságra nevelésben. Avagy miért kellene egy tininek megmentenie a Földet? *Új pedagógiai Szemle, 10,* 17–42.

TÁMOP. (2012). *A fenntartható fejlődés szempontjai a felsőoktatási minőségirányítás intézményi gyakorlatában,* Research report. Oktatáskutató és Fejlesztő Intézet.

UNECE. (2005). *High-level meeting of environment and education ministries* http://www.unece. org/?id=8452. Accessed 9 Sept 2013.

Vajgerné Bőhm, E. (2010). *A fenntarthatóság pedagógiája a gyakorlatban.* Masters thesis at Pannon University, Veszprém, Modern Filológiai és Társadalomtudományi Kar, Neveléstudományi Intézet.

Varga, A. (2004). *A környezeti nevelés pedagógiai és pszichológiai alapjai* [The pedagogical and psychological basis for environmental education]. PhD dissertation at ELTE University, Budapest.

Vásárhelyi, J. (Ed.). (2010). *Nemzeti Környezeti Nevelési Stratégia.* Budapest: Magyar Környezeti Nevelési Egyesület. http://mkne.hu/NKNS_uj/layout/NKNS_layout.pdf. Accessed 11 Mar 2014.

Vásárhelyi, T., & Victor, A. (Eds.). (1998). *Nemzeti Környezeti Nevelési Stratégia. Alapvetés.* Budapest: Magyar Környezeti Nevelési Egyesület.

Vásárhelyi, T., & Victor, A. (Eds.). (2003). *Nemzeti Környezeti Nevelési Stratégia. Alapvetés.* Budapest: Magyar Környezeti Nevelési Egyesület (Revised strategy).

Victor, A. (2002). *Környezeti nevelés—fenntarthatóságra nevelés.* Nemzeti Környezeti Nevelési Stratégia. Manuscript.

Chapter 13
Education for Sustainable Development in Finland

Mauri K. Åhlberg, Mervi Aineslahti, Annukka Alppi, Lea Houtsonen, Anna Maaria Nuutinen, and Arto Salonen

13.1 Introduction

Finland is an interesting case with regard to Education for Sustainable Development (ESD), because of the high quality of Finnish Schools (as evidenced by their PISA performance), their teachers and the high esteem of Finnish teacher education and educational research. This high quality is based partly on long-term educational research, including the popularity of the 'teachers as researchers' movement in teacher education (Åhlberg 1988, 1989, 1990, 1998, 2005a, b, 2012, 2013a, b, c). In Finland the leading Department of Teacher Education is based at the University of Helsinki, where, in addition, the only Professorship of ESD in Finland can be found.

M.K. Åhlberg (✉)
Department of Teacher Education, University of Helsinki, Helsinki, Finland
e-mail: mauri.ahlberg@helsinki.fi

M. Aineslahti
Sorrila School, Valkeakoski, Finland
e-mail: mervi.aineslahti@gmail.com

A. Alppi
Mahnala Environmental School, Hämeenkyrö, Finland
e-mail: annukka.alpi@hameenkyro.fi

L. Houtsonen
Finnish National Board of Education, Helsinki, Finland
e-mail: lea.houtsonen@oph.fi

A.M. Nuutinen
City of Espoo, Espoo, Finland
e-mail: marinuutinen@gmail.com

A. Salonen
Faculty of Welfare and Human Functioning, Helsinki Metropolia University of Applied Sciences, Helsinki, Finland
e-mail: arto.salonen@gmail.com

© Springer International Publishing Switzerland 2015   221
R. Jucker, R. Mathar (eds.), *Schooling for Sustainable Development in Europe*,
Schooling for Sustainable Development 6, DOI 10.1007/978-3-319-09549-3_13

Professor Mauri Åhlberg has been in this position from its inception in 2004 and has supervised research and development in ESD in Finland.

Most of the theoretical and empirical work has been associated with the Finnish Environment and Schools Initiative (ENSI) school network, starting in 1997. In Finland many key ideas of the ENSI programmes, such as innovating action research in schools, have been adopted. Good collaboration with the Ministry of Education and the National Board of Education have been essential. Since 2000, Lea Houtsonen has been involved in the administration and research of ENSI related educational research. Two doctoral theses of innovative ENSI schools have been completed (Ahoranta 2004; Aineslahti 2009) and one is underway (Alppi 2013). Nuutinen (2013a, b) and Siirilä (2013) collected data from the ENSI related UNU IAS RCE Espoo region (described later in this chapter).

In Finland, the importance of biodiversity education as an essential part of ESD was understood early on (see Åhlberg 1988), based on the importance of biodiversity and ecosystem services emphasised by all four UN World Summits for Environment and Sustainable Development (1972–2012) (see Åhlberg 2013e). Kaasinen (2009)—as a member of the Research & Development Group of Åhlberg's department—published her doctoral dissertation on species recognition in schools and universities, as an important part of biodiversity education. Åhlberg (2012, 2013) and Alppi and Åhlberg (2012) described and recommended the use of the interactive NatureGate approach (http://www.naturegate.net) for the identification of local species (see below).

The quality tools we have developed and tested include improved concept mapping. They were tested in Ahoranta's (2004) doctoral dissertation based on 6 years of design experiments. They were presented in publications by Åhlberg and Ahoranta (2005), Åhlberg and Kaivola (2006) and Kaivola and Åhlberg (2007) as tools to promote ESD both in schools and universities. The research and development on these quality tools have been internationally acknowledged, as evidenced by several refereed research articles (e.g. Åhlberg and Ahoranta 2002; Åhlberg et al. 2003, 2005a, b; Immonen-Orpana and Åhlberg 2010; Siirilä and Åhlberg 2012) and a research methods textbook (Wheeldon and Åhlberg 2012). But—as Salonen (2010) demonstrates in his ground-breaking doctoral dissertation for ESD—it is possible also to manage without it.

In Finland as in many other European countries there are many projects in EE and ESD. But many of them seem like fireworks. They look beautiful, but seem to have no lasting impact. They are not adding anything to each other, or building upon each other. In this chapter we describe several research and development projects that try to overcome these limitations.

We have tested cumulative collaborative knowledge building since the year 2000. The original ideas and experiments of collaborative knowledge building were tested and developed with school children in short term studies (Bereiter 2002). We wanted to test whether this method could be developed into a more cumulative version comparable to real scientific knowledge creation. The approach, including the Knowledge Forum® software platform (Åhlberg et al. 2001; Myllari et al. 2010), has now been employed in several doctoral thesis (e.g. Aineslahti 2009; Salonen 2010).

13.2 Mahnala Environmental School: ESD in an (Extra) ordinary School

Mahnala Environmental School (near Tampere) has grades 1–6 (pupils aged from 7 to 12). It follows the common curriculum of Finnish schools guided by the National Board of Education 2004 (OPH 2004). Mahnala Environmental School has demonstrated that an ordinary Finnish comprehensive school can become a centre for EE and ESD. ESD became an integrated part of school activities in 2004 (Table 13.1) (Fig. 13.1).

13.2.1 ESD at Mahnala Environmental School

Using ESD, different school subjects and educational objectives can be integrated into a coherent general education and school organisational culture. School organisational culture includes the way everyday life is organised and lived in the school. According to the Finnish Core Curriculum 2004 (OPH 2004), ESD has to be

Table 13.1 Overview of the most important milestones in the history of Mahnala Environmental School

1990	A kind of nature school was created by the former principal, an ornithologist. He had a licence to ring birds and he started the tradition of ringing birds with third grade pupils. Field studies and gardening were started. Collaboration with local citizens was also started
1996	Classroom teachers, including Annukka Alppi, brought ideas of modern Environmental Education (EE) to Mahnala Environmental School, including Global Learning and Observations to Benefit the Environment (GLOBE)
2002	Ms. Annukka Alppi joined a research group and started university level collaboration with Mauri Åhlberg, and at the same time joined national and international school R&D projects such as ENSI (originally supported by the Organisation for Economic Co-operation and Development (OECD), School Development through Environmental Education (SEED), Sustainability Education in European Primary Schools (SEEPS)
2005	The school developed its own model of ESD, including its own local curriculum and organisational culture to support it
2008	The school was selected to be part of the Development Centre for EE in the Tampere Region, areas of focus included early years' education, teaching and in-service training
2008	The Green Year medal, a national award, was presented to Mahnala Environmental School for their use of the schoolyard as a learning environment
2008	Participated in the international ENSI SUPPORT project and its conference in Espoo, Hanasaari
2008	Since 2008 the NatureGate® Online service (http://www.naturegate.net/) for learning about local species and their sustainability has been tested and used at the school
2010	Environmental Rose Award from National Society of EE
2012	Participation at ENSI CoDeS (collaboration of schools and communities for sustainable development) working conference, Vienna, Austria

Fig. 13.1 Pupils identifying a wild flower with the NatureGate interactive identification tool for local species (Photo: Annukka Alpi), www.naturegate.net (accessed 17 Apr 2014)

integrated in all school subjects and school organisational culture has to promote learning for wellbeing, a healthy environment and sustainable development.

The immediate surroundings of the school are the central source of inspiration for teaching and learning at Mahnala Environmental School. An important shared aims of the teachers of the school was to cooperate with the people of the school region. There was the will to promote environmental awareness and actions for conserving and improving the environment, conserving local biodiversity, and using it sustainably aiming at the wellbeing of the local people.

There are many possibilities for informal networking. Mahnala Environmental School is a rural school and collaboration with local farmers became very important. Harvesting in local farms, local forests, gardens, local nature have become important curricular topics as well as providing organisational activities to promote learning for sustainability, a good environment and a good life.

Rural life is part of the cultural heritage of the municipality of Hämeenkyrö, where Mahnala Environmental School is located. It is important for pupils to learn to understand, to conserve and to develop the local cultural heritage. In spring time the pupils of the Mahnala Environmental School take part in sowing seeds and planting at local farms. In the autumn they take part in harvesting. They have also followed the lives of farm animals such as chicken and sheep.

It has become a tradition at Mahnala Environmental School to incubate hen eggs in spring. In April 2012 pupils incubated eggs of the local provincial chicken breed (Häme-Tavastia). This rare, ancient breed had only 200 chickens left.

After successful incubation at the school, 13 new chickens were added to the flock. In this way pupils learnt to conserve biological cultural heritage and local biodiversity.

There are many lakes in Finland, some near Mahnala Environmental School. The school collaborates with local fishermen and nature entrepreneurs in arranging fishing when the school year starts in August/September. Later, in winter, when the lakes are covered by ice, pupils learn winter fishing from local fishermen, such as using nets under the ice or ice fishing through holes in ice. Pupils observe many of the 27 local fish species and acquire knowledge about them during these events. All the Finnish fish species can be identified using NatureGate online service (2014).

In Mahnala Environmental School pupils have learned both about wild nature and nature locally transformed by humans (local human history and culture). The local curriculum guide book details the learning objectives for each grade.

There has been an active collaboration between Mahnala Environmental School and another ENSI school in the nearby city of Valkeakoski, headed by Mervi Aineslahti. Pupils of city schools have the possibility to learn more deeply about farming and forestry and other aspects of country life when they collaborate with a rural school and vice versa.

13.3   Sorrila School: A Journey of Sustainable School Development

13.3.1   Theoretical Background

Mervi Aineslahti wrote her doctoral dissertation (Aineslahti 2009) about her work at and development of her school (Sorrila School). In her thinking, EE and ESD are concepts which complement each other. However, ESD has a wider meaning and content than EE. This is how Sorrila School understands the three main aspects of sustainable development as applied to school education:

- The ecological aspect contains things like: nature conservation and biodiversity, learning to identify local species as an obligatory part of the Finnish core curriculum for grades 1–6.
- The economic aspect: Sorting and recycling, consumption, supplies, material resources and maintenance. Basic principles of sustainable production, such as cyclical economy: waste from earlier processes become raw materials for the next processes so that all materials get cycled and/or recycled.
- The socio-cultural aspect: school climate and wellbeing, traditions, feeling of community and empowerment, active learning.

The aim at Sorrila School has been to reach for the balance between humankind and nature. In many cases EE has been understood in a narrow way, meaning only activities in the ecological area. Schools may have worked only with nice little 'nature games' or sorting waste. However, the complex concept of ESD necessitates interaction on many levels and between different parts. It means seeing situations connected and as parts of systems.

For this school, ESD is seen as being based on positive thinking and sharing knowledge and energy for the wellbeing of the whole community: schools, homes, the municipality and NGOs. Sustainability can empower people. It gives them a feeling that together people can achieve more.

Learning in this socio-cultural framework is seen as social interaction. It is situated and always in connection with real life conditions. Therefore, local and regional networking and human interaction play a crucial role. Learning is also very closely connected to emotions, not to forget learning from elders and more experienced people (apprenticeships) as well as peers. *Transformative or expansive* learning is meaningful to the learner. Learning may change both the learner and the environment (Beach 1999; Engeström et al. 2002). The environment contains both physical and psychological elements as well as natural and artificial features (Vygotsky 1978; Dillon 2006). Dillon argues: "The central premise is that each time a learner encounters a new situation, the socio-cultural features of that situation are read afresh: learner and context interact and transform each other." (2006: 71)

Dillon calls this a "pedagogy of connection" (ibid.). It is a way to visualise integrative learning. It is finding connections between different subject matters, school and real life, present and past. ESD is a good approach for finding connections. Integrative learning, originally called integrating learning, has been introduced by Åhlberg (1990, 2005a, b). It was adapted to Dillon and Åhlberg (2006). According to his "integrating learning theory" everyone has to take responsibility for his/her own learning in this complex world. It can be accomplished by being open-minded and able to see connections and integrate different subject matters and disciplines (2005a, b, originally in 1998).

Shallcross and Wals (2006) use the concept of a whole school approach. Teachers, pupils and school staff work together in order to develop the school into a sustainable school. Because human beings are part of many other systems, a school as a community of many human beings is even more complicated and therefore a challenging area for development. Learning does not take place only in schools and teachers are not the only professionals. A local community and parents give the school extra resources which empower both the school and the local inhabitants. Networking and cooperation form a win-win-situation. All participants benefit from such learning for sustainability. Pupils encourage their parents and local authorities to adopt more sustainable actions and ways to live. Positive examples make others curious and willing to be part of the 'the circle of positive thinking'.

According to Aineslahtis (2009) school development and sustainable development have similar features. School development can be connected with learning in general and especially with lifelong learning. Each person's experiences of good practices in the past can be used as building material for school development.

Not only the teachers' and the school staff's, but also the pupils', their parents' and the local community's ideas and knowledge should be used to make the school an arena for *collaborative transformative learning*. This means creating new ideas and knowledge for the whole community and acting in cooperation with each other.

13.3.2 Practical Experiences

Sorrila School is a large primary school situated in Valkeakoski, Southern Finland. There are over 400 pupils and grades from the preschool to 6th grade. Over the years, the school has taken part in several national and international school development projects, all connected to ESD. Being part of the Finnish ENSI-schools network has given it many opportunities to network and develop through shared experiences.

The first Comenius-project (2003–2006) was called *Exploring and Interacting with Our Environment* (http://www.peda.net/veraja/valkeakoski/sorrila/comenius1, accessed 9 April 2014). The school had partners from the UK, Austria, Hungary, Lithuania and another Finnish school, Mahnala Environmental school (see above), which also is an ENSI-school and a close partner over the years. The following project was aimed at developing learning environments. It was titled *Encounters*. During all these years 'Aunt Green' has been an important role model for the students and staff members (Fig. 13.2).

In the *Encounters* project, Sorrila School had a partner from Espoo, Keinumäki school, also a partner of the ENSI-network. The coordinating teacher, Anna Maria Nuutinen, Keinumäki school (Espoo, Greater Helsinki Region), is also a member of Prof. Åhlberg's R&D Group for ESD, and she was very influential in the creation of UNU IAS RCE Espoo (see below).

13.3.3 Conclusions

The change process that relies on sustainable school development leads the school along a road of positive renewal. It is not a series of individual projects but an ongoing process. Widening the concept of sustainability into social dimensions, the wellbeing of all the members of the school community as well as that of nature, its ecosystem services, of conservation and sustainable use of biodiversity, go hand in hand. Mervi Aineslahti and Sorrila school have also tested NatureGate online services at http://www.naturegate.net for species identification. According to the Finnish National Core Curriculum 2004 (OPH 2004) it is obligatory to learn to identify local species during grades 1–6.

Fig. 13.2 The Green flag
and Aunt Green, Dr. Mervi
Aineslahti in Sorrila school

13.4   From the Encounters Project to the UNU-IAS RCE Espoo

13.4.1   The Encounters School Project in Keinumäki School (Espoo)

The background of the *Encounters* School Project included the UN DESD (2005–2014), the international ENSI-Project with its SUPPORT-Partnership and Participation for a Sustainable Tomorrow which began in 2007.

The purpose of the *Encounters* Project was to find ways in which sustainable development and methods could become rooted in the school's daily activities. The project developed methods which involve networking with local stakeholders, authorities as well as researchers and experts and the selection of those pedagogical methods which support social interaction and participation and which enrich the working methods within the school's learning environment.

The goal of the *Encounters* Project in Keinumäki School was to ensure that sustainable development and ESD take place in the school's daily life. To reach the objective the school planned and carried out five learning units/packages, which take into account the following dimensions of sustainable development:

(a) The ecological dimension: conserving and sustainable use of biodiversity, ecosystem functions and interaction, environmental exploration, human responsibility and influence, environment-friendly practices in everyday life, relationship with the nature.
(b) The economic dimension: energy and water saving, sorting and recycling, consumption, equity, procurement.
(c) The social dimension: wellbeing, relationships, respect, safety, equality, participation and influence, finding the joy of learning, critical and innovative thinking, empowering and discovery.
(d) The cultural dimension: our own traditions and roots, multiculturalism, justice, tolerance, science and art. Old cultivars of plants and old animal breeds ought to be included also in this cultural dimension, because they are part of local biodiversity, i.e. part of biological cultural heritage.

The units aid children in noticing the socio/cultural history of the area. They learn to understand the relationships between people and nature; to see how the activities of various age groups affect nature; how people learn differently and how to approach differences. Finally they learn to understand how sustainable ways of life make a difference in building the future.

The five learning packages are:

1. Ancient Espoo helps the human senses to empathise with nature, its moods and changes in different seasons. Nature has been respected and appreciated in ancient times. The package combines art, music and history (mythical stories) into a fascinating entity. It allows students to become familiar with their own roots, and cultural history. Also, in other cultural areas, students can identify with the prehistory of their native country. The units cover the cultural sustainability of the region and consist of six workshops.
2. Medieval Espoo helps to understand how human beings are dependent on nature, soil, water and forest. This package familiarises the students with medieval times, ordinary life and special occasions. Medieval people were strongly linked to the surrounding environment, land and forest. Students learn about medieval occupations in different workshops.
3. The Nuuksio National Park supports the relationship with the wild nature. It is important to have one's own experiences in nature creating the basis for future relationships. This learning unit includes a comprehensive experience of Nuuksio's nature during different seasons.
4. Mapping one's own local environment demonstrates how empirical, inquire-based learning promotes understanding of interactions and the functioning of the ecosystem. The environmental learning package contains instructions for both teachers and students for experimental learning. The package includes

instructions for air, soil and water testing as well as for the exploration of organisms. The aim is that pupils obtain information through their own investigations in the local environment and learn by questioning. It is important to learn systems thinking, to think critically and innovatively. The Nature Gate online tool for species identification has been used (http://www.naturegate.net).

5. Keinumäki School as a learning environment covers everyday practices in school and supports sustainable lifestyles. The learning package aims to support students' wellbeing, participation, influence and activities in their own school, at home and in society. Cultural sustainability includes traditional human cultural history, traditions and multicultural aspects and issues of fairness and tolerance, but also conservation of old plant cultivars and animal breeds to conserve biodiversity.

13.4.2 From the School Project to the UNU IAS RCE Espoo

Espoo is the first Finnish municipality to have been approved as a part of the UNU's global RCE network (Regional Centres of Expertise on ESD). The network of RCEs worldwide aims to constitute a Global Learning Space for Sustainable Development. RCEs aspire to achieve the goals of the UN Decade on ESD (UN DESD 2005–2014), by translating its global objectives into the context of the local communities in which they operate.

13.4.3 RCE Espoo Goals 2011–2014

While both the UN DESD (2005–2014) and UN Decade on Biodiversity (2011–2020) are taking place, it is important to learn to identify local species, local biodiversity and to use it sustainably. For the future brochure of UNU IAS RCE Espoo, the NatureGate R&D Group suggested the following points to promote wellbeing and sustainable development:

1. Using NatureGate® http://www.naturegate.net we may learn to identify local species, and to use them sustainably The sustainability aspect is based on the following reasoning chain: We cannot use any species for food, if we are not sure that it is not toxic. If it is edible, it is not wise to use it too much, in which case the species may become extinct.
2. Using NatureGate® we may map species of Espoo digitally, such as mushrooms, invasive species, berries, and wild flowers. At the same time we can monitor effects of climate change to distributions of organisms.
3. Using NatureGate® pupils at school learn to identify local species better and earlier.

13.5 An Ecosocial Approach in Education

13.5.1 Good Life as a Common Goal

A rapid growth of the population and more materialistic ways of life have transferred planet Earth into the era of the Anthropocene (Zalasiewicz et al. 2011). Human activity has become the most important impact factor on Earth (e.g. Cook et al. 2013). The importance of critical thinking has increased because we are more responsible for future generations than ever before in the human history.

Everybody is looking for a good life. In order to have a good life we focus on material things by buying a smarter phone, a faster computer, a bigger TV, or a bigger apartment (Myers 2000). However, while focusing on material things our subjective well-being might be reduced because of a weaker ability to enjoy positive emotions and experiences of everyday life (Quoidbach et al. 2010).

We tend to forget how simple immaterial things such as experiences of solidarity, harmony, affection and friendship keep us happy and satisfied. During the last decades individualism has increased everywhere and the pursuit of the so-called good life has been based on economic growth and on the power of new technologies (Hofstede et al. 2010). Individualism correlates with an accumulation of material goods. However, it is an unsustainable way to achieve subjective wellbeing.

If our relationship with other people, history and nature is instrumental, we may lose the joy of life, happiness and feeling of being supported by the community (Marglin 2008). According to the World Economic Forum inequality is the biggest risk for the near future (Howell 2013). Due to global markets our interconnections and interdependence are global. A hierarchy of ecological, social and economic interests forms the basis of cultural evolution. Different interests and values can be prioritised in such a way that the chances for a good life do not decrease, but rather increase. This consensus is represented in a new educational concept, called eco-social education, in order to have sustainable and resilient societies.

13.5.2 Systems Thinking and a (Scientifically) Holistic Worldview

The Earth forms a closed and self-sufficient system. In a closed system each part of the system must be in order, so that it can work smoothly. Human beings are part of nature's ecosystem. Nature is the foundation of material and immaterial human development. Systems thinking is a key to understanding the complex and dynamic world around us (e.g. Åhlberg 1988; Senge et al. 2008).

If our goal would be the well-being of human kind, our first concern would be the ecological one. The survival of humankind is fully dependent on ecosystem services such as pollinating of plants by insects, water treatment services, waste decomposition, the UV radiation-protective ozone layer, natural pest control, and fruitful soil.

The economy is an ecosocial process. The ecological base for markets are healthy ecosystem services and a sustainable use of natural resources. In a sustainable society renewable resources cannot be used faster than they regenerate, pollution and wastes cannot be generated faster than they decay and are rendered, and harmful and non-renewable resources cannot be used at all (Sterman 2012: 41). Social justice of markets is based on decent work along every step of the product chain of goods and services (Salonen 2013).

The current stage of the Anthropocene extends all over the Earth. Our responsibility has also increased as the impact of human actions has spread all over the world. Systems thinking helps us to recognise that societies cannot be constructed on short-term economic requirements because the crossing of the planetary boundaries means, in the end, destruction of our economy. Therefore there is a hierarchy between ecological, social and economic elements of culture which form a framework for human activity. First, we have to secure the conditions for life on planet Earth. According to systems thinking ecosystems, societies and markets are in a hierarchical relationship to each other (e.g. Giddings et al. 2002):

1. The viability of ecosystems and the sustainable use of natural resources determine success and possibilities of society and economy.
2. Implementation of human rights (justice, equality, democracy, cultural diversity) determines success of the economy.
3. Markets are an instrument to achieve well-being, not an end in itself.

The above hierarchy is the foundation of the ecosocial approach in education. It does not deny the possibility of economic growth but it determines two conditions for that growth: taking care of the ecological boundaries, and a profound respect for human rights. If natural resources, ecosystem services, and human beings are merely instruments for market growth, our debt towards future generations will grow, and ethical values will be undermined (e.g. Stiglitz et al. 2009).

13.6 Ecosocial Approach and Sustainable Society

In the industrial age the pursuit of material prosperity was prioritised more highly than the nurturing of harmony and common wealth. Ecosocially educated people have a wider life orientation. They understand that human beings are part of the fragile planetary entity. They question consumption and an ownership-oriented lifestyle in the pursuit of the good life. Instead of the old paradigm of well-being they recognise planetary boundaries and replace the materialistic goals of life by cultural elements that produce long-term satisfaction, enhance the quality of life, and provide experiences of happiness. They are aware of the fact that once people have met their basic needs, their well-being is connected to immaterial capital

(Kahneman et al. 2006; Kahneman and Deaton 2010). Immaterial capital can grow forever without any boundaries. It includes knowledge, self-expression, freedom, affection and participation (Max-Neef 1992). This means that we understand the value of cooperative relationships and generosity (Rees 2010). Pursuit of happiness leads us to adopt a more community based life orientation (Graham 2011).

Ecosocial education is about learning to understand the balance between freedom and responsibility of humans (Salonen and Åhlberg 2012). A core insight is that we share our common planet with more and more people every day. This sharing should be as equal as possible because the deepest essence of development occurs when rich and poor get closer to each other (see Wilkinson and Pickett 2010). This leads to a more inclusive society. The companies have the right to make a profit for their owners as a reward for their contribution, but at the same time they have to play an active and responsible role in society whose manpower, infrastructure and natural resources they use (Salonen 2010: 54–60).

The transition towards a sustainable society is possible if we have ecosocially educated citizens. This transition is based on cooperation and community rather than competition between individuals. Individualism involves prioritising one's own interests and material goals in life. In contrast, community-oriented people take into account immaterial aspects of well-being and they tend to take care of the opportunities available for future generations (Kasser 2011: 207).

The overall aim of education is a civilised human being, who takes care of herself and her culture, the globe as a whole as well as the possibilities of future generations (Salonen and Åhlberg 2012). Such people are able to apply social and ecological information to a wide range of world situations. They also display the ability to imagine the predicaments of many types of people and future generations, and they have the ability to think reflectively (Nussbaum 2010). These skills and competences are the basis on which moral values grow (Nussbaum 2010). People with an ecosocial orientation are seeking a good life for everyone, everywhere and forever. It is a massive challenge, but their values do not allow less.

13.7   ESD from Viewpoint of the Finnish National Board of Education

13.7.1   The Finnish National Board of Education

The Finnish National Board of Education (FNBE) is responsible for constructing national core curricula for preschool education, for basic education and for general upper secondary education. The FNBE also guides and supports the implementation of the national core curricula which means the drawing up of the main lines of municipal and school based curricula.

Both of these processes—construction of the national core curricula and the drawing up of the main lines of local curricula in schools—are unique in Finland.

Both can be described as collaborative learning processes where it is possible for students to learn together from shared experience, from the experiences in other fields of society, and experiences in other countries. Research and practice are combined. In this process the curriculum developers try to create common understandings on the changes and needs of our society in the global world, and the needs of individuals in this society, and on how to express these in the best ways.

With regard to the creation of new curricula, as well as their evaluation and testing, the informal Finnish ENSI network and its individual members have been frequently used. For example, Mari Nuutinen has been a member of at least one of the groups for new curricula. From time to time Arto Salonen and Mauri Åhlberg have been invited to provide new ideas for the processes.

The teachers have very central role in these processes, not only in their own schools, but also in their municipalities and in national processes. Teachers are regarded as the best experts for teaching and learning, and their input both in curriculum development as well as in the development of the whole education system is highly valued. It has to be remembered that all teachers have university degrees and Master's level degrees including a Master's thesis.

Between 2012 and 2016 the FNBE will renew all national core curricula, and, based on that, municipalities and schools will conceptualise what these curricula mean at the local level and renew their local curricula. In this curriculum process it has to be better explained how we understand sustainability as the main goal of education, and how we include values connected to sustainable development to the value basis and working culture of the schools. The new curriculum shall support the learning and well-being of the children and help them gradually grow in their ability to understand the global world and to take responsibility for its sustainable future (Halinen 2012).

In autumn 2013 a profound reform of national core curriculum for general education was started. Sustainable development was to be emphasised. When preparing the reform the competences of global citizens were described in a preliminary fashion. Competences for leading a sustainable lifestyle are part of these. Sustainable lifestyles are based on a wise use of ecosystem services.

Progressive inquiry learning methods allow pupils to gain in-depth understanding of natural, built-up and social environments. They need to know how nature works, how human activities change the environment and in what ways the environment can be nurtured. It is important for pupils to become aware of their own dependence on the environment as well as of the consequences of their own actions and how these are linked to environmental problems. Guiding children and young people towards ecological thinking calls for familiarisation with local nature and environments. According to current and future curricula for Finnish comprehensive schools, learning to identify local species is obligatory. In Finland we have a large country, plenty of useful herbs, berries and mushrooms. It is wise to learn to identify the most common local ones, to conserve and use sustainably local biodiversity. In upper grades, pupils move from observations of their own local environments to global problems and their potential solutions.

ESD at school provides children and young people with opportunities to think critically and to participate in decision-making processes, involving responsibility and collaboration. Joining networks of like-minded people, social media and other groups brings about confidence in the possibility of change and may lead to structured social participation. Schools play a key role in guiding learners towards constructive criticism and innovative social action (Houtsonen and Jääskeläinen 2011).

In the current curriculum for 7–16 year olds, there are seven integrative cross-curricular themes, one of which is Responsibility for the Environment, Well-being and a Sustainable Future. The FNBE evaluated the implementation of these cross-curricular themes using a questionnaire. There were questions about knowledge, attitudes, values and actions. In total, 1,198 pupils from the 9th grade answered and the results were presented by Uitto (2012). Here we can only highlight some of the most important results. On average pupils have acquired knowledge of those issues related to sustainable development, such as sorting, recycling, energy-saving, where earlier research has established that teachers think they are important (Åhlberg 2005a, b). Newer important issues, such as preferring vegetables over meat products, were not very popular when the inquiry was undertaken in 2010. There was no question on conserving and sustainable use of biodiversity, although it is one on the key elements of local sustainable development according to the four UN World Summits on the Environment and Sustainable Development 1972–2012 (e.g. Åhlberg 2012, 2013a, b, c, d). For future curricula Åhlberg (2012) has suggested more emphasis on species identification as a stepping stone for biodiversity conservation and sustainable use. Use of NatureGate http://www.natutegate.net was suggested to be tested for that purpose.

13.8 Overall Conclusion and Outlook for Post-DESD

The above represents the results of our theoretical and practical work in ESD during the DESD (2005–2014). We have trusted the original ENSI-spirit of local innovative schools and teachers. Each of the presented school experiments and innovations for sustainable development is based on trust in well-educated teachers and their collaboration with university researchers and national school administration. The Finnish system encourages innovative teachers to develop their work and their schools to promote a sustainable future. After 2014 (and after the DESD) each school, teacher and researcher will continue to promote all the three main aspects of sustainable development.

When the DESD has finished, another important UN Decade continues, the UN Decade on Biodiversity (2011–2020). Global biosphere is the sum of local biodiversities. The patented system and method of NatureGate® will be developed to support local biodiversity and sustainability education even better. In Finland new school curricula are currently created. As far as we know, a sustainable future will be one of the core values of the new curricula coming into force in 2016.

Humankind faces bigger threats and problems than never before in its history, such as faster climate change than ever, and a corresponding loss of biodiversity and free ecosystem services. This means, among other things, that there will be food shortages in future. At the same time, humankind has built the internet, which is probably the biggest infrastructure created by humankind. This allows quick learning for those who want to learn by using it. Search engines such as Google provide much up-to-date information quickly. Using Facebook groups, relevant knowledge building and learning for sustainability, a healthy environment and a good life can be promoted. In Finland many of us are eager to test these new options.

References

Åhlberg. (1988). Kasvatustavoitteiden teoreettinen kehikko ja sen empiiristä koettelua. Helsingin yliopiston kasvatustieteen laitos. [A framework for educational objectives and its empirical testing]. *Tutkimuksia*, 117.

Åhlberg, M. (1989). Environmental educators need conceptual innovations and scientific ontology and epistemology. In V. Meisalo & H. Kuitunen (Eds.), *Innovations in the science and technology education*. Proceedings of the second Nordic conference on science and technology education. Heinola, 8–11 Aug 1989. National Board of General Education. Information Bulletin, 2, pp. 280–286.

Åhlberg, M. (1990). Kasvattajille sopivien tutkimusmenetelmien ja-instrumenttien teoreettiset perusteet, tutkiminen ja kehittäminen elinikäisen kasvatuksen ja oppimisen näkökulmasta: KST-projektin tutkimussuunnitelma. [Research methods and instruments which are suitable for educators—theoretical underpinnings, research and development from the viewpoint of lifelong education and learning. A research program.] University of Joensuu. Research Reports of the Faculty of Education, No. 31.

Åhlberg, M. (1998). Education for sustainability, good environment and good life. In M. Åhlberg & W. Leal Filho (Eds.), *Environmental education for sustainability: Good environment, good life* (pp. 25–43). Frankfurt am Main: Peter Lang.

Åhlberg. (2005a). Integrating education for sustainable development. In W. Leal Filho (Ed.), *Handbook of sustainability research* (pp. 477–504). Frankfurt am Main: Peter Lang.

Åhlberg, M. (2005b). Eheyttävän ympäristökasvatuksen teoriasta (1997–2004) kestävää kehitystä edistävän kasvatuksen teoriaan (2005–2014). [From a theory of integrating environmental education (1997–2004) to the theory of education for sustainable development (2005–2014)]. In L. Houtsonen & M. Åhlberg (Eds.), *Kestävän kehityksen edistäminen oppilaitoksissa* (pp. 158–175). Helsinki: Opetushallitus.

Åhlberg, M. (2012). Biologian opettaja oppilaittensa oppimisen arvioijana. [Biology teacher as an evaluator of her/his pupils' learning]. In P. Kärnä, L. Houtsonen, & T. Tähkä (Eds.), *Development challenges of science education* (pp. 49–67). Helsinki: Opetushallitus. Koulutuksen seurantaraportit; 10.

Åhlberg, M. (2013a). Concept mapping as an empowering method to promote learning, thinking, teaching and research. *Journal for Educators, Teachers and Trainers, 4*(1), 25–35.

Åhlberg, M. (2013b). Biodiversity. In S. O. Idowu, N. Capaldi, L. Zu, & A. Das Gupta (Eds.), *Encyclopedia of corporate social responsibility* (pp. 160–173). Berlin: Springer-Verlag.

Åhlberg, M. (2013c). RCE Espoo: NatureGate promoting species identification for conservation and sustainable use of biodiversity. In U. Payyappallimana & Z. Fadeeva (Eds.), *Innovation in local and global learning systems for sustainability* (pp. 84–89). Yokohama: United Nations University, Institute of Advanced Studies, Regional Centers of Expertise.

Åhlberg, M. (2013d). Comparing knowledge forum and facebook for dialogue for truth, good life, good environment and sustainability. What kind of learning is promoted? (manuscript).

Åhlberg, M. (2013e). Sustainable development and education for sustainable development. A contribution for EU CoDeS project (manuscript).

Åhlberg, M., & Ahoranta, V. (2002). Two improved educational theory based tools to monitor and promote quality of geographical education and learning. *International Research in Geographical and Environmental Education, 11*(2), 119–137.

Åhlberg, M., & Ahoranta, V. (2005). Menetelmiä YK:n Kestävän kehityksen kasvatuksen vuosikymmenelle (2005–2014) [Methods for the UN decade od education for sustainable development 2005–2014]. In L. Houtsonen & M. Åhlberg (Eds.), *Kestävän kehityksen edistäminen oppilaitoksissa* (pp. 129–157). Helsinki: Opetushallitus.

Åhlberg, M., & Kaivola, T. (2006). Approaches to learning and teaching environmental and geographical education for sustainability. In J. C.-K. Lee & M. Williams (Eds.), *Environmental and geographical education for sustainability* (pp. 79–93). New York: Nova.

Åhlberg, M., Kaasinen, A., Kaivola, T., & Houtsonen, L. (2001). Collaborative knowledge building to promote in-service teacher training in environmental education. *Journal of Information Technology for Teacher Education, 10*(3), 227–238.

Åhlberg, M., Turja, L., & Robinson, J. (2003). Educational research and development to promote sustainable development in the city of Helsinki: Helping the accessible Helsinki Programme 2001–2011 to achieve its goals. *International Journal of Environment and Sustainable Development, 2*(2), 197–209.

Åhlberg, M., Äänismaa, P., & Dillon, P. (2005). Education for sustainable living: Integrating theory, practice, design and development. *Scandinavian Journal of Educational Research, 49*(2), 167–186.

Ahoranta, V. (2004) Oppimisen laatu peruskoulun vuosiluokilla 4–6 yleisdidaktiikan näkökulmasta käsitekarttojen ja vee-heuristiikkojen avulla tutkittuna. [Quality of learning in a primary school grades 4–6 from the viewpoint of general didactics using concept maps and vee-heuristics]. *Doctoral dissertation.* Joensuun yliopiston kasvatustieteellisiä julkaisuja, no 99.

Aineslahti, M. (2009). Matka koulun kestävän kehityksen maisemassa. [Journey in the landscape of sustainable school development]. *Doctoral dissertation.* University of Helsinki, Faculty of Behavioural Sciences, Department of Applied Sciences of Education. *Tutkimuksia* 295.

Alppi, A. (2013). A manuscript for a doctoral dissertation about and for rural education for sustainable development (manuscript).

Alppi, A., & Åhlberg, M. (2012). Learning from local and global collaborations. In J. Murray, G. Cawthorne, C. Dey, & C. Andrew (Eds.), *Enough for all forever* (pp. 305–317). Champaign: Common Ground Publishing LLC.

Beach, K. (1999). Consequential transitions. A sociocultural expedition beyond transfer in education. *Review of Research in Education, 24*, 101–139.

Bereiter, C. (2002). *Education and mind in the knowledge age.* Mahwah: Lawrence Erlbaum.

Cook, J., Nuccitelli, D., Green, S., Richardson, M., Winkler, B., Painting, R., Way, R., Jacobs, P., & Skuce, A. (2013). Quantifying the consensus on anthropogenic global warming in the scientific literature. *Environmental Research Letters, 8*(2), 1–7.

Dillon, P. (2006). Creativity, integrativism and a pedagogy of connection. *International Journal of Thinking Skills and Creativity, 1*(2), 69–83.

Dillon, P., & Åhlberg, M. (2006). Integrativism as a theoretical and organisational framework for e-learning and practitioner research. *Technology, Pedagogy and Education, 16*(1), 7–30.

Engeström, Y., Engeström, R., & Suntio, A. (2002). Can a school community learn to master its own future? An activity-theoretical study of expansive learning among middle school teachers. In G. Claxton & G. Wells (Eds.), *Learning for life in the 21st century* (pp. 211–224). Oxford: Blackwell Publishers.

Giddings, B., Hopwood, B., & O'Brien, G. (2002). Environment, economy and society: Fitting them together into sustainable development. *Sustainable Development, 10*(4), 187–196.

Graham, C. (2011). *The pursuit of happiness. An economy of well-being.* Washington: Brookings Institution Press.

Halinen, I. (2012). *Opening remarks at the international conference "sustainable development. Culture. Education" reorientation of teacher education towards sustainability through theory and practice, May 22–25 (2012).* Savonlinna: University of Eastern Finland.

Hofstede, G., Hofstede, G. J., & Minkov, M. (2010). *Cultures and organizations: Software of the mind. Intercultural cooperation and its importance for survival.* London: McGraw-Hill.

Houtsonen, L., & Åhlberg, M. (Eds.). (2011). *Kestävän kehityksen edistäminen oppilaitoksissa. [Promoting sustainable development in school organizations].* Helsinki: Opetushallitus.

Houtsonen, L., & Jääskeläinen, L. (2011). Sustainable lifestyle. In L. Jääskeläinen & T. Repo (Eds.), *Schools reaching out to a global world: What competences do global citizens need?* (p. 34). Helsinki: Finnish National Board of Education.

Howell, L. (2013). Global risks 2013. An initiative of the risk response network (8th ed.). Cologny/ Geneva: World Economic Forum. http://www3.weforum.org/docs/WEF_GlobalRisks_ Report_2013.pdf. Accessed 28 Mar 2014.

Immonen-Orpana, P., & Åhlberg, M. (2010). Collaborative Learning by Developing (LbD) using concept maps and Vee diagrams. In P. Torres & R. Marriott (Eds.), *Handbook of research on collaborative learning using concept mapping* (pp. 215–235). Hershey: IGI Global.

Kaasinen. (2009). Kasvilajien tunnistaminen, oppiminen ja opettaminen yleissivistävän koulutuksen näkökulmasta. [Plant species recognition, learning and teaching from viewpoint of general education]. Faculty of Behavioral Sciences. Department of Applied Sciences of Education. Research Report 306.

Kahneman, D., & Deaton, A. (2010). High income improves evaluation of life but not emotional well-being. *Proceedings of the National Academy of Sciences, 107*(38), 16489–16493.

Kahneman, D., Krueger, A., Schkade, D., Schwarz, N., & Stone, A. (2006). Would you be happier if you were richer? A focusing illusion. *Science, 312*(5782), 1908–1910.

Kaivola, T., & Åhlberg, M. (2007). Theoretical underpinnings of education for sustainable development. In T. Kaivola & L. Rohweder (Eds.), *Towards sustainable development in higher education—Reflections* (pp. 42–48). Helsinki: Ministry of Education. Department of Education and Science. http://www.minedu.fi/export/sites/default/OPM/Julkaisut/2007/liitteet/opm06. pdf?lang=en. Accessed 28 Mar 2014.

Kasser, T. (2011). Cultural values and the well-being of future generations: A cross-national study. *Journal of Cross-Cultural Psychology, 42*(2), 206–215.

Marglin, S. (2008). *The dismal science. How thinking like an economist undermines community.* London: Harvard University.

Max-Neef, M. (1992). Development and human needs. In P. Ekins & M. Max-Neef (Eds.), *Real-life economics: Understanding wealth creation* (pp. 197–213). New York: Routledge.

Myers, D. (2000). The funds, friends, and faith of happy people. *American Psychologist, 55*(1), 56–67.

Myllari, J., Åhlberg, M., & Dillon, P. (2010). The dynamics of an on-line knowledge building community. A five-year longitudinal study. *British Journal of Educational Technology, 41*(3), 365–387.

NatureGate (2014). NatureGate online services. http://www.naturegate.net. Accessed 28 Mar 2014.

Nussbaum, M. (2010). *Not for profit. Why democracy needs the humanities.* Princeton: Princeton University.

Nuutinen, M. (2013a). RCE Espoo: The encounters project. In U. Payyappallimana & Z. Fadeeva (Eds.), *Innovation in local and global learning systems for sustainability* (pp. 96–101). Yokohama: United Nations University, Institute of Advanced Studies, Regional Centers of Expertise.

Nuutinen, M. (2013b). Interviews and observations of what is happening in UNU IAS RCE Espoo (manuscript).

OPH (National Board of Education). (2004). *National core curriculum for basic education 2004.* Helsinki: National Board of Education. http://www.oph.fi/english/publications/2009/national_ core_curricula_for_basic_Education. Accessed 28 Mar 2014.

Quoidbach, J., Dunn, E., Petrides, K. V., & Mikolajczak, M. (2010). Money giveth, money taketh away: The dual effect of wealth on happiness. *Psychological Science, 21*(6), 759–763.

Rees, W. (2010). What's blocking sustainability? Human nature, cognition, and denial. *Sustainability: Science, Practice, and Policy, 6*(2), 13–25.

Salonen, A. (2010). Sustainable development and its promotion in a welfare society in a global age. *Doctoral dissertation in Finnish*. Helsinki: University Press.

Salonen, A. (2013). Responsible consumption. In S. Idowu, N. Capaldi, L. Zu, & A. Das Gupta (Eds.), *Encyclopedia of corporate social responsibility* (pp. 2048–2055). Berlin: Springer.

Salonen, A., & Åhlberg, M. (2012). The path towards planetary responsibility—Expanding the domain of human responsibility is a fundamental goal for life-long learning in a high-consumption society. *Journal of Sustainable Development, 5*(8), 13–26.

Senge, P., Smith, B., Kruschwitz, N., Laur, J., & Schley, S. (2008). *The necessary revolution: How individuals and organizations are working together to create a sustainable world*. London: Doubleday.

Shallcross, T., & Wals, A. (2006). Introduction: Mind your Es, Ds and Ss: Clarifying some terms. In T. Shallcross, J. Robinson, P. Pace, & A. Wals (Eds.), *Creating sustainable environments in our schools* (pp. 3–9). Stoke-on-Trent: Trentham Books.

Siirilä, J. (2013) A manuscript of a doctoral dissertation of education for sustainable development, especially in UNU IAS RCE Espoo region (manuscript).

Siirilä, J., & Åhlberg, M. (2012). A research methodological study of concept mapping to foster shared understanding to promote sustainable development in the UNU-IAS RCE Espoo, Finland. In A. Cañas, J. D. Novak, & J. Vanhear (Eds.), *Concept maps: Theory, methodology, technology. Proceedings of the Fifth International Conference on Concept Mapping* (Vol. 1, pp. 1–8). Valetta: Veritas Press.

Sterman, J. (2012). Sustaining sustainability: Creating a systems science in a fragmented academy and a polarized world. In M. Weinstein & E. Turner (Eds.), *Sustainability science* (pp. 21–58). New York: Springer.

Stiglitz, J., Sen A., & Fitoussi, J.-P. (2009). The measurement of economic performance and social progress revisited. Commission on the measurement of economic performance and social progress. www.stiglitz-sen-fitoussi.fr/documents/overview-eng.pdf. Accessed 28 Mar 2014.

Uitto, A. (2012). Vastuu ympäristöstä, hyvinvoinnista ja kestävästä tulevaisuudesta. In E. Niemi (Ed.), *Aihekokonaisuuksien tavoitteiden toteutumisen seuranta-arviointi 2010* (pp. 155–183). Helsinki: OPH.

Vygotsky, L. S. (1978). *Mind in society. The development of higher psychological processes*. London: Harvard University Press.

Wheeldon, J., & Åhlberg, M. (2012). *Visualizing social science research: Maps, methods, & meaning*. Thousand Oaks: SAGE.

Wilkinson, R., & Pickett, K. (2010). *The spirit level. Why equality is better for everyone*. London: Penguin.

Zalasiewicz, J., Williams, M., Haywood, A., & Ellis, M. (2011). The Anthropocene: A new epoch of geological time? *Philosophical Transactions of the Royal Society, 369*(1938), 835–841.

Chapter 14
Learning from ESD Projects During the UN Decade in Norway

Mari Ugland Andresen, Nina Høgmo, and Astrid Sandås

14.1   Introduction

Despite the long tradition in Norwegian schools for outdoor education, the transition to ESD has been challenging. Using our experience from work with international and national ESD projects, this chapter aims to reflect on lessons learnt during the UN Decade. All of the authors worked at the national level in the education sector in Norway. It should be emphasized that the views and claims expressed in this chapter are based on our understandings and experiences from our work with ESD in Norway. To describe the progress of ESD in Norway during the DESD, we first give a brief insight into the traditional human-nature relationship in Norway, then we provide a short introduction to the overall framework for ESD, including the national strategy for ESD and the formal curriculum, and then we discuss the implementation of ESD in Norway.

M.U. Andresen (✉)
Norwegian Public Roads Administration, Fløenbakken 436, 5009 Bergen, Norway
e-mail: mariuandresen@gmail.com

N. Høgmo
Buskerud County Municipality, Ruduegen 520, 3540 Nesbyen, Norway
e-mail: ninahogmo@hotmail.com

A. Sandås
Vestenga 2, 1413 Tårnåsen, Norway
e-mail: asa1@live.no

© Springer International Publishing Switzerland 2015   241
R. Jucker, R. Mathar (eds.), *Schooling for Sustainable Development in Europe*,
Schooling for Sustainable Development 6, DOI 10.1007/978-3-319-09549-3_14

14.2 Background

The poetic epic Edda, a collection of old Norse poems from the 1,200 to 1,300, is regarded as a compressed programme for the early Norwegian culture. The poem promotes an ecological system of knowledge which serves as a sort of textbook of survival (Witoszek 1998). The utilitarian ethos that came out of this period survived for centuries, and is still characteristic for Norwegian society and national identity. The Polish scholar Witoszek puts it as follows: "There is an extensive semiotic immersion from nature into the Norwegian perception of the Norwegian self" (Witoszek 1998: 18, translated by the authors). Arctic explorers such as Roald Amundsen and Fridtjof Nansen were, and still are, regarded as brave men with knowledge as well as understanding and love for nature. Together with the "father of the deep ecology", the philosopher Arne Næss, and Gro Harlem Brundtland, after whom the Brundtland Report (formally known as the report of the World Commission on Environment and Development (WCED) 1987) was named, these people are often held up as symbols of Norway and Norwegians. Furthermore, natural heritage has a great symbolic value for Norwegians. For example, the works of Norway's most famous artists like Henrik Ibsen, Edvard Munch and Edvard Grieg, are filled with symbols and inspiration from nature.

Being close to nature and knowing how to carry out outdoor recreation have always been important in Norwegian schools. Consequently, we have a long tradition for outdoor-education, and many countries have looked to Norway for inspiration in this respect. Unsurprisingly, Norway has welcomed international initiatives that involve sustainable development, including the UN Decade. Despite challenges with implementation in schools, which we will return to later, we argue that the overall framework for education in Norway is supportive of ESD. This includes the National Strategy for ESD and the formal curriculum, both of which give schools and others a clear mandate for ESD. We will begin with the vision and goals of ESD in Norway, as outlined in the national strategy.

14.3 The Norwegian Framework for ESD

14.3.1 National Strategy for ESD

The National Strategy for ESD *Knowledge for a Common Future* sets an ambitious vision for ESD: "an education system that contributes to sustainable development" (The Norwegian Ministry of Education and Research 2012: 5, translated by the authors). Education for sustainable development aims to:

- Develop children's and young people's competences so they can contribute to sustainable development in different areas in nature and society.
- Provide a framework for education for sustainable development for kindergartens and schools so that they may gain competence in ESD.

- Stimulate network development and collaborative relations between kindergartens, schools, relevant agencies, NGOs and research institutions at national, regional, and local levels.
- Promote participation in international forums for exchange of experience and raise the quality in teaching for sustainable development both in Norway and in other countries (The Norwegian Ministry of Education and Research 2012: 5, translated by the authors).

A recurring theme in the Strategy is the collaboration between schools and external stakeholders. Furthermore, it defines the role and responsibilities of the stakeholders involved. In Norway, the municipalities are the school owners responsible for primary and lower secondary schools, and a significant responsibility is placed on the school owner.

14.3.2 ESD in the Formal Curriculum

The collective objectives and principles for teaching in primary and secondary schools are found in the national curriculum. The latest curriculum reform was carried out in 2006, and it was titled the Knowledge Promotion (KP-06) (Norwegian Ministry of Education and Research 2006). The reform covered primary, lower secondary and upper secondary education. The KP-06 was the first pure competence-based curriculum in Norway, and the reform placed an increased focus on basic skills and knowledge promotion through outcome-based learning. The curriculum comprises the Core Curriculum which is the same for all school levels, subject curricula for various school levels, as well as an overall quality framework. The Core Curriculum was developed in the 1994 curriculum, but was kept in the KP-06. In addition the Directorate of Education and Training provides regulations for individual assessment and for how schools should distribute teaching hours over subjects. Although ESD is not mentioned in the curriculum documents, SD is mentioned throughout KP-06. In particular, there is a substantial integration of SD in the Core Curriculum. The Core Curriculum distinguishes six different 'human beings'; the spiritual human being, the creative human being, the working human being, the liberally-educated human being, the social human being, the environmentally aware human being and the integrated human being. All these "beings" contain a high level of SD- and ESD-thinking. Here are some examples from the Norwegian Core Curriculum:

The interplay between economy, ecology and technology must make unique demands, scientific and ethical, on our age, if we are to ensure sustainable development [. . .] Education must therefore provide a broad awareness of the interconnections in nature and about the interplay between humans and their habitat. (The Norwegian Ministry of Church and Education 1994)

Our way of life and our form of society have profound, threatening effects on the environment [. . .] It increases the need for more knowledge, more holistic knowledge, and for more conscious ecological, ethical and political decisions made by individuals, and by society as a whole [. . .] (The Norwegian Ministry of Church and Education 1994)

With the introduction of the Knowledge Promotion, teachers were given the freedom to choose teaching methods. In addition, each municipality was given the responsibility and right to develop their own local curriculum based on the learning objectives set out in the national curriculum. It is required that the Core Curriculum is integrated in the local curricula developed in schools as well as in the individual teaching plans in all subjects. This opportunity to develop local learning objectives tailored to the individual child, the school, and the community gives great opportunities for ESD.

14.3.3   Short Background to Early Work for ESD in Norway

Network for Environmental Education

Norway participated in a network called Environment and School Initiatives (ENSI) that was started by the Organisation for Economic Development and Co-operation (OECD) in 1986. The goal of ENSI was to find out whether ESD is an appropriate instrument for school development and effective learning at the pupil, teacher and school level. The following were the focus areas:

- At the pupil level, the network has been concerned with developing the pupils' abilities to take the initiative, co-operate, formulate issues, collect information, construct knowledge, and draw conclusions.
- At the teacher level the focus has been on the teachers' abilities to construct opportunities for active pupil learning, lead educational activities, co-operate across disciplinary boundaries, and co-operate with stakeholders outside of the school.
- At the school level, the project has examined the impacts of this kind of educational practice on the way schools organise their educational programmes (Norris et al. 1993).

The following conclusions emerged from the ENSI project in Norway:

- Pupils were inspired by the co-operation between the schools and stakeholders outside of the school.
- Teachers were inspired by the co-operation with their colleagues in other disciplines and by networking co-operation between schools.
- Co-operation between schools and research institutions appeared to be of mutual benefit and seemed to promote pupils' learning.
- The pupils were adept at collecting information, drawing conclusions and communicating their findings to local decision-makers and the press.
- Teachers found that it was productive to work in interdisciplinary ways.
- Schools can participate in the work of Local Agenda 21 and contribute to the development towards SD (Sandås and Benedict 1993).

We learnt that, as long as schools receive support and guidance, it was not difficult to change school practice along the lines recommended internationally. When the project period was over, schools sometimes reverted to their previous practices. The challenge was to establish structures that could provide continuous support to schools (Sandås and Benedict 1993). To establish continual support structures and to give schools access to scientific competence, work begun in 1997 to develop national programmes on ESD themes. The programmes consisted of a set of quality-assured activities and background material, while the development of individual school projects was the responsibility of each school.

During the course of a 4-year period, approximately 100 schools per year participated in the project Vanda, where pupils investigated local lakes (Knutsen and Van Marion 1997), and 80 schools per year participated in stream investigations in the project Bekkis. The Coast project attracted almost 400 participating schools, and there were about 150 schools per year involved in the energy project Meis. An additional 54 schools were involved in measuring solar radiation in 1994. An evaluation of these programmes concluded that:

- They provided the school with support.
- They led to interdisciplinary and active learning processes in the schools.
- The pupils considered the tasks to be meaningful and they were proud to be part of a larger project.
- The schools can be a contributing partner in the work for SD. (Engesæter et al. 2002)

The programmes were later combined to create a simpler and more internally consistent system. The result is the current internet-based programme Miljolare. no (accessed 3 April 2014). The network functions as a meeting place for schools, research institutions and public management, and it provides continual support to schools. The purpose is to give examples of methods, suggest how the education can be organised and give schools access to updated and quality-assured information. Another purpose is to quality assure the products of the educational activities and to build competence in this kind of education for all of the participants: pupils, teachers and other interested parties.

Figure 14.1 shows a conceptual model for Miljolare.no.

Miljolare.no has been developed as a tool for ESD, and it has been used by several projects both nationally and internationally. The international learning activities on Miljolare.no are available at Sustain.no.

It is this concept and the Miljolare.no web tool that has been developed and tested in the SUPPORT project.

14.3.4 "SUPPORT: Partnership and Participation for a Sustainable Tomorrow" (2007–2010)

The overall objective of the Comenius network project SUPPORT: Partnership and participation for a sustainable tomorrow was to promote and enhance the quality of ESD. Students and teachers must acquire a range of skills and competencies to

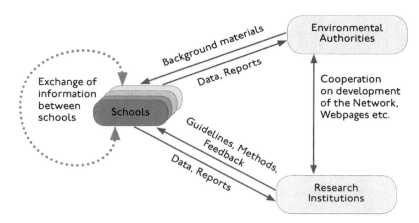

Fig. 14.1 The conceptual model of the internet based programme Miljolare.no gives an overview of the actors, the services and the quality management as well as their interlinking (Sandås et al. 2010: 23)

become responsible citizens, citizens not only able to take care of their social and natural environments, but also to understand how the economic systems interact with both the social and natural environment. The required competencies include the ability to cooperate and to act autonomously as well as language skills and digital competence. In SUPPORT, key features of ESD were partnership and participation. The *partnership* part acknowledges that schools and pupils benefit from reaching out and forming networks and partnerships, collaborating with a range of other stakeholders in society on sustainability issues. The *participation* part refer to ESD benefits gained by pupils and schools engaging directly and actively with real sustainability issues in quests for more sustainable solutions. SUPPORT used information and communication technology (ICT) to facilitate such a collaborative network, linking schools, researchers and local communities.

The challenge was to develop an online instrument that could take into account the features of ESD on the one hand, and avoid a traditional ready-made easy-to-use teaching package on the other hand. The aim was to develop an instrument that would allow schools and students to create their own learning situation with enough space for local freedom and creativity. The challenge was to support an "emancipatory" approach rather than to be "instrumental". To meet this need, a school campaign was developed on the topic of sustainable transport. This campaign was called "CO_2n-nect: CO_2 on the way to school". In this campaign a large number of schools, pupils, parents and communities were able to work with SD in the field of climate change and transport. This model was developed on the basis of the model showed in Fig. 14.1.

14.3.5 CO₂ on the Way to School

The pupils participated by investigating their own CO_2 emissions on the way to school, entering the data into an international database, before analysing and comparing results and doing local project work. We argue that this was more of an instrumental part of the campaign. The emancipatory part of the campaign was that schools were encouraged to collaborate and create partnerships through the web-based network with other schools, researchers or organisations in their community. The aim was to increase schools' competencies to deliver quality ESD and to provide ICT-based tools including guidelines, links, a CO_2 transport emissions calculator and opportunities for partnership. One aspect of the campaign was to generate information useful to research and management concerning transport and climate issues and to generate innovative ideas for sustainable transport. The activity is an example of how the SUPPORT project has reached out to people outside of the project, and of how partnerships can be established when employing existing local resources in combination with the opportunities that a central, established project can offer.

The evaluation of CO_2nnect covered the educational aspects of CO_2nnect, with an emphasis on how CO_2nnect has functioned as an educational innovation or prototype for mainstreaming ESD in education systems (Benedict 2010). The evaluation assesses attainment of both CO_2nnect's operational goals and the long-term goal of "improving the understanding and practice of ESD" (ibid.: 9). Information sources included the CO_2nnect website, information and feedback from the SUPPORT partners and an online evaluation filled out by 207 participating teachers.

Participating teachers commented that pupils were particularly motivated by opportunities to partner globally with research bodies and other schools to make a useful joint contribution. They were also motivated by opportunities to participate in the local democratic debate on sustainability issues in their communities. ICT was viewed as an effective approach to develop such collaborative relationships and joint efforts. The issues inquiry approach and high degree of pupil initiative and autonomy were also well received. Teachers reported that CO_2nnect contributed significantly to pupils' achievement of diverse ESD learning outcomes including understanding, skills and abilities, awareness, attitudes and values.

Learning outcomes for pupils and schools, as reported by teachers, consistently correlated with the intensity of collaboration between the school and stakeholders outside the school. Improvement in several areas of skills, abilities, personal attitudes, and values tended to be more challenging to achieve and more dependent on the degree of collaboration than outcomes related to understanding of complex issues and awareness-raising. The CO_2nnect tool facilitates such collaboration by:

1. Creating a learning arena for school-school and school-research collaboration on the internet.
2. Providing a framework of scientific activity and school guidance to help schools construct local learning arenas anchored in school-community collaboration.

CO_2nnect shows that ICT-based tools have a potential to provide shared global opportunities and motivation for ESD. CO_2nnect produced several kinds of positive outcomes for schools, but CO_2nnect and similar ICT-based tools may need to be coupled with other kinds of school development initiatives to reach their full potential: good tools alone may not be sufficient. Integration of such tools into existing programmes of school support, guidance and development could multiply the impact on long term school development and capacity building. We suggest that in Norway, the responsibility for managing such collaboration and implementing such tools should be given to the municipalities.

14.3.6 The Nordic ESD Project "Extreme Weather"

The Extreme Weather project was initiated as a follow-up to the SUPPORT project. Mitigation and adaptation to climate-change related extreme weather incidents was recognized as a relevant topic, and one that required a local focus where schools could be important stakeholders. Vulnerability and robustness to meet the consequences of climate change in a local community is linked to how the inhabitants perceive and experience natural hazards, and local knowledge has increasingly been valued as a major factor when planning responses to climate change. In a developmental perspective, it is necessary for people to feel safe and secure to enable a community to attract inhabitants and prevent people from leaving. We saw a great potential in learning and ESD within this theme and considered it likely that involving schools in this might act as a catalyst for debate and knowledge exchange in their local community. The project partners were the Finnish National Board of Education, the University of Copenhagen and the Norwegian Centre for Science Education (lead partner). The project started in 2011 as a 1 year project. Extreme Weather was part financed through the Nordic Councils strategy for SD and the Norwegian Ministry of Environment.

The main aim for the project was to develop and disseminate a learning activity for schools capable of stimulating children and adolescents to take an active role in societal challenges related to extreme weather and the adaptation to climate change. Miljolare.no was used as a platform to facilitate participation and collaboration between schools and other stakeholders in their local communities.

In the learning activity, pupils have had the opportunity to contribute local information regarding extreme weather and natural hazards, including their consequences and the pupils' own personal experiences. This way, pupils have been encouraged to contribute information of relevance to themselves and to the society. Quantitative and qualitative information are both registered with the ICT-tool. This allows schools from all over the Nordic Region to compare and assess each other's results. Researchers as well as local and regional management are target groups for the results. The activity is translated into all Nordic languages, and at the time of writing (November 2013), 1,271 people from the Nordic countries have been interviewed by school children about their thoughts on climate change adaptation and locally preparedness. All the results have been registered by the pupils on the Extreme Weather activity at Miljolare.no, and all results have been made available for other schools, scientists, educational management, and the general populace. The results have provided great opportunities for reflection and learning in subjects like the natural and social sciences, language, mathematics and history. In addition, the data may provide the local management and scientists with valuable informa-tion, which in turn can lead to more SD locally and nationally, and strengthen the learning and collaboration on climate change and extreme weather in the Nordic region.

The Extreme Weather project is another example of how ICT-tools can be used to support collaboration and learning for SD. Results of an evaluation survey among participating teachers showed that the teachers considered the ICT-tool a valuable starting point for learning, reflection, and cross-curricular teaching and learning. Despite a substantial marketing effort, there were few participants from Norway, and none of these worked with the activity in a cross-curricular fashion. The Extreme Weather project was developed as a project for both social and natural sciences, but in Norway only natural science teachers used the learning activity Still, we have examples from other Nordic schools that showed that the "Extreme Weather" project made it possible to develop a school activity that supported teachers and schools in implementing SD projects in line with Local Agenda 21 and the UN's DESD.

14.3.7 The Environmental Toolbox

The environmental toolbox programme (ET) was initiated in 2008, and it has been the main national initiative for implementing ESD practice in Norwegian schools during the UN decade. The target group for the ET programme has been pupils and teachers in primary and secondary education. Schools could apply for funding of up to 50,000 NOK (€6,090 as of 3 April 2014) to implement and share experiences from their local ESD project. To be eligible for funding, the projects must be interdisciplinary, they must include natural and/or social science, and they must be anchored in the national curriculum.

The initiative is funded by the Ministry of Education and Research and the Ministry of the Environment. The Directorate for Education and Training and The Norwegian Directorate for Nature Management have the responsibility for implementing the ET programme, and the Norwegian Centre for Science Education (a national competence centre for natural science education) holds the secretariat for the project.

The secretariat collaborates with a reference group and a resource group. The reference group consists of different stakeholders involved with education and/or environmental issues. The NGOs in the reference group work together with the project secretariat to promote collaboration between schools and other stakeholders in local ESD-projects. The resource group consists of teacher educa-tors working at Norwegian colleges and universities. This group was established in order to connect the ET programme with teacher education and to draw on the teacher educator's expertise.

It is required that the project applications in some way address the development of the pupils' awareness and commitment to the environment and sustainable development. Beside this requirement, the schools are free to choose their theme or topic. Schools are, however, encouraged to use local issues relevant for SD as a starting point. In addition to a budget, the application requires a description of the teaching plan for the project, pupil participation, how SD is included in the project, how the project will be evaluated, which external stakeholders will be involved, and how the schools will collaborate with these.

Many NGOs hold knowledge and experiences valuable to teachers and pupils and, since the very beginning of the ET in 2008, NGOs have played a significant role in the programme. Through the evolution of the programme, many schools have shown an increasing ability to collaborate with local stakeholders and to put SD at the centre of their project. We have also seen that for many participants, the ability to work and teach in an interdisciplinary fashion has developed as their projects proceed.

However, being a national initiative with more than 100 schools with successful applications to follow up, the ET programme has not yet succeeded in the aim of implementing ESD in the majority of schools that have participated in the programme. Instead, many school projects have taken the form of traditional outdoor-education projects. Important qualities in ESD like pupil participation and collaboration with external stakeholders towards a shared goal and addressing the real issues of SD have been challenging for many schools. The collaboration between schools and NGOs has often been limited to NGOs being a source of knowledge and support for teachers. Also interdisciplinary teaching and learning has proved difficult in many projects. The majority of the teachers implementing the local projects have been natural science teachers. Also the majority of the NGOs that have collaborated with schools are mainly concerned with outdoor recreation or nature management. Few projects have been concerned with social themes such as inclusion, democracy, or social justice, and this indicates that many Norwegians tend to understand ESD within the frame of natural science.

At the beginning of the ET programme, expertise in the curricular subjects: social science, food, health, physical education, and natural science were represented in the resource group along with expertise in pedagogy. Despite an increase in the number of experts in the resource group since the project started, by spring 2013, the resource group no longer contained experts on social science or food and health. Considering the "watertight bulkheads" that often are found between different disciplines in the education system, it has proved a challenge to get the intensions of the ET programme communicated to all teachers and disciplines.

14.3.8 The Cross-Sectorial EcoRegion Project

For many years the Baltic Sea region suffered from challenges arising from excessive land use, erosion, and acidification of rivers as well as air and water pollution. In the wake of the growing demand for energy and other environmental and social challenges following the dissolution of the Soviet Union, development in the region was far from sustainable. Despite these signs of unsustainable development, many good practices on local SD have been initiated in the region. The EcoRegion project worked towards an ambitious goal of:

> Turning the Baltic Sea Region into the world's first Eco Region where economic growth goes hand in hand with environmental integrity and social justice. (Schultz-Zehden and Delsa 2011: 6)

The EcoRegion project was initiated in 2007/2008 and implemented between 2009 and 2011.

Experts and organisations from eight key sectors involved in SD—agriculture, education, energy, forest, industry, spatial planning, tourism and transport—and ten regions active in SD processes from all the Baltic Sea region countries were brought together in the project, with the aim of conducting pioneering work with regard to working method, implementation, and policy level.

The Norwegian Centre for Science Education represented the education sector in the project, and focused on how pupils and schools may contribute to SD and how this can be an integral part of the regional work for SD.

One of the outcomes from the project was the development of a database with good examples of local SD. The education sector provided examples of good practices to the database, and it was evident from the documentation of good practices reports how Miljolare.no served as a tool for collaboration between schools and local stakeholders. The database has been open for all and can be reached via the EcoRegion website www.baltic-ecoregion.eu (accessed 3 April 2014).

The experiences of the different sectors and regions from work for SD provided a base for political recommendations at the international level as well as at regional and local levels. The publication *EcoRegion Findings* (2011) presents the partners' reflections on regional work with SD and outlines which future steps are required for furthering SD in the Baltic Sea region. The Norwegian Centre for Science Education has contributed with experience from the education sector's work with ESD in the project. The experiences from the education sector mainly focused on:

- The gap between policy and practice.
- Factors for successful implementation of ESD.
- The need to strengthen collaboration, both horizontally and vertically in the system (Andresen and Høgmo 2011).

One of the strengths of the EcoRegion project was that many different sectors and regions were involved. This gave the project a multifaceted experience base for SD work, and this diversity of experience produced a diversity of approaches to the work with SD. This is addressed in the EcoRegion publication: *Education and Innovation for Sustainable Development* (Schultz-Zehden et al. 2011), which documents a great variety in how the project partners in Eco Region interpreted and addressed ESD in their work. Professionals outside the education sector saw themselves as credible and relevant partners in ESD. In order to ensure horizontal implementation of ESD, the education sector should welcome such diversity and maintain an inclusive approach to other sectors and professions.

14.4 Challenges and Opportunities for ESD in Norwegian Schools: Lessons Learnt

The four projects introduced in this chapter represent a selection of initiatives carried out in Norway during the UN Decade. There are many other successful initiatives for ESD in Norway, yet experience with the projects described here give a relatively good overview of the challenges and opportunities for ESD in Norway.

Although the projects described differ in many ways, they have common learning points in terms of how they have been received by teachers and how they have been carried out in schools and in their local communities:

- Norwegian teachers report an increased pressure to prepare pupils for national testing and exams. There is also a high pressure to focus on teaching and learning of basic skills like numeracy, reading, writing, and ICT. Time constraints and lack of flexibility also prove to be obstacles for the planning and development of ESD projects.
- There is a need to link existing programmes like the ET and Miljolare.no to a whole-school approach. Even though the formal curriculum encourages development of local curricula, many schools find it a challenge to develop the local curricula in line with ESD. Headteachers and teachers have stated that the Core Curriculum (where we find the most substantial integration of SD and ESD) is not used when developing local learning plans (Andresen 2007).
- Interdisciplinary and cross-sectoral collaboration is demanding both for professionals within and outside the school system. In general, overcoming the barriers for horizontal policy integration for ESD appears to be very challenging.
- Both schools and external stakeholders need support and guidance to set up meaningful sustainable projects, especially in the project-initiation phase.
- We have seen a strong correlation between the pupils' learning outcomes and the level of collaboration with stakeholders outside of the school. The ICT-tool Miljolare.no has proven to be a valuable tool for learning and collaboration in the projects described in this chapter. Teachers—who have used the tool *and* collaborated with other local stakeholders in their work with the learning activities/projects—report that the tool is a good starting point, and that it helps strengthen the learning and the motivation for the pupils. However, many participating schools in Norway have not collaborated with external stakeholders in their projects. This is a challenge in the work to develop schools in line with ESD.
- To realise the full potential of the ICT-tool and support school development for ESD along the lines formulated in our National Strategy for ESD, there is a need to develop a system at the local level that can facilitate and sustain the collaboration.
- At the school level, the majority of the local projects have mainly oriented their implementation towards natural science, even if the project designs strongly have encouraged cross-curricular learning.

- In order to advance school development both in the field of education in general and in ESD, a system is needed to ensure that the learning outcomes and evaluations of the projects reach the policy level. This has not yet been established, and this is a challenge that must be overcome if vertical policy integration is to be attained.

14.5 Discussion

Taking a bird's eye perspective on the cold and long country in the North, a sort of paradox becomes evident. On the one hand, Norway has a long history and tradition for including elements of environmental management and natural heritage in the governing and administration of the country. On the other hand, Norway is a significant producer of oil and gas, and the governing of this industry has brought Norway from poverty to becoming one of the wealthiest nations in the twenty-first century. As inhabitants of Norway, we have used this merciful opportunity to leave considerable carbon footprints and develop unsustainable consumption habits. It seems that this paradox has not yet reached the educational authorities, and we find it pertinent to raise the question whether our long traditions for outdoors education might have misled us to believe that we are carrying out ESD. Our lessons learnt show that interdisciplinary learning, systems for collaboration, and a practice-to-policy integration are not obvious consequences of an implementation that is understood and interpreted within the frames and traditions of natural sciences and outdoors education only. If we are to succeed in implementing our national goals, we need to establish a common understanding of the broad and multifaceted content of ESD.

As our societies get more developed and complex, so does our strive for a sustainable future. Challenges for SD are increasingly complex, and demand complex solutions that often are found across traditional disciplines and areas of responsibility.

For the education sector, this requires a local focus. Schools should work locally with topics and issues that are both locally relevant and meaningful for the pupils. The evaluation of the ICT-tool CO_2nnect (Benedict 2010) shows that the learning outcomes for ESD correlate with the level of collaboration. Interdisciplinary teaching and collaboration between schools and other local stakeholders is required to attain an ESD approach. We have seen that this poses new challenges for schools and teachers. Even if we find examples of individual ESD-projects in Norwegian schools, and to a lesser extent some schools that work systematically with ESD, we have a long way to go before we have reached the aim of the UN Decade "to integrate the principles, values, and practices of sustainable development into all aspects of education and learning" (UNESCO 2005: 6).

Norway has a framework with a clear vision and goals outlined in the National Strategy for ESD. Our National Curriculum contains a significant potential for ESD with the Core Curriculum and its structure for developing local curricula. However, from our experience, ESD is not yet sufficiently integrated in schools' local curricula. The Strategy for ESD has not been followed up with an action plan. Nor is ESD part of the evaluation or reporting system in Norway. Consequently, ESD lacks priority and is reduced to an "extracurricular activity" for the most dedicated teachers. The schools alone should not have to carry the whole responsibility for ESD. This should be a shared responsibility between the schools and the governing authorities, and we believe a systematic effort is needed to support schools with the development of local curricula in line with ESD.

For the realisation of ESD, management and governing authorities, including the education sector, have a responsibility to ensure a holistic approach where connections and relations between and across disciplines is an obvious and natural approach. The Norwegian Strategy for ESD was revised in 2012 and gives the municipalities a clear mandate for leading the local work with implementing ESD, including collaboration between schools and external stakeholders. This is a new and demanding task for the municipalities that has not yet reached the level of implementation. In order to fulfil the horizontal collaboration aims in the strategy for ESD, and to ensure that individual ESD-initiatives become sustainable, there is a need to facilitate the establishment of a local system that may assist schools in their ESD work.

14.6   Conclusion

It has been important to work with teachers and schools during the UN decade to identify how Norway can work to improve the level of ESD implementation in schools. While the overall framework for ESD, including the national strategy for ESD and the Core Curriculum give schools a clear mandate and opportunities for ESD, we have seen that the actual implementation in schools is challenging and far from mainstreamed. There is a great need to support the collaboration of the schools and the municipalities in order to build a robust and sustainable structure for ESD. Such a system should ensure that local issues of SD and the pupils' learning become integrated into the overall objectives of both education and regional development for SD. To facilitate the transition from traditional outdoor education and predominantly natural-science based approaches to ESD, we see a need to revise local curricula in line with the Strategy for ESD and the National Core Curriculum. To ensure priority in schools, the national authorities should implement a system of reporting and evaluation.

We have illustrated how ICT can be used to support the implementation in schools. In the future work of ESD, we believe the use of ICT-tools like Miljolare. no can be valuable both as a starting point for collaboration, and as a tool that can sustain and maintain collaborations.

In order to further implementation of ESD in Norway we see a great potential in building on existing work for EE and ESD, including the ET programme. It is our hope that this chapter will contribute to further debate and action on what we consider the most important task for our school system—to educate for a just and sustainable future!

References

Andresen, M. U. (2007). *Exploring environmental policy integration in the formal education sector: A comparative study of education for sustainable development in Scottish and Norwegian primary schools. Master thesis.* University of Edinburgh.

Andresen, M. U., & Høgmo, N. E. (2011). Lessons learned from the education sector. Results from the Baltic 21-EcoRegion project. *EcoRegion Findings, 2011*, 16–17. http://www.baltic-ecoregion.eu/downloads/EcoRegion_Findings.pdf. Accessed 5 May 2014.

Benedict, F. (2010). *Innovation and instruments for education for sustainable development.* Oslo: Norwegian University of Life Science. http://support-edu.org/webfm_send/616. Accessed 5 May 2014.

Engesæter, P., Flygind, S., & Nyhus, L. (2002). Evaluering av Nettverk for miljølære. *Østlandsforskning Rapport, 1.*

Knutsen, A. E., & Van Marion, P. (1997). *The national environmental programme, evaluation of sub-program "water", report nr 1.* Trondheim: NTNU (Norwegian University of Science and Technology).

Norris, N., Posch, P., & Kelley-Lainè, K. (1993). *A review on environmental education policy in Norway.* Paris: OECD.

Sandås, A., & Benedict, F. (1993). *ENSI: Indepth policy review in Norway. Background report for the OECD-expert team.* Oslo: The Norwegian Ministry of Church and Education.

Sandås, A., Benedict, F., Grønstøl, G., Hindson, J., & Renman, Å. (2010). ICT tools to facilitate partnerships for ESD. Booklet developed in the Comenius SUPPORT project. http://support-edu.org/webfm_send/768. Accessed 5 May 2014.

Schultz-Zehden, A., & Delsa, L. (Eds.). (2011). *EcoRegion findings—results from the Baltic 21-EcoRegion project 2009–2011.* Berlin: S.Pro—Sustainable Projects GmbH. http://www.baltic-ecoregion.eu/downloads/EcoRegion_Findings.pdf. Accessed 5 May 2014.

Schultz-Zehden, A., Delsa, L., & Torres, C. (2011). Innovation and education for sustainable development. EcoRegion Perspectives, 3. http://baltic-ecoregion.eu/downloads/final.pdf. Accessed 5 May 2014.

The Norwegian Ministry of Church and Education. (1994). *The national core curriculum.* Oslo: The Norwegian Ministry of Church and Education.

The Norwegian Ministry of Education and Research. (2006). *The national curriculum for knowledge promotion.* Oslo: Norwegian Ministry of Education and Research.

The Norwegian Ministry of Education and Research. (2012). *Kunnskap for en felles framtid. Revidert strategi for utdanning for bærekraftig utvikling 2012–2015.* Oslo: Norwegian Ministry of Education and Research.

UNESCO. (2005). *Framework for the international implementation scheme.* Paris: UNESCO Education Sector. http://unesdoc.unesco.org/images/0014/001486/148654e.pdf. Accessed 14 Apr 2014.

Witoszek, N. (Ed.). (1998). *Norske Naturmytologier.* Oslo: PAX Forlag A/S.

World Commission on Environment and Development (WCED). (1987). Report of the world commission on environment and development: Our common future. http://www.un-documents.net/our-common-future.pdf. Accessed 14 Apr 2014.

Chapter 15
Searching for a Sea Change, One Drip at a Time: Education for Sustainable Development in Denmark

Simon Rolls, Katrine Dahl Madsen, Torben Ingerslev Roug, and Niels Larsen

In order to properly understand Education for Sustainable Development (ESD) in Denmark, it is necessary to trace a development extending considerably further back than 2005 and the launch of the United Nations Decade of Education for Sustainable Development (DESD); beyond even the publication of the Brundtland report in 1987. By doing so, one can identify many of the concepts and principles now considered as comprising ESD within the Danish tradition for Environmental Education (EE).

Internationally, EE is seen by many as a forerunner for ESD; by others as a major ingredient in ESD; or even, by some, as a competitor to ESD. In Denmark, at least at the conceptual level, ESD can be seen as a fairly straightforward development and continuation of a firmly established tradition for research and practice within the field of Environmental Education. As such, Denmark may in some respects be considered to have had a head start in terms of implementing the goals of the DESD. Nevertheless, as the end of the Decade approaches, Danish educational policy regarding the fulfilment of these goals remains largely disjointed, half-hearted

S. Rolls (✉) • K.D. Madsen
Department of Education, Aarhus University, Copenhagen, Denmark
e-mail: siro@edu.au.dk; kdma@edu.au.dk

T.I. Roug
Faculty of Science, University of Copenhagen, Copenhagen, Denmark
e-mail: tiro@science.ku.dk

N. Larsen
Regional Centre of Expertise on ESD, Center for Design, Innovation and Sustainable Transition, Aalborg Universitet, Copenhagen, Denmark
e-mail: niladk5@gmail.com

© Springer International Publishing Switzerland 2015   257
R. Jucker, R. Mathar (eds.), *Schooling for Sustainable Development in Europe*,
Schooling for Sustainable Development 6, DOI 10.1007/978-3-319-09549-3_15

and vague. We therefore begin this chapter by presenting some of the key characteristics of EE in Denmark and its development towards ESD, culminating with a brief examination of possible explanations for the apparent national derailment.[1]

However, ESD does of course have a life away from the policy level. Providing an overview of developments in Danish ESD practice is beyond the scope of this chapter, as many of the most interesting and innovative initiatives within ESD are small-scale, locally-based and decentralised development projects where information within the public realm is frequently scarce or even non-existent. Meanwhile, the establishment in 2010 of a Danish UN Regional Centre of Expertise on Education for Sustainable Development (RCE Denmark) is a step towards pooling these resources. Through a national network of ESD researchers and practitioners, examples have been compiled of a broad range of ESD practice from formal, non-formal and informal learning contexts.

This chapter will present three cases focusing on ESD within the field of formal primary and lower secondary education in Denmark. Three cases can, of course, by no means provide a comprehensive picture of ESD in Danish schools; nor should these cases be considered in any way representative. Each case can, however, be regarded as an example of a promising practice, illustrating different approaches, but with the common objective of increasing and improving ESD in Denmark.

These cases provide evidence that ESD is finding its way into Danish schools, albeit one drip at a time. Meanwhile, they also show that ESD is a concept which can be difficult to grasp within a school organised along strict disciplinary lines. Many of the leading proponents of ESD speak strongly in favour of what is commonly referred to as a 'whole-school approach' (McKeown and Hopkins 2007; Breiting and Schnack 2009)—indeed, it is stated in the International Implementation Scheme for the DESD that:

> The overall goal of the DESD is to integrate the principles, values, and practices of sustainable development into **all** aspects of education and learning. (UNESCO 2005: Annex 1, our emphasis)

To do so requires a sea change in terms of not only the content, but the organisation of schooling. We conclude this chapter by posing the question of what the future might hold for ESD in Denmark: How can we ensure that ESD becomes a deeply rooted aspect of Danish schooling and achieves this goal far beyond 2014 and the end of the DESD?

---

[1] Much of the inspiration for this historical overview stems from the Danish contribution to the report *Climate Change and Sustainable Development: The Response from Education* (Breiting et al. 2009).

15.1   A Brief History of Education for Sustainable Development in Denmark

15.1.1   Environmental Education in Denmark

The widespread incorporation of environmental issues within teaching in the Danish *Folkeskole*[2] mirrors the growing concern about such topics in Denmark in the late 1960s and early 1970s (Breiting and Wickenberg 2010). Previously, the focus had largely been on nature and preservation, but now, pollution and environment were firmly on the agenda (Læssøe 2008).

It is necessary here to draw attention to the strong traditions for pedagogical autonomy within Danish schools as the relative freedom of schoolteachers in terms of curriculum has played and continues to play a significant role in the development of first EE and currently ESD. While recent years have seen a growing trend towards greater accountability within the Folkeskole, with greater focus on standards and testing, the national curriculum in Denmark remains more or less confined to the formulation of overall objectives and a specification of which subjects are to be taught at the various levels of schooling and the central knowledge and skill areas. The objectives for the various subjects are of a relatively general nature with more detailed stipulations of content determined at a local level. As such, schoolteachers in Denmark continue to have a relatively free rein when it comes to determining what and how they teach.

According to Breiting and Wickenberg this pedagogical autonomy goes some way towards accounting for the early inclusion of environmental issues in Danish schools. Teachers who were themselves concerned about growing pollution were able to teach their students about the causes of and possible solutions for pollution without having to wait for a political decision to place such matters on the curriculum (Breiting and Wickenberg 2010: 19).

Whilst this freedom can result in an organic educational system, able to respond quickly to new issues as they arise, it places the responsibility for this responsiveness largely on the shoulders of the individual teacher and can result in lethargy at the policy level. In terms of EE, it also meant that its development in the early stages was largely the result of the efforts of a relatively small proportion of teachers, themselves often actively involved in the environmental movement. There was little to compel other teachers to include environmental issues in their lessons. If EE was to spread beyond a few firebrands trying to convince their colleagues of the importance of their environmental concerns, these issues needed to be incorporated within the overall objectives and curriculum guidelines stipulated by the Ministry of Education.

---

[2] Municipal public schooling covering the entire period of compulsory education, i.e. both primary and lower–secondary although this division is not found in Denmark.

With the new Education Act in 1976, environmental issues were officially recognised as an educational concern for the first time and included in the guidelines for geography and, in particular, biology:

> ...the teaching shall contribute to students' understanding of people's living conditions and opportunities and in that way provide a basis for deciding about local and global environmental problems. (Teaching Guidelines for Biology 1976, Danish Ministry of Education, cited in Breiting and Wickenberg 2010: 20)

This passage also provides evidence of the links between EE and citizenship education in Denmark with its emphasis on increasing understanding and providing a basis for democratic decision making rather than technical or scientific knowledge. While students should, of course, acquire useful knowledge and skills, in Denmark it is generally considered of at least equal importance that students gain experience in forming their own opinions and, on the basis of serious reflection, taking appropriate action.

The same Act also introduced a specific subject *samtidsorientering* (modern/contemporary studies) dealing with 'important contemporary issues', many of which would today be classified as within the scope of ESD.

To complete the picture, it should be noted that teachers engaged in 'peace education', 'global education' and 'development education' were also active during this period, and, although these areas were not directly reflected in the school's curriculum, they could be easily accommodated. As such, it is already possible to see the roots of many aspects of current conceptions of ESD at these early stages.

15.1.2   The Emergence of ESD in Denmark

In the wake of the publication of the Brundtland commission's report *Our Common Future* (United Nations, World Commission on Environment and Development 1987), educating the population with regard to sustainable development was for the first time placed firmly on the national political agenda in Denmark. However, the impact was initially within non-formal youth and adult education ('folkeoplysning').

While the report of the Brundtland Commission kick-started non-formal ESD in Denmark, the UN summit on environment and development held in Rio in 1992 increased the focus on ESD within the school sector. The first fruits came in the form of collaboration between the Baltic countries. In March 2000, this led to the Haga Declaration (Baltic 21 2000), signed by the Ministers of Education from each of the participating countries.[3] The Haga Declaration describes sustainable development as a challenge which will require an integrated approach to economic, environmental and societal development, encompassing a broad range of related

---

[3] Denmark, Estonia, Finland, Germany, Latvia, Lithuania, Norway, Poland, The Russian Federation and Sweden.

issues such as democracy, gender equity and human rights. It also stresses the importance of broad participation and including ESD in all curricula at every level of education, demanding "an educational culture directed towards a more integrative process-oriented and dynamic mode emphasising the importance of critical thinking, and of social learning and a democratic process" (Baltic 21 2000: 2.4).

In 2002, the same year as the DESD was adopted at the UN summit in Johannesburg, the Danish Government approved a national strategy for sustainable development: *A Shared Future—Balanced Development* (Danish Government 2002). It contained less than half a page on education, but is still worth highlighting, partly because the social-democratic government that signed the Haga Declaration had now been replaced by a new liberal-conservative coalition, and partly because it nevertheless contains a number of statements regarding the educational sector's role and objectives for ESD. Despite being largely consistent with the Haga Declaration, it is worth noting that a shift can be observed away from a prioritisation of critical thinking and democratic learning and towards knowledge and individual responsibility.

As a direct prelude to the DESD, Denmark formed a part of the United Nations Economic Commission for Europe's (UNECE) joint European preparations. As such, Denmark is a cosignatory of the common strategic guidelines for the decade which were approved in March 2005 (UNECE 2005). This is an important document—even though only guidelines—as it describes what ESD is about, contains a number of requirements for implementation of ESD at the national level, and a joint timetable for the decade. The document additionally obliges Denmark to deliver assessments of national ESD initiatives to UNECE based on common European criteria. As such, the UNECE strategy document puts pressure on the Danish Government to live up to the agreement. It should be noted that the shift away from critical and democratic citizenship and towards individual knowledge and responsibility which one finds in the Danish strategy for sustainable development is not apparent in the UNECE strategy. Rather the latter stresses the importance of ESD encouraging "systemic, critical and creative thinking and reflection in both local and global contexts" (UNECE 2005: 4) and states that ESD should promote participatory learning, hereby facilitating processes "encouraging dialogue among pupils and students and the authorities and civil society" (UNECE 2005: 5). Contrary to the Danish focus on knowledge, the UNECE strategy thereby recommends "a reorientation away from focusing entirely on providing knowledge towards dealing with problems and identifying possible solutions" (UNECE 2005: 6).

15.1.3   The Danish Strategy for the United Nations Decade of Education for Sustainable Development

Despite Denmark's strong traditions for EE and democratic education and involvement in the compilation of various international ESD strategies, a national strategy for the Decade was not published until almost halfway through, in 2009. Indeed, cynics may wonder whether the strategy would have remained on the backburner was it not for Denmark hosting the COP15 meeting (United Nations Climate Change Conference) later that year.

In the strategy, the overall goals are largely couched in terms familiar from the international documents. For example, in the foreword by then Minister of Education Bertel Haarder, it is stated that

> the strategy shall ensure that children, young people and adults become aware of the concept of sustainable development and learn how to act competently through knowledge and skills (. . .) The aim is to introduce sustainable development in all relevant curricula used in basic education, youth education and teacher training in order to establish a link between natural and social sciences and humanities. (Danish Ministry of Education 2009: 3)

While it is here stated that linking the natural sciences, social sciences and humanities is one of the goals of the strategy, there can be little doubt that the natural sciences are ascribed the most important role. Sustainable development issues are to be added to school curricula when relevant; however existing disciplinary boundaries are maintained preventing a truly interdisciplinary approach. Particularly when presenting concrete initiatives, broader policy priorities for education are apparent. Indeed, it is explicitly stated within the strategy that:

> The strategy takes a point of departure in the following messages:
>
> • Personal responsibility and engagement are important for guiding own actions and behaviour.
> • Democratic decisions should be made on the basis of sound scientific knowledge.
> • The desired economic growth should ideally not damage the opportunities for growth of future generations or other continents. (Danish Ministry of Education 2009: 12)

The first two of these messages are very much in line with overall educational policy at that time. The third message, however, would seem to paraphrase the Brundtland report's definition of sustainable development as "development that meets the needs of the present without compromising the ability of future generations to meet their own needs" (United Nations, World Commission on Environment and Development 1987). However, the use of the somewhat cagey "should ideally" results in a less ambitious objective and seemingly suggests that it may (in non-ideal situations) be necessary to sacrifice the opportunities of future generations or other continents to ensure contemporary economic growth in Denmark.

As the Danish DESD strategy was published in the lead up to Denmark hosting the COP15 meeting in Copenhagen, and considering Danish ESD's firm roots within EE, it is perhaps not surprising that environmental issues, and particularly climate change, overshadow other aspects of ESD. The dominance of climate change within the ESD discourse represents a problem which the strategy itself draws attention to in the section on "strategic efforts":

> During the UN Decade of Education for Sustainable Development, action is to be taken to focus and differentiate in relation to other problems than climate change. (Danish Ministry of Education 2009: 15)

However, precisely which problems or what action are referred to here is not specified. In a press release by the Ministry of Education in conjunction with the launch of the Danish DESD strategy, it was pointed out that ESD in Denmark has focused on a number of issues during recent years, including citizenship and equal opportunities. In 2009, it was stated, it was the turn of climate change to take centre stage (Danish Ministry of Education Press Release 2009). One might ask whether dividing ESD into focus areas in this way does not partly defeat the object of operating with the more holistic approach—accounting for the complexity and interconnectedness of the issues at stake—that a concept of sustainable development would seem to entail. Furthermore, concentrating to such an extent on one particular element, highlighted as a focal point for the 2008/2009 school year, seems somewhat at odds with the launch of what was intended to comprise the Danish ESD strategy over the next 5 years.

15.1.4   A New Government: A New Beginning for ESD?

In November 2011, a new centre-left government was elected in Denmark. At least on paper, this change of government presented new opportunities for Danish ESD with education and environmental issues highlighted as key policy areas from the outset. However, these twin foci have primarily remained parallel with little convergence in the form of ESD.

One area where ESD is present is the Ministry of Education's introduction of the New Nordic School initiative. In a ten-point manifesto, the final goal is that New Nordic Schools should

> through teaching, pedagogical practice and exemplary behaviour in the institution's everyday work and activities make children and young people co-creators of a democratic and sustainable society—socially, culturally, environmentally and economically. (Ny Nordisk Skole, authors' translation)

Schools can apply voluntarily to become New Nordic Schools, and the manifesto does not as such represent policy requirements to incorporate ESD. The subsequent 2013 legislative reform of the Folkeskole focused on improving academic standards, particularly in the core subjects of Danish and Mathematics, increasing professional standards among staff, and greater local autonomy through increased powers for local authorities and school principals. Despite mentioning the goals of teaching children to become active participants in democratic processes and preparing young people for the demands of the world of tomorrow, as well as stressing that the reform's goals are in line with those of New Nordic School, ESD is absent from the reform.

While the increased local autonomy potentially improves opportunities for introducing ambitious whole school approaches to ESD, this is very much dependent on a commitment to such an approach at the local level. Furthermore, the emphasis on strengthening standards within core subjects, among both pupils and teachers, could potentially reinforce barriers to interdisciplinary approaches inherent to ESD.

Another relevant reform is found in the publication of a new statutory order for the Bachelor's Degree Programme in Education in March 2013 (Danish Ministry of Science, Innovation and Higher Education 2013). Here, sustainability is included as a learning objective for student teachers within a number of subjects. For example the subjects Biology, Physics/Chemistry, Geography and Nature Science all have the stated objective that students gain knowledge about action competence and sustainability in relation to mankind's interaction with nature and technology. Home Economics and Social Studies also include sustainability dimensions, while Woodwork and Needlecraft both have sustainability as one of four overall competence areas for prospective teachers.

15.1.5   Reflections on the Impact of the DESD in Denmark

The UN DESD has to some extent placed education for sustainable development on the educational agenda in Denmark and has resulted in a number of collaborative projects within both formal and non-formal education, some of them made possible by the establishment of RCE Denmark. At school level *sustainable development* has since 2009 formed part of the curriculum in a number of subjects; however, without unfolding the ideas of *education* for sustainable development, as described as part of the UN Decade of ESD, nor reflecting the international theory development within the field(s) of EE/ESD. At the moment, no explicit strategy exists on a national level to further ESD in Denmark beyond the decade.

As noted above, ESD is largely compatible with the Danish tradition for environmental education, formed around the concepts of action competence, participation, 'hands on' learning and critical sense (Schnack 2003; Breiting and Wickenberg 2010). However, these ideas are yet to find widespread acceptance in schools, and it is an ongoing challenge to entrench ESD in schools and to broaden sustainability and climate change from relevant issues for natural science lessons to a vital part of teaching in all subjects and at all levels of education. There is also a risk that ESD is reduced to a question of behavioural change, due to a strong desire to 'get the message across', rather than to focus on strengthening the students' ability for critical reflection, as stressed in a critical pedagogical conceptualisation of ESD (Breiting et al. 2009; Schnack 2003).

15.2   Three Promising Practices

Whilst development within ESD at the national policy level has been slow, there are a number of examples of local initiatives at the practice level which can perhaps offer ideas and inspiration for approaching and furthering ESD in Denmark and beyond.

The three cases presented here are all of an experimental character and formed as short-term development projects addressing ESD. They are each inspired to some extent by the UN Decade of Education for Sustainable Development, but are not part of a cohesive national strategy. The first project was co-initiated by the Green Flag organization (FEE) and a Teacher Training College and concerns teacher education, while the two remaining projects were initiated by Danish universities: one as a research and development project within primary and lower secondary schools; the other designed by a school activity centre at a Danish university and aimed at lower secondary school classes. Together the cases present various experiences of how to 'do' ESD and highlight some of the challenges and possibilities related to working with ESD in practice.

15.2.1   Case 1: ESD in Teacher Education: Working for Change

Introduction

Between August 2009 and May 2010, six classes at the Teacher Training College (TTC) in Odense, Denmark, took part in a pilot course on ESD. The classes represented different subjects: geography, biology, physics, and chemistry. All in all there were 70 participating students and five lecturers involved in the project.

The idea for the course on ESD in teacher education stemmed from the UN DESD and Denmark's status as host for the COP15 meeting in Copenhagen. The pilot course was in line with the Eco-Schools programme, normally intended for primary and lower secondary schools, adapting the concept of Eco-Schools to teacher training institutions. It was created in collaboration with several partners including the Foundation for Environmental Education (FEE International) and the head office for Green Flag, Green Schools in Denmark; the local municipality of Odense; and other Danish teacher training colleges. Funding for the project came from the Government and the Danish Outdoor Council.

Case Description

Training of primary and lower secondary teachers is in Denmark offered at teacher training colleges/centres of higher education. The programme takes 4 years and leads to the award of the Bachelor of Education.

The aim of this project was to develop teaching and didactics for student teachers so as to enable them to teach education for sustainable development and environmental education. The normative aims and teaching design form part of the didactical approach, as well as the ways in which pupils participate and the outcomes of the learning process. As part of the project, the student teachers had the opportunity to develop learning materials and cooperate with pupils in the schools they visited for their practical training.

The design combined not only subject knowledge, but also didactic knowledge and more practical experiments which were applied in schools. Finally, students at the teacher training college evaluated their projects. The basic idea was that the students would be able to *do something* also related to *local environmental policy* and *actions, investigate,* do *something practical* or *experimental, promote and communicate about results in different media* and *make and implement teaching plans* (Kristensen 2010).

According to FEE, Eco-Schools and the concept of Green Flag, Green Schools have the following learning and teaching approach:

> Eco-Schools is a programme for environmental management, certification, and sustainable development education for schools. Its holistic, participatory approach and combination of learning and action make it an ideal way for schools to embark on a meaningful path for improving the environments of schools and their local communities, and for influencing the lives of young people, school staff, families, local authorities, NGOs, and more. (FEE: What is Eco-Schools?)

Through a number of steps, the Eco-Schools programme transforms educational institutions into environmentally-oriented schools. These steps include: forming an Eco-School committee, reviewing the organisation, establishing an action plan, incorporating environmental issues within the curriculum, collaboration with external partners, monitoring, and assessing.

The main experiences gained from this project are *firstly* related to the students' acceptance of a well-defined project approach, i.e. Green Flag, Green School; *secondly* to how it was integrated within a well-defined and, to a certain extent, well-structured teacher training system; *thirdly* to the way student teachers from the TTC participated in the project; and *fourthly* to the way student teachers have transformed the principle of ESD into their own school practice. In the following description, some of these central issues and experiences will be discussed.

Challenges

The experiences from the 1 year project are complex and diverse. Overall, it seems the student teachers found it exhilarating and relevant, not only in relation to their training as schoolteachers, but also on a personal level and as democratic citizens.

The first challenge in the project was to ensure initial motivation among the student teachers. Students at the TTC found the concept and approach of Green Flag, Green School rather 'top down' in nature. As mentioned earlier, the approach of the Eco-Schools concept includes different steps towards 'greening' the school and it can therefore appear a ready-made and somewhat rigid concept. Participation and involvement were discussed in different ways, because students did not feel their opinions were properly incorporated. Some students at the TTC found the concept too focused on environmental issues and with little room for discussion. Others found the concept of sustainability, incorporating social, economic, political, cultural and ecological dimensions, overwhelming.

As the project continued, the student teachers found the process positive, but at the same time also frustrating, as the amount of time available for the project was very limited and it was not possible to work for extended periods in the classes. The need for greater influence and a more open approach were among the students' critiques (Kristensen 2010).

One reason could be the relationship between the Eco-Schools concept and its integration within higher education. The Eco-Schools programme has advantages, but also disadvantages. First of all, in many ways it is extremely straightforward to grasp for primary school classes. The programme consists of different steps which are easily manageable and obtainable—not just for the pupils, but also for the busy teacher with little time for preparation who nevertheless wants to work with sustainability. The programme and the steps involved are visible and action-oriented, and involve not only pupils and teachers, but also the school management and the school as a whole. In a compulsory school context this may be workable. But for teacher training colleges and their students it can represent more of a problem if the concept is too 'closed'. Teacher training and education in Denmark is based on principles of enquiry, participation, critical reflection, and collaboration, and is reflective in relation to theory and practice. Deeper and more open reflections on sustainability can be difficult when student teachers are confronted with a ready-made concept like the Eco-Schools programme.

Another experience for the student teachers was related to implementation among pupils in the classroom. The most difficult part was to handle the participatory approach. Much research has reflected on participation in ESD and, especially in Denmark, on the development of "action competence" (Jensen 2002; Jensen and Schnack 1997).

Different kinds of competencies and different types of knowledge can be developed: content knowledge, which refers to theories, principles and concepts in the different scientific disciplines involved; pedagogical and didactic knowledge refers to teachers' general knowledge about the mediation process and methods; pedagogical content knowledge concerns the type of knowledge which is unique to teachers and the way teachers relate pedagogical knowledge to subject matter (Cochran 1997). In all these types of knowledge generation processes, the students' participation is central if they are to develop action competence. Especially the last two types of knowledge relate to the way students conduct their product presentation. Some of the different themes were about energy and waste, others were about climate change, while others again were about pupils' opinions and values regarding the environment and sustainable development. The main problem was to integrate scientific-environmental knowledge and pedagogical-didactic knowledge and skills to action competence related to sustainability. Even though it was possible to use a large variety of media, most of the products consisted of a PowerPoint presentation, and these were not always based on a deeper analysis of the actual problem considered in the case. On the other hand, the student teachers responded critically, for example stating: "green is not just good because it's green; we have to find real and authentic reasons for acting in a sustainable way. Not just because we tell each other it's sustainable or green". Critical reflections were an essential part of the whole process.

Opportunities

In some of the classes, student teachers worked with central concepts such as sustainable development, interdisciplinarity and action competence. These concepts were employed in a project-related theme called sustainable consumption and included a whole school approach. The student teachers were particularly excited about the opportunity to develop their didactic skills related to the concept of action competence, and this resulted in a more complex perspective on sustainable development.

Most of all, the aspect of working with important themes, while also relating them to life outside school, motivated the TTC students. At the same time, they discovered that it was possible to involve pupils and get them interested and excited in the classroom, which makes the student teachers aware of the possibilities of integrating ESD in their future teaching.

A portfolio tool and a formative evaluation process were the overall evaluation methods used in the project. While there were some critical remarks regarding the short time available for the project and the difficulty of incorporating the project in the ordinary teacher training programme, the general impression from the students' point of view was positive and they stated that it was exciting to plan, implement and evaluate their theme and case in front of the pupils in the schools. The organisers had positive experiences with cooperation, not only with schools and other teacher training colleges, but also with the local authorities.

Students suggested that the TTC should form a committee for sustainable development. As one student said: "The most important thing we experienced in this project: you should go for change." The TTC has since established this committee and continues to integrate more subjects and strengthen the participatory approach. The main aim is to focus on the development of action competence and the integration of subjects and pedagogy related to sustainability in a more reflexive manner.

15.2.2   Case 2: Experiences of ESD as Didactical Practice: Teachers and Researchers Collaborating on ESD

Introduction

This case will present teachers' and researchers' reflections on working with ESD in a school setting, based on interviews with participants collaborating on an

ESD research and development project. The project involved researchers from the Department of Education at Aarhus University and teachers from four primary and lower secondary schools in Denmark during the period 2008–2009. Randomly selected teachers were invited to participate by the researchers. The project was formed as parallel courses, jointly designed by the teachers and researchers involved and carried out in science, social studies and as a cross-curricular theme. The project had twin objectives: (a) to develop ESD as didactical practice in various schools and (b) to generate knowledge of *how* to work with ESD in practice (Breiting and Schnack 2009).

Case Description

The collaboration was based on principles of participation on two levels: both among researchers and teachers, and among teachers and pupils. As the researchers invited the teachers to participate in the development project, their intention was never to offer a 'ready-made ESD package', but to introduce various ways of doing ESD in schools, and to support the teachers' own development of ESD teaching. As such, the project was formed around ideas of dialogue, participation, experimentation and ownership. The researchers did introduce the teachers to certain key principles and issues, partly inspired by the UN definition of ESD, and partly drawing on existing Danish theory development regarding environmental education and action competence. Finally, the didactical ideas were framed within a Danish democratic tradition. The idea was to introduce ESD as a 'perspective through which to understand the world', rather than a toolbox or an extra-curricular activity. As such, the researchers on the one hand intended to introduce key principles and ideas of ESD, and on the other hand wished to work with open ended processes where the result is never known in advance, linked to the participatory approach—a tricky balancing act, as stressed by the researchers.

Examples of ESD as Didactical Practice

In a third grade class at one of the schools, a story about a boy from Guatemala was used as an example of differences between Denmark and Guatemala in terms of living conditions, everyday life, opportunities, and dependence on natural resources. As part of the lesson, the pupils created small paper dolls picturing the boy as they imagined him. The teaching focused, among other things, on the pupils' ability to put themselves in another person's shoes, and thereby on developing an understanding that, just as our present situation is the product of historical development, the future will be shaped by our actions today.

Another example of a concrete project to initiate ESD teaching is a timeline concerning the use of a chosen natural resource. In this case, the pupils carry out interviews with elderly and middle-aged people in their local community. They ask them how the chosen natural resource was used in the past, how it is used in the present and how it will be used in two futures—one being the future they *expect*, one being the future they *hope for*. Back in class, the pupils illustrate the points from the interviews on four different posters: past, present, and the two future scenarios. The students then discuss and reflect on the issues raised on the posters. Thus, the approach involves aspects of episodic memory, visualisation, reflexivity and conceptual development on ESD.

Challenges

Interviews with teachers indicate that they are unfamiliar with the concept of ESD. They know of the concept of sustainable development, but are unsure of how to approach sustainable development as part of their teaching practice, both in general and in particular in relation to their various subjects. Common for the teachers who participated in this project is an initial and personal interest in sustainable development and environmental issues which they find important to integrate into school. ESD as it was presented by the researchers, however, seemed unfamiliar to the teachers. This includes new dimensions, perspectives and methods compared to existing teaching practice, as stressed by one of the teachers: "...it is not an established route, sustainability in school...it is most definitely not an asphalt road". The complexity of the concept, which is stressed by science teachers in particular, also makes ESD difficult to approach, as the elements of insecurity, risk and conflicting interests are difficult to integrate within the existing science curriculum.

A major challenge was how to get started with the ESD lesson plans, and the teachers' uncertainty as to whether they were on the 'right track'. To transform what for the teachers seemed 'lofty ideas' into concrete teaching practice was a challenge. To start with, the researchers wanted the teachers to come up with ideas for ESD lessons themselves, based partly on existing curricula, partly on their existing teaching experiences, and partly on the key issues and principles of ESD as presented by the researchers at a preliminary workshop. To tackle the teachers' uncertainty, however, the researchers ended up presenting a number of examples of how to approach ESD as didactic practice. This uncertainty points to the need for substantial teacher support in the initial phase of developing and implementing ESD lesson plans, and development projects like this therefore include elements of in-service training for participating teachers.

Possibilities

Although the teachers describe the collaboration process as quite demanding, they also describe it as: "an eye opener", "turned my way of thinking upside down", "rewarding", "developing", "fun", "creating space for discussion in a busy day filled with practical demands" and "it is nice that the researchers confront our reality in school".

One of the main aims of the project was to create space for innovation and reflexivity around ESD for researchers, teachers and pupils. As part of this, the teachers were asked to write down their thoughts, concerns and considerations in a diary during the period of teaching ESD. The teachers found this quite difficult, partly due to lack of time, partly due to this being an unusual process for them. However, the method is a way to create space for reflexivity in a busy day, and to support the teachers' learning processes within ESD. As such, development projects like this support space for reflection in busy everyday teaching practice, and potentially work as in-service teacher training in ESD.

One of the lessons from the work with pupils in the third grade is that even the smallest children are able to reflect and reason rationally about the basic conditions that a sustainable development must take into account. A main point here is that no child is too young to work with education for sustainable development.

Interviews indicate that teachers' sense of ownership of the teaching sessions in ESD is significant, but as long as no supportive structures exist at the school, the work with ESD is difficult to maintain, as Wickenberg (2004) also found in his research on ESD in Swedish schools. Rooting ESD within teams with a certain inherent continuity seems to be a promising supportive structure to create space and time for development of ESD at schools. Teachers point to changes in their teams and subjects as the main obstacle to the continuity of the ESD courses. The process was most successful when a team of teachers collaborated on a lesson plan within ESD, each contributing from their subject, offering both time and expertise. In this case, teachers recounted interested students, fruitful learning processes, hands on learning, reflexivity and creativity. Changes in teams and subjects, meanwhile, resulted in the cessation of ESD teaching. Subsequently, the teachers told how they introduced ESD as "small drips".

15.2.3   Case 3: Building Bridges in the Educational System: University SD Research as ESD Cases in School

Introduction

In the autumn of 2010/2011, the school outreach centre at the Faculty of Science, University of Copenhagen (Skoletjenesten) conducted a pilot project on a hands-on pupil workshop on sustainability in urban development involving mobile ICT as an essential tool. The school outreach centre is part of the communications department at the Faculty of Science and functions as a service introducing hands-on scientific research to approximately 15,000 visiting pupils from lower secondary school classes each year.

During the 2-month pilot period, 24 classes (approximately 500 pupils) participated in the workshop "The flooded city". The target group for this project was pupils aged 13–16. Participating classes were mainly from the Greater Copenhagen area, but classes from the whole of Denmark attended and teacher feedback was positive.

Case Description

Objectives

The main objective was to address sustainability issues relevant to the pupils and encourage them to develop their action competence. A secondary objective was to introduce mobile media in an informal learning context. This was to assess how smartphones can support primary learning objectives by working in parallel in a virtual and physical space. A third and final objective was to investigate whether it is possible to take a quick and dirty approach to areas within the field of ESD and still properly engage with the individual pupil, enabling the development of action competence. There is a demand for such quick and dirty approaches since the amount of time available in, e.g. geography, leaves little time to work with the different topics on the curriculum. However, a quick technical fix necessitates certain compromises and we will consider the dilemmas in relation to maintaining a participatory approach below.

Process

Before Before a visit to the specific activity, teachers receive email information on how the facilitators expect the class to be prepared for the activities on the day of the workshop. The teachers may have some specific objectives that integrate well into the curriculum.

During The pupils are welcomed at the university campus and introduced to the facilities and workshop by a student facilitator.

They are then sent in groups to a nearby location where they start by analysing what they see as problems; for example, too much traffic noise, no daylight, no green areas, or no room for cycling/sustainable transportation. When the groups have created their interpretation of the space they are working in, they start to work on finding a location-specific solution; for example, proposals for sustainable transportation, or for efficient management of precipitation. Pupils document their work with photos, texts and videos uploaded directly to a central server from their smartphones. The groups can be located by teachers and facilitators due to the GPS tracking system built into the phones.

After approximately 75 min, groups upload their final video presentation, filmed on location using the smartphone, before returning to the university. When they return, a slideshow of all their pictures is displayed on the projector, and they are asked to recreate their solution in the lab, before the groups share and watch the video presentations. The class votes for the best solution to be entered into a competition, judged by a panel consisting of researchers within the field of urban development alongside representatives from the local municipality. While this concept does not offer school pupils a dialogue with the local authorities, they are aware that their video presentations may potentially reach the eyes and ears of local policymakers.

The facilitator wraps up the day by commenting on the groups' work and illustrating the similarities to the methods and products of professional urban developers. The idea is to simulate, and perhaps even affect, real life urban development.

After The class have online access to all the material they have collected and produced at the workshop. This is used as a tool of reflection on both process and product back in the classroom. It is up to the individual teacher how much they focus on ESD in relation to the basic geography curriculum. Pupils are asked to reflect on the issues raised during the workshop in relation to their own local areas, thereby transposing the knowledge they have acquired to different contexts.

Opportunities and Challenges

Experiences have been drawn from the online evaluation form that all teachers receive after a visit to Skoletjenesten, regardless of topic; from interviews with teachers; informal conversations during the workshop; pupil evaluation surveys; and observations. The quality and complexity of the pupils' ESD solutions and products have also been taken into consideration.

Opportunities

Based on the teacher evaluations, teachers find it valuable to be provided access to technology not otherwise available to them—here it is worth noting the significant progress in mobile ICT technology and its availability since 2010 when the project was piloted. They also appreciate the availability of a ready-made learning resource. They are positive regarding the equal focus on process and product, with pupils playing an active role in their own learning process and working in groups largely without adult intervention. The opportunity to introduce current research topics within their teaching and create a link to higher education is another positive aspect highlighted by teachers. The mobile technology seems in some cases to encourage participation—several teachers mentioned that they experienced pupils who seldom participate in classroom discussions playing a very visible role in the video presentations. According to the teachers, the cases generally seemed to engage the pupils in trying to create solutions for a more sustainable future.

In a short questionnaire, the pupils highlighted what they considered the best aspects of the workshop. Here, they focused on the opportunity to be creative in finding solutions to real world problems, and the freedom to decide amongst themselves on the best course of action. In general they also liked working with smartphones. However, many of the pupils found it difficult to define exactly what the sustainability issues were if asked immediately after the session ended, and there were considerable variations in terms of the quality and complexity of their proposed solutions. Being able to watch their own and others' solutions was nevertheless frequently mentioned by pupils as a valuable source of learning.

Based on their experiences during the workshops, the facilitators highlight the possibility of using university research as a platform for working with ESD as a major impetus for teachers, with the informal discussions between teacher and facilitator valued highly by both parties.

Challenges

Based on the teacher evaluations, it can be difficult to find the time to prepare the pupils properly; some teachers found that pupils had difficulty grasping the concept of sustainable development and struggled to explain it to them, even within the context of a specific case. The teachers found it easier to incorporate cases on precipitation management within their geography lessons than cases concerning sustainable transportation.

Some of the pupils, meanwhile, expressed their difficulty in relating to the cases as the context was too far removed from their own local communities.

For facilitators, the main challenge was a lack of preparation among classes. Despite communicating certain expectations to teachers prior to the workshop, one in four classes were found to be insufficiently prepared. This resulted in considerable disruption to the planned workshop programme as pupils were not equipped to properly consider the complex issues involved in SD. Another issue was that the participating teachers' understanding of ESD was often strongly influenced by their backgrounds within the natural sciences, resulting in a strong focus on environmental issues at the expense of social and economic aspects.

The workshop is based on the pupils finding their own solutions to real world problems through critical reflection and democratic processes among their peers. This places much of the responsibility for learning with the pupils themselves and limits the influence of both teacher and facilitator in ensuring that the proposed solutions are aligned with sustainability values. Indeed, with its democratic approach, pupils may agree upon a non-sustainable solution, as long as the arguments for this solution are based on critical reflection and discussion. Teachers may tackle these issues during both preparation and follow-up lessons; however, as shown above, preparation was frequently inadequate for this purpose, and teachers themselves often found it difficult to fully grasp and communicate the principles of sustainable development.

At the time of writing (2013), the smartphones bought for the project in 2010 still work, but cannot keep up with the new applications the pupils use in their everyday lives. In response, the workshop has been developed to use more video-based inspiration hidden in the physical locality as QR codes to be scanned by pupils (for examples, see www.youtube.com/byundervand). In this way, the physical work on modelling solutions is prioritised rather than the time-consuming use of ICT for documentation purposes. It all comes down to the key issue: To figure out how to stimulate and challenge pupils sufficiently with a relevant paradox or dilemma so that they choose to work with the problem within an ESD context. This remains the primary objective; ICT is merely a learning tool which, for pupils, constitutes a natural part of their everyday lives.

Another important question remains: is a quick and dirty approach to ESD related topics possible—supporting student participation while fitting within the constraints of the school timetable? Since the organisational learning structure of the Danish public school system is predominantly divided into subject-specific classes, teachers only have a limited amount of time available to spend on topics such as cultural landscape and urban development within the geography curriculum. Mirroring the dilemma at the core of this chapter, ESD proponents may ask: Can such approaches function as a supplement to all-encompassing whole-school approaches such as eco-schools, or do such attempts to work within the existing structures in fact undermine more radical approaches?

15.3 Conclusion

What can we learn from these three cases? They illustrate that ESD is a relatively unfamiliar concept for teachers and student teachers in Denmark, despite many of the principles of ESD resembling the more well-known ideas of environmental education and democratic citizenship education in Danish schools. The cross disciplinary nature of ESD challenges teachers' practices, as well as the organisational structure of schooling, due to their formation along strict disciplinary lines. This disciplinary structure represents the primary challenge to ESD within the established education system. Nevertheless, the three cases presented here provide examples of promising practices of ESD implementation within schools. From these cases we can also learn that collaboration among central actors within the education system enables the sharing of ideas, reflections, methods and solutions within ESD as didactical practice. Collaboration challenges researchers, teachers and student teachers in their understandings and practices—the collaboration facilitates mutual learning processes in an explorative and experimental space.

One important question which remains is: How could ESD be rooted within the education system in a longer term perspective? Traditionally, the Danish curriculum leaves space for relative autonomy for teachers to decide the more precise content of each class, as stressed at the beginning of this chapter. Consequently, for teachers who might have an interest, it is possible to work with ESD in various ways, as we saw in the three cases. However, the relatively loose framework of the Danish curriculum also includes the risk that little teaching is actually carried out within the field of ESD, primarily led by a few firebrands. This risk may be lessened by the recent introduction of ESD into legislation on teacher training—indeed, monitoring how sustainability issues are reflected in the teacher training curriculum and subsequently implemented by these student teachers in their classrooms represents a potentially fertile area for future research.

However, policy reforms of public schooling send a less clear message and the continuing focus on improving academic standards within the core subjects may weaken ESD with its focus on interdisciplinarity and lack of clearly measurable learning outcomes.

Whether the aim is to introduce ESD as "small drips" within the existing education framework or bring about a sea change by reorienting the school system towards ESD in a whole school approach, supporting structures which create space for development and reflections on ESD as didactical practice seem to be crucial if ESD should find a more permanent place in schools. On a school level, organising teacher teams for cross-curricular collaboration, and cooperation between schools and external partners, such as the local community, businesses, universities and teacher colleges, is one way to do this. On a national level, this requires strategies which address issues of sustainable development in innovative and participatory ways within all types of education.

References

Baltic 21 (2000). Haga declaration. http://cbss.idynamic.lv/component/option,com_attachments/id,45/task,download/. Accessed 26 Feb 2014.

Breiting, S., & Schnack, K. (2009). *Uddannelse for Bæredygtig Udvikling i danske skoler: Erfaringer fra de første TUBU skoler i Tiåret for UBU*. The Danish School of Education, University of Aarhus.

Breiting, S., & Wickenberg, P. (2010). The progressive development of environmental education in Sweden and Denmark. *Environmental Education Research, 16*(1), 9–37. doi:10.1080/13504620903533221.

Breiting, S., Læssøe, J., Rolls, S., & Schnack, K. (2009). Climate change and sustainable development: The response from education: Report from Denmark. In: J. Læssøe et al. (Ed.), *Climate change and sustainable development: The response from education: National reports* (pp. 147–191). Copenhagen: International Alliance of Leading Educational Institutions. http://dpu.dk/Everest/Publications/Om%20DPU/Institutter/Institut%20for%20didaktik/20091207121704/CurrentVersion/DEN.pdf. Accessed 26 Feb 2014.

Cochran, K. F. (1997). Pedagogical content knowledge: Teachers' integration of subject matter, pedagogy, students, and learning environments. Research matters – to the science teacher. 9702 (14 Jan 1997). Online Document of the National Association for Research in Science Teaching. http://www.narst.org/publications/research/pck.cfm. Accessed 26 Feb 2014.

Danish Government. (2002). A shared future: Denmark's national strategy for sustainable development. Copenhagen: Danish Environmental Protection Agency. http://www2.mst.dk/Udgiv/publications/2002/87-7972-279-2/pdf/87-7972-259-8.pdf. Accessed 26 Feb 2014.

Danish Ministry of Education. (2009). Education for sustainable development: A strategy for the United Nations decade 2005–2014. Copenhagen: Danish Ministry of Education. http://www.uvm.dk/Service/Publikationer/Publikationer/Tvaergaaende/2009/Education-for-Sustainable-Development. Accessed 26 Feb 2014.

Danish Ministry of Education Press Release (2009, March 4). Undervisningsministeriet opruster på klimaområdet. http://uvm.dk/Uddannelser/Folkeskolen/Fakta-om-folkeskolen/Nyheder-om-folkeskolen/~/UVM-DK/Content/News/Udd/Videre/2009/Marts/090304-Undervisningsministeriet-opruster-paa-klimaraadet. Accessed 26 Feb 2014.

Danish Ministry of Science, Innovation and Higher Education. (2013). Bekendtgørelse om uddannelsen til professionsbachelor som lærer i folkeskolen. https://www.retsinformation.dk/Forms/R0710.aspx?id=145748. Accessed 26 Feb 2014.

FEE (2014) What is Eco-Schools? http://www.fee-international.org/en/Menu/Programmes/Eco-Schools. Accessed 26 Feb 2014.

Jensen, B. B. (2002). Knowledge, action and pro-environmental behaviour in environmental education. *Environmental Education Research, 8*(3), 325–334. doi:10.1080/13504620220145474.

Jensen, B. B., & Schnack, K. (1997). The action competence approach in environmental education. *Environmental Education Research, 3*(2), 163–178. doi:10.1080/1350462970030205.

Kristensen, P. (2010). *Afrapportering af pilotprojekt Grøn Flag Grøn Læreruddannelse: læreruddannelsen på Fyn.* Odense: UC Lillebælt.

Læssøe, J. (2008). Uddannelse for bæredygtig udvikling i Danmark: hvad bliver det til? Global Økologi, June 2008, pp. 8–11. http://old.ecocouncil.dk/global/global_okologi_2008/nr2_2008_8_11.pdf. Accessed 26 Feb 2014.

McKeown, R., & Hopkins, C. (2007). Moving beyond the EE and ESD disciplinary debate in formal education. *Journal of Education for Sustainable Development, 1*(1), 17–26. doi:10.1177/097340820700100107.

Ny Nordisk Skole. (2012). Mål, Manifest og Dogmer for Ny Nordisk Skole. http://nynordiskskole.dk/Om-Ny-Nordisk-Skole/Hvad-er-Ny-Nordisk-Skole/Maal-Manifest-og-Dogmer-for-Ny-Nordisk-Skole. Accessed 26 Feb 2014.

Schnack, K. (2003). Action competence as an educational ideal. In D. Trueit et al. (Eds.), *The internationalization of curriculum studies* (pp. 271–291). New York: Peter Lang.

UNECE. (2005). UNECE strategy for education for sustainable development. http://www.unece.org/fileadmin/DAM/env/documents/2005/cep/ac.13/cep.ac.13.2005.3.rev.1.e.pdf. Accessed 26 Feb 2014.

UNESCO. (2005). Report by the Director-General on the United Nations decade of education for sustainable development: International implementation scheme and UNESCO's contribution to the implementation of the decade: Draft international implementation scheme for the United Nations Decade of Education for Sustainable Development (2005–2014). UNESCO. http://unesdoc.unesco.org/images/0014/001403/140372e.pdf. Accessed 26 Feb 2014.

United Nations, World Commission on Environment and Development. (1987). Our common future: Report of the World Commission on environment and development. Published as Annex to General Assembly document A/42/427. http://www.un-documents.net/wced-ocf.htm. Accessed 26 Feb 2014.

Wickenberg, P. (2004). Norm supporting structures in environmental education and education for sustainable development. In P. Wickenberg et al. (Eds.), *Learning to change our world? Swedish research on education & sustainable development* (pp. 103–127). Lund: Studentlitteratur.

Chapter 16
Education for Sustainable Development in Flanders: The UN Decade and Beyond

Katrien Van Poeck, Jürgen Loones, and Ingrid Claus

16.1 Introduction

In this chapter we address the implementation of the UN-Decade of Education for Sustainable Development (ESD) as well as the prospects for future ESD policymaking in Flanders, the northern, Dutch-speaking part of Belgium.[1] The implementation of the Decade in Flanders is highly shaped by the Strategy for ESD developed by the United Nations Economic Commission for Europe (UNECE). At the high-level meeting of Environment and Education Ministries in Vilnius in 2005, Flanders committed itself to the implementation of the Strategy. An 'ESD consultation platform' was created in response to UNECE's appeal to install a coordination mechanism for the ESD Strategy so as to stimulate implementation, information exchange, and partnerships. The platform is composed of representatives of diverse public administrations on different levels, including ministers' political advisors, and non-state actors such as NGOs, unions, institutes for higher education, school systems within compulsory education and strategic advisory councils. The environmental education (EE) unit of the Flemish government, which was responsible for stimulating networking and capacity building concerning ESD, coordinated the development of an Implementation Plan for ESD in Flanders. One of the policy objectives put forward in this Plan, as well as in the UNECE Strategy, was the need for scientific research on ESD practices and

---

[1] This chapter reflects the authors' personal opinions and in no way represents any official stand or opinion of the Flemish government.

K. Van Poeck (✉)
Laboratory for Education and Society, University of Leuven, Leuven, Belgium
e-mail: katrien.vanpoeck@ppw.kuleuven.be

J. Loones • I. Claus
Environmental Education Unit of the Flemish government, Brussels, Belgium
e-mail: jurgen.loones@lne.vlaanderen.be; ingrid.claus@lne.vlaanderen.be

© Springer International Publishing Switzerland 2015
R. Jucker, R. Mathar (eds.), *Schooling for Sustainable Development in Europe*,
Schooling for Sustainable Development 6, DOI 10.1007/978-3-319-09549-3_16

policymaking. Since it is a major assignment of the EE unit to serve as a centre of expertise on EE and ESD, a policy advisor was entrusted with this task (incidentally one of the authors of this chapter). The rationale behind this decision was that by assigning a policy advisor to conduct the research herself, the resulting knowledge development would simultaneously strengthen the unit. Moreover, the decision was supposed to contribute to reducing the 'theory-practice gap' and the idea was that research outcomes could regularly inform ESD policy and practices. Considering that rigorous and critical academic research on ESD beyond a mere practice-based orientation is required, the policy advisor started doctoral studies at the University of Leuven.

In this chapter, we present some conclusions of this doctoral research project (Van Poeck 2013), focusing on the implementation of the UN Decade of ESD in Flanders. Thereby, we also address the relation between research and policymaking within the context outlined above. Next, we present an analysis of a meeting of the ESD consultation platform during which Flemish ESD policy beyond the Decade was discussed. We conclude the chapter by going into two future ESD initiatives and addressing how both the outcomes of the doctoral research and the particular setting in which research and policymaking took place have influenced these policy intentions.

16.2 The UN Decade of ESD in Flanders

As the above mentioned doctoral research revealed, the implementation of the UN Decade of ESD in Flanders is inextricably intertwined with three broader developments in environmental and educational policy: the increasing impact of ESD policy and discourse on EE, the framing of social and political problems as learning problems, and ecological modernisation (see also Van Poeck 2013; Van Poeck et al. 2013). More precisely, our analysis showed how these developments give shape to the boundaries of a particular policy regime and, thus, affect what is possible and acceptable within Flemish ESD policy and, consequently, within educational practices.

16.2.1 The Increasing Impact of ESD Policy and Discourse on EE

The increasing impact of ESD on EE is a policy-driven tendency (Jickling and Wals 2007; Nomura and Abe 2009), highly affected by decisions made in international institutions. Although 'sustainable development' is omnipresent in policy discourses, the concept remains largely contested (Bruyninckx 2006; Gunder 2006; Jickling and Wals 2007). Ever since the publication of the

Brundtland Report (WCED 1987), sustainable development (SD) is increasingly affecting environmental education policy and practice. The UN Decade is part of this evolution. As a result, a reallocation of resources (funds and personnel[2]) from EE to ESD occurred in Flanders, and EE policymaking was increasingly influenced by strategies put forward by international institutions (Van Poeck et al. 2013).

However, the fact that ESD is becoming more and more established in EE does not imply that the relation between both concepts is clear for everybody. Several authors argue that a multitude of different perspectives exists simultaneously (e.g. Hesselink et al. 2000; Reid and Scott 2006) and in our case study we also observed efforts made so as to clarify the 'difficult concept' of ESD and to develop a shared understanding of how it relates to EE and other educational fields (Van Poeck et al. 2013). For instance, a task force composed of members of the ESD consultation platform was established as an attempt to create a common conception of ESD resulting in the brochure *ESD: Flag and Cargo* (Flemish government 2011). Furthermore, opinions concerning the desirability of ESD as a new focal point for environmental education are sharply divided in scholarly literature (e.g. Jickling 1994; Sauvé 1996, 1999; Gough and Scott 1999; Huckle 1999; Smyth 1999; Foster 2001; Scott 2002; Sauvé and Berryman 2005; Selby 2006; Chapman 2007; Jickling and Wals 2007; Bajaj and Chiu 2009; Mogensen and Schnack 2010). In Flanders, too, diverse actors emphasised the value of 'basic nature education' and stressed that not the entire EE sector should be reoriented towards ESD. A key policy actor regarded the UN Decade of ESD as 'a trigger for reflection' on current EE practices. In particular, discussions arose concerning divergent interpretations of SD as well as about the tension between an instrumental approach to education aiming at promoting predetermined 'sustainable' behaviour and a perspective that highlights 'pedagogical and emancipatory' values such as critical thinking and autonomous decision making (Van Poeck et al. 2013). The latter is closely related to a second development we take into account.

---

[2] The implementation of the Flemish ESD Implementation Plan largely depended on the redistribution of funds within the existing budgets of several departments. Initially, actions were mainly financed with funding for EE. Later on, however, other collaborating partners started to contribute, they too falling back on the reallocation of existing means. As to the deployment and division of personnel, the situation was very similar. New recruitment for ESD failed to occur but the DESD brought about changes in the tasks and responsibilities of existing personnel. In the EE Unit, two policy advisors were deployed to coordinate the ESD consultation platform and to study and foster ESD as an important trend in EE policy and practice. In other policy areas (e.g. the Department of Education and Training, the Tourism Flanders Brussels Agency, the Flemish Department of Foreign Affairs, the Department of Culture, Youth, Sport and Media and the Agency for Socio-Cultural Work for Young People and Adults) staff members were not full-time seconded but spent time on the promotion, coordination and implementation of ESD in their policy area.

16.2.2   The Framing of Social and Political Problems as Learning Problems

In contemporary society, we face a tendency to frame social and political problems as learning problems (Biesta 2004; Simons and Masschelein 2006, 2009). 'Learning' emerges as a solution for numerous problems, that is, individual learners are expected to acquire the 'proper' knowledge, insights, skills, and attitudes so as to 'learn' to adapt their behaviour to what is considered desirable and make themselves competent to deal with societal challenges. Learning policy and experts in education are deployed to resolve social problems, and educational policy and reforms are designed to change people's behaviour, attitude, and mentality in a particular, preconceived way. Thus, the responsibility for social problems is increasingly delegated to individual people (Finger and Asùn 2001; Simons and Masschelein 2006).

This tendency applies to SD in particular. Ever since the relationship between people and their natural environment has been conceived as problematic, appeals have been made to education in order to tackle the evolving ecological problems such as urban children's increased alienation of nature, problems of nature conservation, the environmental crisis and issues of (over- and under-) development (Postma 2004). A field of educational theory and practice evolved from nature education over conservation education and EE toward ESD and is characterised by the prevalence of a conception of education as an instrument to tackle ecological challenges (Van Poeck 2013). Policymakers assume that the pursuit of SD requires a continuous learning process of groups and individuals. Agenda 21, the plan of action adopted at the United Nations (UN) Conference on Environment and Development in 1992, considers learning as indispensable for reaching SD (UNCED 1992) and the UN Decade of ESD endorsed this framing of SD as a learning problem. An ecologically sound and sustainable society emerges then as a challenge that can be met by applying the proper learning strategies and, thus, education becomes first and foremost a matter of socialisation, that is, of the acquisition of particular knowledge, skills, or competencies.

Also in our analysis of Flemish ESD policymaking, SD issues predominantly emerged as matters of individual learning and the aims of EE and ESD were almost exclusively defined in terms of individual dispositions (Van Poeck et al. 2013). For instance, curriculum objectives translated SD into a set of 'key competencies' individual pupils should achieve and the brochure *ESD: Flag and Cargo* as well as policy documents regarding the relation between ESD and a green economy emphasise the transfer of knowledge and values, green skills, and competencies such as systems thinking.

16.2.3   Ecological Modernisation

A third development we want to address is the increasing hegemony of the discourse of ecological modernisation in today's Western societies. Hajer describes "discourse" as a specific ensemble of ideas, concepts, and categorisations that are produced, reproduced, and transformed in a particular set of practices and through which meaning is given to physical and social realities. (1995: 44)

Basically, the discourse of ecological modernisation assumes that existing political, economic, and social institutions can internalise care for the environment (Hajer 1995). Although the structural character of the ecological crisis (i.e. having its roots in the core institutions of modern society) is acknowledged (Mol and Spaargaren 2000), a fundamental transformation of these societal structures does not appear here as a prerequisite for tackling this crisis. A fundamental assumption is the possibility of reconciling economic growth and the solution of ecological problems. This implies that everyone is expected to do their bit, that consumers should buy and promote techno-scientific innovations, and that citizens and NGOs are regarded as potential allies rather than adversaries (Læssøe 2010). Within this discourse, the environmental challenge emerges as a management problem rather than a political issue and as a "positive-sum-game": "there would be no fundamental obstructions to an environmentally sound organisation of society, if only every individual, firm, or country, would participate" (Hajer 1995: 26). Thus, ecological modernisation brought about a reconsideration, i.e. a narrowing of participatory practices (Læssøe 2007, 2010), focusing on consensus building and marginalising conflicts or contestation concerning values, political ideology, and the ever-present tension between private and collective interests.

In our analysis of ESD policymaking in Flanders (Van Poeck et al. 2013), the influence of an ecological modernisation perspective is reflected in the discourse about a green economy where economic prosperity can go 'hand-in-hand' with ecological sustainability as well as in the prevailing conception of SD as a 'balance' between ecological, social, and economic concerns. However, we also observed that both issues were the subject of discussion among the actors involved. Furthermore, a reconsideration of participatory practices came to the fore in the role of the ESD consultation platform (i.e. contributing to the implementation of policy), the way it was composed of a variety of actors collaborating as a coalition of ESD advocates, and the persistent criticism about the lack of involvement of 'all relevant stakeholders'. Yet, here too, contestation emerged as well, particularly as to the desirability to build alliances (and, thus, consensus) with partners from business circles.

16.2.4   The Boundaries of a Policy Regime

Our analysis of ESD policymaking in Flanders shows how the three above mentioned developments gave shape to the boundaries of a particular governmental regime. Referring to Foucault's "governmentality" perspective, Ferreira argues that within such a regime "a range of semi-normative prescriptions (...) work to include, exclude and govern what it is acceptable (possible) to think and what it is acceptable (possible) to do" (Ferreira 2009: 612). It sets the contours of what is "sayable", "seeable", "thinkable", and "possible" (Masschelein and Simons 2003; Simons and Masschelein 2010: 512). As such, this regime not only affects what becomes (im)possible in both EE/ESD policymaking and practices in the field, but also how we can (or should) think and speak about it. That is, policymakers as well as practitioners and participants are somehow expected to be willing and able to see EE and ESD, think and speak about it and act toward it in a very particular way. The policy-driven emphasis on ESD promotes a consensual understanding of SD in favour of neoliberal economic and political thought and pushes into the background arguments for fundamental social change. The framing of SD as a learning problem contributes to individualisation and the attribution of the responsibility for unsustainability to individuals. The discourse of ecological modernisation also marginalises appeals for fundamental social and political transformations and reduces the space for conflict and contestation.

We showed above how this regime affects ESD policymaking in Flanders. Nevertheless, our case study also revealed that this regime does not completely determine Flemish ESD policy. We observed, for instance, that the prevailing competencies-oriented, instrumental conception of education has been the subject of discussion as well as the connection between ESD and the green economy, and the need to build alliances with partners from business circles. Furthermore, whereas stakeholders' roles in the consultation platform have been set rather formalistically and consensus-oriented in terms of representation, consultation, and implementation of the UN Decade of ESD, participants themselves referred to the platform as creating a space for a valuable dialogue, where differences could sometimes be articulated and could work as a trigger for reflection on existing educational practices and policy.

During our doctoral research we not only analysed ESD policymaking in Flanders but we also studied seven diverse educational practices: the project 'Environmental Performance at School' ('Milieuzorg Op School'), the Transition Towns movement, an environmental education centre, a Community Supported Agriculture initiative, a transition arena for a climate neutral city, a regional centre for action, culture, and youth, and an organisation that offers workshops to promote ecological behaviour change (see also Van Poeck 2013). These case studies, too, reveal how the elaborated social and political developments and the way they trickle-down in international and Flemish ESD policy contribute to the establishment of a regime that co-constructs what becomes (im)possible in concrete practices and how we can (or should) think and speak about these practices. Our analysis showed how the increased focus on the consensual catch-all term 'sustainable development' reduced the space for contestation and controversy within the examined EE practices. Furthermore, we found that the framing of SD as a learning problem brought about an emphasis on socialisation and qualification within the examined EE practices. Finally, we repeatedly observed how the prevailing discourse of ecological modernisation encouraged practitioners to see SD issues and to think and speak about them in terms of 'collaboration' between 'allies' and of 'managing' ecological problems. Yet, just like in our analysis of ESD policymaking, we found that the studied practices were not fully determined by this regime and that, at particular moments, something different emerged. For instance, we witnessed practices where complex and contested SD issues and the often conflicting values, interests, and knowledge claims inherent in them were thoroughly explored and discussed. We also observed educators explicitly questioning the consensual account of SD implied in the Triple-P perspective (the balance between People, Planet, and Profit).

Our analysis of ESD policymaking and practices within the context of the aforementioned social and political developments thus allowed us to understand how the bounds of a particular governmental regime are legitimised and maintained as well as counteracted (Duyvendak and Uitermark 2005; Ferreira 2009). As such, our inquiry aimed at contributing to what Ferreira calls "unsettling the taken-for-granted" (2009: 607) in EE and ESD. It illuminates how certain orthodoxies have become 'normal' and 'obvious' and, all the same, how these orthodoxies assume the possibility of infringement and subversion (Duyvendak and Uitermark 2005; Ferreira 2009).

16.2.5   At the Crossroads of Research and Policymaking

As Masschelein and Simons (2003) emphasise, such a governmental regime cannot be understood as a 'system' that can be changed methodically in a preconceived direction. Rather, it generates effects by *appealing* to people. As such, the boundaries set by the three developments we described above do not *force* policymakers and practitioners into a particular way of seeing, speaking, thinking, and acting. It is 'merely' appealing for it. And indeed, as we argued, at particular moments both the policymaking process and the practices we examined do resist this appeal. By describing this we want to invite and inspire the reader to be attentive to different ways of seeing, speaking, thinking, and acting. After all, Masschelein and Simons argue, "resistance can take the form of simply doing different things" (2003: 87; our translation).

Considering the focus and context of our research project, one might have expected the report of this inquiry to contain a number of recommendations so as to stimulate an 'adequate' ESD policy in Flanders that fosters 'desirable' EE or ESD practices. For several reasons, we did not formulate such recommendations. First, we deliberately wanted to avoid a position where researchers—as experts—could easily derive from their analyses guidelines for a 'better' ESD policy. As Latour argues, taking the position of an expert reinforces the problematic demarcation between science and politics:

> Basically, the expert (no matter how sympathetic and modest he might be) always reinforces the impossible Demarcation, attempting as he does to conceal from the public the kitchen of science in the making and to protect scientists from the interests and passions of the public. And the *worst* is that the cover of expertise is just sufficiently solid to allow politicians to hide behind experts' advice so that they do not have to decide by themselves and for themselves. (2010: 166; our translation)

Therefore, the researcher did not approach the policymaking process and the cases she studied from an evaluative perspective and subsequently put forward instructions for policymakers and practitioners based on 'valid' knowledge and expertise. Rather, as indicated above, she aimed to describe policymaking and educational practices as an invitation to see, think, and speak about it in such a way that, she hopes, might inspire policymakers and practitioners to experiment with different practices and to reflect upon them. After a presentation of the research for civil servants, one of the participants explained how the insights and concepts we developed can indeed inspire policymakers and practitioners and stimulate reflection among them as they 'provide words' to think and speak about their experiences with EE and ESD. Such encounters are encouraging as to these—perhaps optimistic—aspirations. It indicates how presenting this research can incite a dialogue about and search for how educational practices can be understood and given shape in the light of SD issues and how, indeed, 'something else' might become possible. Researchers can thus contribute to this endeavour and incite reflection and dialogue among practitioners, policymakers, and scientists. Yet, it requires time and effort to engage in such a common search and experimentation. A second, more pragmatic, reason for not suggesting more practical recommendations—or better: considerations—for policymaking is precisely the lack of time and space for dialogue, reflection, and experimentation with professionals within the scope of the presented doctoral project. However, as we will explain in the remainder of this chapter, we do aim to take up this challenge in the future.

16.3   Beyond the UN Decade of ESD

In this final section, we address some prospects for future ESD policy in Flanders. First, we briefly analyse a meeting of the ESD consultation platform during which Flemish ESD policy after the end of the UN Decade was discussed in general. Next, we go into two concrete initiatives that are already being prepared, thereby explaining how the lessons learnt from the doctoral research have influenced these intentions.

16.3.1   Discussing Future ESD Policy in Flanders

During a meeting of the ESD consultation platform in April 2013,[3] the participants discussed prospects for Flanders' ESD policy following the DESD. As we will show, this discussion aligned well with the aforementioned tension between, on the

---

[3] Our analysis of this meeting is based on a document analysis (preparatory notes and report) and participant observation (we attended the meeting as policy advisors).

one hand, an attempt to manage systematically the policy process in a well-defined direction and, on the other hand, the need and willingness to experiment with and reflect upon concrete practices.

In preparation of this discussion, the members of the platform had been asked to answer a questionnaire about Flanders' future ESD policy. One of the policy advisors then analysed the participants' opinions, preferences, priorities, and intentions and reported on it in a note that was distributed before the meeting. One of the tensions highlighted in this document is the contradistinction between promoting ESD through pursuing formal policy measures (e.g. embedding tangible targets in official policy documents or obtaining structural funding within the Budget of the Flemish government) and ad hoc cooperation and experimentation with concrete practices based on voluntariness and commitment. Another, related tension concerns the choice between pursuing predetermined results or fostering reflexivity and dialogue about what can be considered valuable educational processes within concrete practices.

In the discussion during the platform meeting, divergent stances were voiced relating to these tensions. Some participants argued for a rather managerial approach and emphasised the importance of systematic policy measures and formal agreements:

> We have to receive recognition. We should make an appeal to the Flemish educational sector and the minister and ask them to commit themselves to the target of realising sustainable schools and to set a deadline for it. Let's get inspired by the "sustainable schools framework" in the UK. (report of the platform meeting; our translation)

Their concerns and proposals thereby mainly reflected an ecological modernisation perspective as well as the framing of SD as a matter of individual learning. They advocated, again, the involvement of 'all relevant partners', collaboration with partners from business circles, efforts to promote competencies for SD, etc. Others, however, contested this approach and argued for dialogue and reflection about the role of education in the light of SD issues and for creating a space for experimentation with open-ended educational processes:

> If we focus too much on targets, we will be blind for the process which has to be open-ended. Otherwise, we will miss opportunities because we won't see what happens. This discords with a logic of planning. Is there a space for the process? Moreover: for inspiration, for passion, for people that move others? We cannot plan or predict such things. And it is insufficiently implied in an understanding of education in terms of competencies. (report of the platform meeting; our translation)

These participants implicitly challenged the boundaries of the above mentioned governmental regime, e.g. by valuing passion, commitment, enthusiasm, and inspiration over competencies and expertise, and by arguing for an approach to education as a distinct domain, separated from managerial problem-solving:

> In the end it is all about the question: what kind of a world do we want? Emancipation: that is the major power of education. (...) We have to avoid ending up as an instrument for a green economy, or for a particular society. Education should remain critical. (report of the platform meeting; our translation)

In the preparatory note the policy advisor proposed five strands for future ESD policy in Flanders focusing on experimentation, reflection, and dialogue rather than on formal policy measures:

> We preferred concrete cases as it is within such practices that the "how" and "what" of ESD comes to the fore. We want to encourage, support and value pilot projects and acknowledge the passion and commitment at hand. Therefore, we will gather people who develop similar initiatives and start a dialogue. (report of the platform meeting; our translation)

Pilot projects in diverse educational settings (such as teacher training, vocational training, primary education, non-formal education) and coaching were put forward as important ESD policy initiatives in the future.

16.3.2 A Closer Look at Two Initiatives

Finally, we end this chapter by illuminating the above elaborated approach to ESD policy with a closer look at two concrete initiatives that are already being prepared or implemented.

First, the EE unit will organise a series of conferences and symposia addressing the role of education in the light of SD. Thereby, we will particularly address 'hot items' in (Flemish) SD policy today, such as green economy, sustainable technologies, and transition management. The aim is to explore these issues, to present different (and often irreconcilable) perspectives at hand, and to stimulate reflection on the concomitant roles attributed to education. For instance, an EE conference was organised focusing on green economy. Different conceptions of a green economy were presented and debated and, afterwards, we addressed the role and purposes of education in this respect. A member of the UNECE Steering Committee on ESD went into international policymaking about ESD and green economy and an educational scientist presented a critical analysis of the policy discourse, particularly of the instrumental perspective on education and the omni-present focus on individual competencies. With this conference, we wanted to promote dialogue and reflection, and thereby we deliberately aimed to challenge and discuss the boundaries of the governmental regime that came to the fore in the doctoral research. Particularly, we made an effort to move beyond a narrow focus on individual learning and beyond the discourse of ecological modernisation that strongly affect the international policy on ESD and green economy (Van Poeck et al. 2013).

A second initiative is a pilot project in an EE centre of the Flemish government. In the context of the ESD implementation plan, educators of this centre have organised training events and workshops about 'systems thinking', a concept that received major attention in the brochure *ESD: Flag and Cargo* and that is often mentioned as an important competence people should achieve through ESD. However, the educators involved were uneasy with the competencies-focused way these workshops were set up and with the emphasis on rather abstract, didactical content (a brief theoretical introduction in systems thinking combined with an overview of exercises and teaching methods). The policy advisor that conducted the doctoral research collaborated with them so as to reconsider these training events. This brought about an interesting dialogue and a common search for a way of dealing with the complexities and uncertainties of SD issues without falling into an abstract teaching of the 'right' competencies. Although the attention for complexity and uncertainty has certainly been enhanced by the increased influence of ESD on EE policy and practice, this common search brought about a renewed interest in the merits of EE's long tradition and particularly in the existing body of EE literature. Sauvé (2005) described 15 currents in EE. The "bioregionalist current" inspired us to develop a pilot project starting from the geographical environment of the EE centre and the complexities and linkages at hand. Rather than focusing on didactics and competencies, we will experiment with educational activities that encourage participants to "see a place from the point of view of natural and social systems, whose dynamic relations contribute to creating a sense of 'living place' rooted in natural as much as cultural history" (Sauvé 2005: 21–22). The project will be set up as a 'collective inquiry' (see also Simons and Decuypere 2013), resulting in a dynamic and interactive exposition: a collective and committed search for 'what matters' in this bioregion and for diverse (new) ways of giving shape to the 'living together' of people and nature in this particular place. Thereby, we aim to invite participants to 'explore' and to 'map' the region and the concerns, tensions, commitments, paradoxes, passions, etc. emerging there.

By presenting these practices to policymakers and practitioners, we hope to inspire other educators, to foster reflection and dialogue and to broaden the scope of what is "sayable", "seeable", "thinkable", and "possible" (Masschelein and Simons 2003; Simons and Masschelein 2010: 512) in the fields of EE and ESD.

References

Bajaj, M., & Chiu, B. (2009). Education for sustainable development as peace education. *Peace and Change, 34*(4), 441–455.

Biesta, G. (2004). Democracy—A problem for education or an educational problem? In T. Englund (Ed.), *Five professors on education and democracy* (pp. 89–109). Örebro: Örebro University.

Bruyninckx, H. (2006). Sustainable development: The institutionalization of a contested policy concept. In M. M. Betsill, K. Hochstetler, & D. Stevis (Eds.), *Palgrave advances in international environmental politics* (pp. 265–298). New York: Palgrave Macmillan.

Chapman, D. (2007). *Environmental education/education for sustainability: What is the difference?* New Zealand Association for Environmental Education. http://www.nzaee.org.nz/index.asp?pageID=2145880172. Accessed 7 Sept 2009.

Duyvendak, J. W., & Uitermark, J. (2005). De opbouwwerker als architect van de publieke sfeer. *Beleid and Maatschappij, 32*, 76–89.

Ferreira, J. (2009). Unsettling orthodoxies: Education for the environment/for sustainability. *Environmental Education Research, 15*(5), 607–620.

Finger, M., & Asún, J. M. (2001). *Adult education at the crossroads. Learning our way out.* London/New York: Zed Books.

Flemish Government. (2011). *Education for sustainable development: Flag and cargo.* Brussels: Flemish Government.

Foster, J. (2001). Education as sustainability. *Environmental Education Research, 7*(2), 153–165.

Gough, S., & Scott, W. (1999). Education and training for sustainable tourism: Problems, possibilities and cautious first steps. *Canadian Journal of Environmental Education, 4*(1), 193–212.

Gunder, M. (2006). Sustainability. Planning's saving grace or road to perdition? *Journal of Planning Education and Research, 26*(2), 208–221.

Hajer, M. (1995). *The politics of environmental discourse. Ecological modernization and the policy process.* New York: Oxford University Press.

Hesselink, F., van Kempen, P. P., & Wals, A. (2000). *ESDebate. International debate on education for sustainable development.* Gland/Cambridge: IUCN.

Huckle, J. (1999). Locating environmental education between modern capitalism and postmodern socialism: A reply to Lucie Sauvé. *Canadian Journal of Environmental Education, 4*(1), 36–45.

Jickling, B. (1994). Why I don't want my children to be educated for sustainable development: Sustainable belief. *The Trumpeter, 11*(3), 2–8.

Jickling, B., & Wals, A. E. J. (2007). Globalization and environmental education: Looking beyond sustainable development. *Journal of Curriculum Studies, 40*(1), 1–21.

Læssøe, J. (2007). Participation and sustainable development: The post-ecologist transformation of citizen involvement in Denmark. *Environmental Politics, 16*(2), 231–250.

Læssøe, J. (2010). Education for sustainable development, participation and socio-cultural change. *Environmental Education Research, 16*(2), 39–57.

Latour, B. (2010). *Cogitamus. Six lettres sur les humanités scientifiques.* Paris: La Découverte.

Masschelein, J., & Simons, M. (2003). *Globale immuniteit. Een kleine cartografie van de Europese ruimte voor onderwijs.* Leuven: Acco.

Mogensen, F., & Schnack, K. (2010). The action competence approach and the 'new' discourses of education for sustainable development, competence and quality criteria. *Environmental Education Research, 16*(2), 59–74.

Mol, A. P. J., & Spaargaren, G. (2000). Ecological modernisation theory in debate: A review. *Environmental Politics, 9*(1), 17–49.

Nomura, K., & Abe, O. (2009). The education for sustainable development movement in Japan: A political perspective. *Environmental Education Research, 15*(4), 483–496.

Postma, D. W. (2004). *Because we are human. A Philosophical inquiry into discourses of environmental education from the perspective of sustainable development and man's caring responsibility.* Leuven/Nijmegen: Katholieke Universiteit Leuven/Radboud Universiteit Nijmegen.

Reid, A., & Scott, W. (2006). Researching education and the environment: Retrospect and prospect. *Environmental Education Research, 12*(2/3), 571–587.

Sauvé, L. (1996). Environmental education and sustainable development: A further appraisal. *Canadian Journal of Environmental Education, 1*(1), 7–34.

Sauvé, L. (1999). Environmental education between modernity and postmodernity: Searching for an integrating educational framework. *Canadian Journal of Environmental Education, 4*(1), 9–35.

Sauvé, L. (2005). Currents in environmental education: Mapping a complex and evolving pedagogical field. *Canadian Journal of Environmental Education, 10,* 11–37.

Sauvé, L., & Berryman, T. (2005). Challenging a "closing circle": Alternative research agendas for the ESD decade. *Applied Environmental Education and Communication, 4*(3), 229–232.

Scott, W. (2002). Education and sustainable development: Challenges, responsibilities, and frames of mind. *The Trumpeter, 18*(1), 1–12.

Selby, D. (2006). The firm and shaky ground of education for sustainable development. *Journal of Geography in Higher Education, 30*(2), 351–365.

Simons, M., & Decuypere, M. (2013, September 10–13). *Collective experimentation: Field notes on making University*. European conference on educational research, Istanbul.

Simons, M., & Masschelein, J. (2006). The learning society and governmentality: An introduction. *Educational Philosophy and Theory, 38*(4), 417–430.

Simons, M., & Masschelein, J. (2009). The public and its university: Beyond learning for civic employability. *European Educational Research Journal, 8*(2), 204–217.

Simons, M., & Masschelein, J. (2010). Hatred of democracy … and of the public role of education? Introduction to the special issue on Jacques Rancière. *Educational Philosophy and Theory, 42*(5/6), 509–522.

Smyth, J. (1999). Is there a future for education consistent with Agenda 21? *Canadian Journal of Environmental Education, 4*(1), 69–83.

United Nations Conference on Environment and Development (UNCED). (1992). *Agenda 21, the United Nations programme of action from Rio*. New York: United Nations.

Van Poeck, K. (2013). *Education as a response to sustainability issues. Practices of environmental education in the context of the United Nations Decade of education for sustainable development*. PhD dissertation, Leuven: Katholieke Universiteit.

Van Poeck, K., Vandenabeele, J., & Bruyninckx, H. (2013). Taking stock of the UN decade of education for sustainable development: The policymaking process in Flanders. *Environmental Education Research*. doi:10.1080/13504622.2013.836622.

World Commission on Environment and Development (WCED). (1987). *Our common future*. Oxford: Oxford University Press.

Chapter 17
Education for Sustainable Development in France

Michel Ricard and Maryvonne Dussaux

17.1 Teaching in Schools

In France, the National Education Ministry, and, to a lesser extent, the Ministry of Agriculture, are responsible for primary and secondary education. Over 12 million students attend primary, middle, and secondary schools in our country, with nearly one million teaching, auxiliary, administrative and service staff.

ESD has led to some major changes, implemented within a highly-centralised system, which is very complex to manage due to its size and diversity, as well as a culture that does not particularly favour openness to outside partners, dialogue, interdisciplinary approaches, and the uncertainty intrinsic to SD.

17.1.1 Implementing ESD in the School System

The first steps towards teaching about the environment in the French education system were initiated in the 1970s, following the Stockholm conference. In 1977 it included the publication of a "directive on educating schoolchildren about the environment", which led to the gradual implementation of projects on societal issues in schools. Unfortunately, this directive was not fully applied and reached only a minority of schoolchildren and a small number of motivated teaching staff whose actions were not officially recognised by an institution that focused mainly on implementing traditional, single-subject programmes.

M. Ricard (✉)
ENSEGID, Institut Polytechnique de Bordeaux, 1 Allée Daguin, 33607 Pressac, France
e-mail: michel.ricard@wanadoo.fr

M. Dussaux
Université Paris Est, Place du 8 mai 1945, 93 203 Saint Denis, France
e-mail: maryvonne.dussaux@u-pec.fr

© Springer International Publishing Switzerland 2015
R. Jucker, R. Mathar (eds.), *Schooling for Sustainable Development in Europe*,
Schooling for Sustainable Development 6, DOI 10.1007/978-3-319-09549-3_17

Following the Johannesburg summit, France introduced a policy on education about the environment and SD to meet the needs of our society. Its implementation, from 2002 to the present, has been marked by several milestones:

- The Prime Minister's appointment, in 2002, of a policy officer in charge of developing ESD, and a report produced by two national education inspectors on the extent to which the environment and SD were included in school programmes;
- The introduction, in 2003, of a National SD Strategy and the attachment of the "Environment Charter" to the French constitution in 2005. This defined a proactive government policy backed by institutional stakeholders in education, supported by partnerships involving central government, local authorities, and civil society.
- The launch of the United Nations Decade on ESD 2005–2014 and the setting up of the French Commission for the Decade in 2005. This contributed largely to reinforcing the national and European dynamic for implementing ESD.
- The national conference on SD in 2007 brought together all the stakeholders in French society. The educational aspects of this "Grenelle Environment Forum" assessed the actions on ESD implemented since 2002 and issued a certain number of recommendations. They included, in particular, the fact that ESD must provide elements for students to reflect on their own values and become aware of their individual and collective responsibilities in order to make fully-informed choices and commitments.

In 2009, ESD was enshrined in a law on the "national commitment for the environment", thus becoming mandatory in every field of initial and continuing education.

In each of these various stages, therefore, the national education authorities have tried to find the best ways to meet the challenges of SD by introducing revised school curricula and new teaching methods.

As it is currently implemented, the ambition of ESD is to involve all students throughout their school careers, by meeting a triple objective, with scientific, civic, and professional implications. Those have gradually become clearer as reflection on adapting courses and teaching practices has progressed.

Scientific Objective

To educate responsible citizens on the basis of scientific knowledge that is confirmed and updated, using a global approach to these highly diverse issues. This requires the consideration of a broad range of constantly reviewed topics, including pollution and risks, climate change, biodiversity, health, and natural resources, as well as urban development and transport, not to mention inequalities of all kinds.

Civic Objective

ESD is defined as a major component in civic education. The key focus is to train students in critical thinking, as well as offering learning suited to them rather than dogmatic teaching. Projects and case studies are complementary to classroom teaching and offer students an opportunity to "learn by doing", via individual or group work, connected to the area where they live and featuring meetings with local partners and stakeholders.

Professional Objective

ESD must prepare students to take SD imperatives into account in their work, irrespective of their business sector or level of responsibility. It must also give them an overview of the many careers likely to develop in the environmental field and SD over the next few decades and provide them with the resources to enter these professions.

The introduction of ESD in the school system, focusing on these three objectives, consisted of three stages preceded by an experimental phase:

2002–2004: an 18-month experimental phase:
In 2002, an inter-ministerial commission was set up, with general inspectors from the national and agricultural education authorities, as well as representatives of the Ministry for SD, higher education, and youth. This commission proposed launching experimental ESD in 85 schools. The results, assessed by experts from the ministries involved, led to the issue of a ministerial circular in July 2004, making ESD mandatory in all schools (Ministère de l'éducation nationale 2004; Ricard 2004a, b).

2004–2007: Introduction of ESD in schools:
This phase established the principles of cross-cutting education and mobilised the school authorities to implement ESD in all schools starting in September 2004. From that time on, all students were, therefore, gradually offered suitable and consistent ESD throughout their primary and secondary education (Ricard 2007).

This had two major consequences: first, integrating ESD in the curriculum gave official status to the work of teachers who had often been obliged to teach about the environment and SD on a voluntary basis; secondly, the roll-out of ESD eliminated a number of pedagogical and institutional blockages resulting from single-discipline approaches.

To assist in implementing these new programmes and the associated teaching methods, supporting materials were made available on the Eduscol site (www.eduscol.fr), accessible by level and subject schools (Ministère de l'éducation nationale 2006).

2007–2010: Introduction of ESD in school curricula:
The second phase consisted of integrating SD topics and issues in the teaching curricula, setting up academic committees on ESD, and promoting a global approach to SD in primary, middle, and secondary schools (E3D, see below). Academic plans for supporting ESD were implemented, in conjunction with local authorities (responsible for education at these levels) schools (Ministère de l'éducation nationale 2007; Ricard 2009).

2011: Stepping up the ESD approach
The third phase reinforced previous efforts: greater emphasis on integrating SD issues in teaching programmes, proliferation of comprehensive approaches in schools, teacher and staff training schools (Ministère de l'éducation nationale 2011; Ricard 2012).

17.1.2   The Current Situation

The educational issues and principles of ESD are now integrated in primary and secondary school curricula, for all students on general, technological, and professional courses, coupled with a continuity of teaching throughout their school career that enables students to acquire the SD-related knowledge and skills they need as future citizens (Eduscol 2012; Ministère de l'éducation nationale 2013a).

ESD is managed as a school project, within a framework to ensure its consistency on two levels: on the one hand, between the course material and the various school work plans; on the other hand, among the various activities organised in partnership between schools and local stakeholders.

This third phase was based on several major axes implemented both inside and outside schools:

Governance and Control of ESD

The general implementation of ESD required the mobilisation of all levels in the education system (national, regional, and local) as well as the entire education community (students, teachers, management staff, teacher trainers and inspectors), and all the various partners.

On a regional level, this process was coordinated by the academic committee on ESD, chaired by the president of the regional education authority. This committee maintains the momentum and monitors the general implementation of ESD; furthermore, the academic committee is in charge of coordination among the various partners in ESD: government services, local authorities, approved associations, public establishments, research centres, and enterprises. The various bodies of inspectors provide active support to teachers, assisting them towards full integration of ESD in their teaching programmes.

New Programmes, the Common Base of Knowledge and Skills, and ESD

Following a general curriculum review, ESD now forms an integral part of the school programme. The new primary and middle-school programmes integrate SD issues via the common base of knowledge and skills, particularly aspects concerning "scientific and technical culture", "humanist culture", "social and civic skills", and "autonomy and initiatives".

The new secondary-school programme also devotes considerable space to SD, not only in general, but also in technological and professional courses. The certification requirements are undergoing an in-depth review to integrate SD issues, especially in training for the building, energy, and chemical industries.

Suitable Teaching Resources

As ESD implies new scientific, ethical, and teaching approaches, it must be based on suitable resources from scientific and expert sources, which may or may not be state-controlled. These resources are available from the Centre National de Documentation Pédagogique (CNDP—National Teaching Resource Centre, (www.cndp.fr)) and its regional centres, including, in particular, Amiens (www.crdp.ac-amiens.fr), appointed as the national ESD resource centre CCRDP Amiens (2013).

The "E3D" Project

Schools are invited to apply for recognition as a "School with a global approach to SD-E3D" for an SD project involving teaching, daily school life, structural management and maintenance of the building, as well as cooperation with external partners. This Agenda 21 approach in schools facilitates the integration of the situation in the local area, while bringing an educational dimension to the SD policies in these areas. The school management appoints a teacher responsible for ESD.

One particular case in the E3D approach is that of "security precaution plans and risk education" (EDUSCOL 2009). The issue of risk lends itself to a variety of educational projects for SD, well-suited to an interdisciplinary treatment due to their cross-cutting, systemic, and civic aspects.

ESD and Cross-Cutting Education

In the framework of the Millennium Development Goals and the national strategy for SD, the complementarity between education projects for development, international solidarity, and SD actions has been reinforced. Indeed, education for international solidarity and development offers students keys to a clearer understanding of major imbalances on a planetary scale. By contributing to their comprehension of environmental, economic, social, and cultural interdependence on a global scale, it encourages them to reflect on ways of remedying the situation.

From this perspective, the new programmes explicitly integrate issues related to divisions in the modern world, in terms of geopolitics, economics, social issues, demographics, energy, and food; furthermore, schools are encouraged to develop all sorts of field projects to promote the implementation of practical partnerships to handle these wide-ranging issues.

ESD and Partnerships

Academic policy on ESD relies on cooperation with local stakeholders acting in favour of SD policies: including government services, local authorities, associations, public bodies, research centres and enterprises. These partnerships facilitate the cooperation of stakeholders and different disciplines, as well as promoting awareness of the outside world and grounding, via practical activities, in topics important to the area around the school. This is especially interesting in the case of long-term projects that offer students the opportunity to reflect on the short-, medium- and long-term consequences of these actions.

Publicising ESD Activities and Sharing Success Stories

Schools and education authorities are encouraged to publicise and promote their projects and actions, this both internally and publicly, with regard to government services, responsible local authorities, students' parents, and other partners, by sharing their experiences, promoting team work, decompartmentalising initiatives, and boosting a common ESD culture.

Academic committees contribute to communication on actions and documents produced by the teams, both to educational authorities and the national ESD resource centre (CRDP Amiens: www. crdp.ac-amiens.fr). National SD events (national SD week, maritime days, etc.) are ideal occasions for publicising and giving recognition to work carried out in primary and secondary schools. Schools are also invited to take part in major European and international events focusing on SD issues.

17.1.3   ESD and Staff Training

Successful mainstreaming of ESD requires the implementation of a plan for training various groups of personnel: teachers, heads of staff, and, more generally, all the supervisory and administrative staff.

Until 2008, training for primary and secondary teachers was provided on Master 1 level in "Institut Universitaire de Formation des Maîtres—IUFM" (Teacher Training University), which focused mainly on subject-based training. Due to their relative autonomy, they had a very uneven approach to the cross-cutting methods required by ESD.

During a 4-year transition period, from 2009 to 2013 (Ministère de l'éducation nationale 2013b), teacher recruitment levels progressively changed from Master 1 to Master 2 (5 years of higher education) with several objectives:

- To raise the qualification level of newly-recruited teachers;
- To integrate primary-school teacher training into the BMD (Bachelor/Master/ Doctorate) system;
- To maintain other career paths open for students who are not recruited as teachers;
- To prepare students gradually for their profession before the competitive recruitment examinations by offering classroom observation and practical experience;
- To offer mechanisms to encourage and promote social advancement.

The aim of the June 2013 Framework Act for Fundamental School Reform was to create "the conditions for raising the level of all students and reducing inequalities". It defined new arrangements for initial and in-service training for teaching and education professions and aimed to update teaching methods:

- Its provisions included the introduction of "Advanced Schools of Teaching and Education" (Écoles Supérieures du Professorat et de l'Éducation—ESPE), integrated into universities from September 2013.
- ESPEs will provide initial training for all teachers and education staff, from pre-school to higher education, and contribute to their continuing education.
- ESPEs will be responsible for developing and promoting innovative teaching methods and the use of digital technologies.
- The new initial training courses will enable students wishing to take up teaching careers to acquire all the necessary skills and benefit from a gradual introduction to the profession. This course features a strong professional dimension and includes an introduction to research.
- The competitive exam will take place at the end of the first year of the Master's course. During their second year of the Master's degree, successful students will take a teaching course, including classroom experience in a school or educational establishment. These students will then acquire the status of trainee civil servants and receive appropriate remuneration.
- At the end of this course, students will be awarded a Master's degree with a specialisation in teaching, education, and training (Métiers de l'Enseignement, de l'Éducation et de la Formation (MEEF)).

It is too soon to know how SD will be integrated into teacher training. This will depend on the mobilisation of the instructors responsible for introducing these new qualifications in faculties and departments UFRs or ESPEs.

The general framework outlined by the Ministry of Higher Education includes mandatory training in cross-cutting teaching methods. While SD is not specifically included in the topics to be covered, it will still be possible to raise awareness of twenty first-century issues and provide an introduction to SD-related topics: such as citizenship, health, risk and international solidarity. A more in-depth treatment may be offered in the context of in-service training, also within the purview of the ESPE. The Ministry of Education departmental memo issued on 24 July 2013, which specified the criteria for approval of establishments adopting an SD approach, certainly emphasised the need for staff training. Programmes to support the implementation of these projects should be designed to foster the expected innovation in teaching methods.

In schools and ESPEs, teachers can base their work on the Fundamental School Reform Act, approved on 8 July 2013. This introduces education on the environment throughout a child's school career and is intended to provide "food for thought on key environmental issues, such as air quality, climate change, managing natural resources and energy, and preserving biodiversity", as well as "raising awareness of environmentally-responsible behaviour and know-how that will enable us to preserve our planet".

17.1.4   ESD and Digital Technologies in the School Environment

Since September 2004, the National Education Ministry has made ESD one of its priorities in primary and secondary schools and the programmes focus increasingly on the use of digital technologies via Information and Communication Technology (ICT) and Digital Work Area (DWA). In 2010, the Ministry launched a five-part development plan for the use of digital technologies in schools: facilitate access to high-quality digital resources; train and support teachers in schools; mainstream the use of DWA; build partnerships with local authorities; and train students in the responsible use of ICT (computer and internet certificate). Two websites are particularly useful sources of digital resources for ICTs: the Ministry of Education EDUSCOL site (eduscol.education.fr) and the CRDP Amiens site (www.cndp.fr/crdp-amiens), which makes accessible a large quantity of digital resources.

The circular dated 24 October 2011 concerning the third phase of mainstreaming of ESD is based on three major orientations: strengthening of governance, enlarging partnership, and improving dissemination of information. It also refers to several key concepts:

- integrating the realities of the immediate area in cooperation with local stakeholders;
- considering different viewpoints, issues, and time-and space-scales;
- identifying innovative teaching projects in ESD and communicating them to teachers;
- assisting teachers to progress in the use of digital teaching tools.

On 13 December 2012, the Minister of Education announced a series of measures concerning ICT in education, called "Bringing school into the digital era":

- Education is to be provided within the information society, this implies that students must be taught in digital information literacy;
- A public service for ICT (Digital Education Council) is to be created, providing teachers with digital resources produced by operators of the Ministry (public body) to diversify their teaching;
- Initial teacher training is to be reorganised, with a stronger focus on ICT.

At the Ministry level, the General Directorate for School Education (DGESCO) is responsible for matters related to ICT: encouraging teaching practices using ICT; developing school equipment; creating networks; teacher training; helping creation, production and distribution of multimedia resources; and supporting the product and services industry.

At regional level, education authorities are responsible for implementing national directives and policies; they give impetus to the development of ICT, coordinate the different levels of teaching and establish partnerships with local and regional authorities, companies, and other administrations and organisations.

Overall ICT policy in education both implements measures to increase internet/ICT use and supervise all ICT sections of the SCEREN/CNDP (National Centre for Educational Documentation and its network) and the National Distance Learning Centre (CNED: www.cned.fr).

All these initiatives take ESD into account with a permanent reference to the global approach of all schools called E3D. This approach is considering three levels of action by all different stakeholders: the classroom level, the school level, and the integration of the school inside its societal territory.

17.2   Higher Education

There are approximately 2,300,000 students in higher education in France, or six times fewer than in schools. The situation in higher education is totally different from the highly-centralised primary and secondary system: while universities have a relatively autonomous status, their funding comes mainly from the government, most of the staff are civil servants, and qualifications are essentially regulated by the state.

17.2.1   Initial Training

As it is the case in a number of countries, particularly in Europe, many attempts to integrate SD into the various sectors of higher education have not produced particularly conclusive results. Despite the clear interest of the Ministry of Higher

Education, the many obstacles to the integration of SD include the absence of a roadmap that would facilitate progress in this area. Furthermore, there are often difficulties and obstacles when initiatives go beyond the academic framework, with the aim of truly integrating SD in the curriculum, even if ESD is part of the teachers' work plan.

French universities have all adopted the BMD system and most of the courses dealing with SD are on a Masters level. Non-vocational Masters courses in France are usually based on an academic, single-discipline approach.

Professional courses on ESD and SD are offered in such institutions as universities, Grandes Écoles (elite higher-education establishments), engineering schools and business schools. These courses have proliferated in recent years, mainly in areas relating to business and management, raising the recurrent issue of integration into working life for their graduates. There are now several dozen establishments each producing over 2,000 graduates per year.

Most of these courses are in the second and third cycles (professional master, research master, advanced master) in either initial or continuing education, aimed at a huge variety of professions. Apparently, however, recruitment in enterprises—and, to a lesser extent, in local authorities—is limited, as departments responsible for SD are often small and people are generally recruited in-house for their professional experience.

17.2.2   Continuing Education

When referring to higher education, it is necessary to replace the debate inside the European frame and to consider both initial and continuous training. The European Union requires graduates able to fulfil the needs of the scientific community and the employment market. In this context, it appears that only 22 % of the French population from 25 to 64 years holds a higher education diploma compared to 34 % in Japan and 37 % in USA.

In France, there is a will in favour of greater access to University not only with the goal to increase enrolment but also to address social and cultural inequality with the help of initial and continuous lifelong training, both general and vocational. Indeed, to make effective lifelong training is one of the key objectives of the European Union Council.

In France, only universities are allowed to deliver national degrees acquired through continuing education activities. They are also the only ones to allow non-graduate individuals to access higher education by obtaining a special diploma granting access to university called DAEU and "Capacité en Droit". Furthermore, the French University is increasingly practising VAE (Validation of the Acquisition of Experience) and VAP (the accreditation of acquired skills). This system allows a national education diploma to be obtained on the basis of a professional experience dossier to follow diverse internal and external training sessions and to get higher education qualifications.

However, French universities are still under-represented in professional training activities with less than 20 % of the public concerned. This implies that they have to evolve to developing more suitable actions to present and future expectations of the society, particularly concerning SD.

The Conference of Presidents of Universities (CPU) adopted a charter on "lifelong learning" (LLL) in 2008, including proposals intended to reinforce existing provisions:

- Integrate LLL in overall university policy.
- Offer teaching and learning to a diversified student population and adapt the curriculum to raise the level of student participation and attract adults back to university.
- Provide suitable guidance and advisory services. Guidance must focus on a global approach (lifelong). University departments must be capable of participating in this public lifelong guidance service: advice and guidance must be offered both to students and adults, from the youngest through to seniors.
- Accreditation of work experience. Universities are very keen on developing accreditation of work experience (AWE). Development of this approach should continue, on the basis of skills that are transposable to the entire higher education system.
- Integrate LLL in a quality approach combining training per se, inside and outside the educational establishment, as well as assessing staff activities, to take them into account in career evaluations.
- Strengthen relations between research, teaching, and innovation. Continuing education has always been recognised as a key entry point for existing and future partnerships.
- Develop partnerships on a local, regional, national, and international level and propose attractive, relevant programs with a priority: the need to develop local partnerships to change these systems and respond to the issue of developing these qualifications. Enhanced recognition and positioning of universities in the bodies responsible for in-service training (e.g.: EUA 2008) is likely to foster the development of continuing education in universities.

Alongside the proposals in this charter, universities are working on a certain number of actions aimed at eliminating the gap between initial training and continuing education. This will enable universities to play a comprehensive role as a public service in lifelong learning, encouraging adults to return to study part-time, developing sandwich courses and AWE, clarifying and improving the governance of vocational training, and clarifying the skills and roles of the various stakeholders.

The current economic crisis has merely reinforced this necessary consolidation of public higher education. Unemployment and an ageing population make it essential to optimise everyone's talents and broaden access to higher education. Integrating lifelong learning, research, innovation, and knowledge-transfers, as is possible in higher education, is a challenge to be met while, at the same time, maintaining the principles of autonomy, university freedom and social equity.

17.2.3 The Green Plan Reference System

Article 55 of the Grenelle's law of 3 August 2009 requires all higher education establishments to set up a SD approach (targeting economic, social and environmental factors) under the name "Green Plan". The initial expression of intent has now been embodied in the Green Plan scheme.

Since 2010, on the government's request, universities and Grandes Ecoles have developed a joint strategy aimed at implementing a common reference system, the Green Plan, as a guidance tool for integrating SD into all higher-education establishments (CPU-CGE 2012, 2013).

This national reference is a consolidated document shared by all French universities and Grandes Écoles. It contains all of the data related to its implementation and addresses nine challenges: sustainable production and consumption; a knowledge society; governance; climate change and energy; transport and sustainable mobility; biodiversity and natural resources; public health and risk reduction; demography, immigration and social inclusion; SD and international challenges.

This document also refers to nine priority themes that incorporate the nine challenges of the Green Plan:

- Sustainable production and consumption;
- Building a knowledge society based on education, training and research
- Governance
- Climate changing and energy
- Sustainable transportation and mobility
- Sustainable management of biodiversity and natural resources
- Public health, risk prevention and management
- Demography, immigration and social inclusion
- International challenges related to sustainable development and poverty
- Every establishment prepares, implements, and assesses its strategy on a campus-wide basis and this strategy has to be understood as a first step to get a green labelling:
- Produce a review, including an analysis of the current situation and a diagnosis of strong and weak points;
- Define an SD strategy consistent with its overall policy;
- Draft and implement an action plan;
- Assess and develop a continuous improvement process to ensure further progress.

Solutions must necessarily be based on the following required information: performance indicators (qualitative or quantitative) and factual documents, outlining actions taken and their results: charters, procedures, measurement tools, budgets, reports, and printouts. The reference system is a summary document, reviewing actions taken and results achieved within a defined time period, in the context of commitments, in the framework of a strategy, and consistent with the managerial approach of each establishment. Up to now, results are variable according to establishments but the search for a common labelling in the next 2 years should bring more concrete and visible results.

17.2.4 ICT, E-Learning and Digital Universities

The French Ministry of Higher Education and Research has, since 2004, promoted a policy of pooling national educational digital content by major subject areas. Thus, the seven Thematic Digital Universities (UNT) were born and developed, covering seven major themes: engineering science and technology, basic sciences, economics and management, humanities and social sciences, languages and culture, law and political science, health sciences and sport. A seventh interdisciplinary university was created in 2005: the digital university on environment and sustainable development—UVED (2012):

- UVED's mission is to promote the development, production, distribution and provision, for all students and teachers, of a coherent set of tools and useful digital learning resources in training and lessons. UVED neither register students nor awards degrees but works towards the establishment of a national digital educational heritage of quality service to the entire university community, identifying existing resources and organizing the production of new educational content.
- UVED encourages and supports the creation of new digital resources related to the different fields of environment and SD. Calls for proposals are launched annually by UVED to encourage the production of relevant and necessary resources, thus complementing the disciplinary offer.
- UVED contributes to the development and dissemination at national and international levels of digital resources, validated technically, scientifically and educationally by the scientific community in the field. UVED encourages sharing of learning, experiences and existing tools between higher education institutions. Promoting, coordinating and supporting the production of pooled and reusable resources, it can create synergies between institutions and thus achieve an economy of provision.
- UVED promotes the implementation at the national level of editorial processes, guaranteeing the quality and sustainability of productions. Promoting an appropriate and relevant indexing, they guarantee efficient access to all available resources. Furthermore, UVED contributes to the evolution of digital pedagogy in higher education. The provision of interactive online resources can support teachers and students.

Along with their policy of producing new digital learning resources, UNT's mission is to provide high visibility of resources of the digital educational heritage of French higher education, at national and international levels.

At present, 2,138 resources are indexed in the search engine. All these resources are accessible through the dedicated UVED portal (http://www.uved.fr/ mutualisation-et-valorisation-des-ressources.html) but also through social networks and partner portals like Eduscol (2013), Thot cursus, Channel U, Facebook, Twitter, Knowtex, In-Deed, France Culture More.

17.3 Research on ESD

Prior to the introduction of the DNUEDD (higher national diploma on ESD), research in this field developed, thanks to a strong research community focused on the "Institute for research on education for environment" (IFREE: www.ifree. asso.fr). Starting in 2006, the "Progress and proposals on ESD" symposium, held at UNESCO headquarters in Paris, launched a new dynamic. Since then, a large number of symposia on this issue have been organised by researchers and teacher-trainers from the IUFM, now known as ESPE.

A major advance was achieved in 2008—both from a scientific point of view as well as in terms of organising a specific research community in the ESD field—with the acceptance of a project supported by National Research Agency (ANR: www.agence-nationale-recherche.fr/). This project brought together a team of researchers working on education on the environment as well as "Urgent Societal Issues" (Questions Socialement Vives—QSV). This research programme, entitled "ESD: support for and obstacles to mainstreaming inside and outside schools", was designed as a study of the process of mainstreaming ESD in and around schools, focusing on issues, limitations, and conditions for its feasibility.

The main results obtained were on two levels: the intelligibility of the mainstreaming process, on one hand, and the production of new education research theory, on the other hand (Dussaux 2011; Lange 2008, 2011; Lebaume and Lange 2008). It was quite clear that supporting factors and obstacles represented the two sides of the parameters characterising the policy and purpose of existing education projects. Thus, one strong, specific education project for formal education in this field consists of viewing schools as a laboratory for new types of individual/community relations and new relationships between nature, science, and society. However, effective practices are more dependent on behavioural and/or environmental objectives, conducted from an uncritical perspective, and customary practices intended to make learning meaningful.

The project also gave impetus to scientific production. Thus, it has given rise to six new Habilitations to conduct Research (HDR), including three specific to this field, the publication of three special issues of international reviews, two composite publications, two symposia, several masters dissertations and doctoral theses; and, above all, international collaborations leading to responses to several national and international calls for programmes, including two on a European level.

Several fundamental research projects have also been completed in the context of programmes led by various stakeholders, including the University of Rouen, the Ecole Normale Supérieure de Cachan (STEF-ENS) (Cachan 2013) and Lyon, French Institute for Education (IFé) (Musset 2010).

17.4 Conclusion

The success of the actions of education and training, which should bring about SD, depends on the responsible behaviour of everyone. Education, seen as a structuralising element for the SD culture, is a relatively ambitious objective given the fact that it is inscribed in a real society project.

Schools and universities are the main actors in that education. They should develop all approaches to reach this goal, but should also be associated in a common approach, in partnership with all the other actors, and should adopt innovative and adaptive modalities.

During the last 10 years, the protocol of common action emanating from the governmental sphere and the progressive awareness of the different actors implemented, on diverse time and space scales, coordinated approaches for education or training for environment and then for SD. Such coordination has already allowed the emergence of numerous dynamic examples in the territory.

The French national strategies for SD claim this ambition and set voluntary objectives in terms of information, education and training. This ambition is spread through several innovations that have been initiated during the last 10 years, more or less related to the UN Decade for ESD.

Regarding schools, important progress has been observed, but even the most recent decisions and innovations do not refer directly to SD—see for example the June 2013 Framework Act for Fundamental School Reform, seeking to create "the conditions for raising the level of all students and reducing inequalities"; the creation of the "Advanced Schools of Teaching and Education". All are opening a promising future as they have defined new arrangements for initial and in-service training for teaching and education professions and aimed to update teaching methods. We moved from an implicit approach of sustainable development in all sectors of education towards a more explicit one that reveals the reality of a global and increasing account of sustainable development.

In higher education, among the many hindrances to infuse sustainability is the fact that, in spite of governments' interests in ESD, specific roadmaps are often lacking to further this infusion in the higher education sector. Moreover, there are still difficulties to go beyond the academic frame to really implement SD in curricula, even if ESD is part of in-service training in most EU member states, with a tendency to go beyond environmental dialogue.

Referring to these conclusions we can consider some general prospects for discussion and reflection with a view to a future strategy beyond 2014:

- Schools and universities need to develop a new common strategy intended to overcome the main obstacles by implementing new pilot approaches and key actions aimed at developing a continuity between these two educational systems;
- Enhanced efforts are needed to increase the use and impact of digital technologies to harness the collective intelligence of teachers and learners to develop new pedagogical approaches with the help of mobile devices and social networks;

- To develop a better technical vocational education and training in future programmes to participate in the implementation of green economy, green skills and other important initiatives necessary for an expected major environmental shift.
- To tackle the problem of a suitable training of teachers so that their competencies in ESD are improved to address the interdisciplinary and holistic nature of ESD and to adapt their mission to the requirements of a quickly evolving society.

References

CCRDP Amiens. (2013). *Pôle national de compétence en Éducation au développement durable*. http://crdp.ac-amiens.fr/edd/index.php/accueil/pole-national-competence-edd. Accessed 19 Mar 2014.

CPU-CGE. (2012). *The green plan national framework*, 1–19. http://www.developpement-durable.gouv.fr/Green-Plan.html. Accessed 19 Mar 2014.

CPU-CGE. (2013). *Le canevas et le référentiel de plan vert des établissements d'enseignement supérieur*. 1–19. http://www.developpement-durable.gouv.fr/Le-canevas-et-le-referentiel-de.html. Accessed 19 Mar 2014.

Dussaux, M. (2011). Former les universitaires en pédagogie. *Recherche & Formation*, 67. http://lectures.revues.org/8267. Accessed 28 Mar 2014.

EDUSCOL. (2009). *L'EDD dans l'établissement du second degre*. http://eduscol.education.fr/cid47461/e3d-etablissements-en-demarche-de-developpement-durable.html#gestion. Accessed 19 Mar 2014.

EDUSCOL. (2012). *Les grandes thématiques du développement durable*. http://eduscol.education.fr/pid23362-cid47860/les-grandes-thematiquesdu-developpement-durable.html. Accessed 19 Mar 2014.

EDUSCOL. (2013). *Enseigner avec le numérique*. http://eduscol.education.fr/pid25718/espaces-numeriques-de-travail-ent.html. Accessed 19 Mar 2014.

EUA (European University Association). (2008). *Charte des Universités européennes pour l'apprentissage tout au long de la vie*. http://www.eurosfaire.prd.fr/7pc/doc/1227801443_eua_charte_fr_ly.pdf. Accessed 19 Mar 2014.

Lange, J.-M. (2008). Education à l'environnement. In A. van Zanten (Ed.), *Dictionnaire de l'éducation* (pp. 279–282). Paris: PUF.

Lange, J.-M. (2011). *Éducation au Développement Durable: Problématique éducative/problèmes de Didactiques*. Mémoire HDR. Université de Rouen.

Lebeaume, J., & Lange, J.-M. (2008, September 18–20). Quelle(s) didactique(s) pour une formation des enseignants 'aux éducations à'? Convoquer la complémentarité et la spécificité des didactiques. *Colloque "Didactiques: Les didactiques et leurs rapports à l'enseignement et à la formation. Quel statut épistémologique de leurs modèles et de leurs résultats?"* Bordeaux 2008.

Ministère de l'éducation nationale. (2004). *Généralisation d'une éducation à l'environnement pour un développement durable: circulaire 2004–110 du 8 juillet 2004*. http://www.education.gouv.fr/bo/2004/28/MENE0400752C.htm. Accessed 19 Mar 2014.

Ministère de l'éducation nationale. (2006). *Décret 11 juillet 2006 sur le socle commun de connaissances et de compétences*. http://www.education.gouv.fr/cid2770/le-socle-commun-de-connaissances-et-de-competences.html. Accessed 19 Mar 2014.

Ministère de l'éducation nationale. (2007). *Seconde phase de généralisation de l'éducation au développement durable: circulaire n° 2007–077 du 29 mars 2007*. http://www.education.gouv.fr/bo/2007/14/MENE0700821C.htm. Accessed 19 Mar 2014.

Ministère de l'éducation nationale. (2011). Troisième phase de généralisation de l'éducation au développement durable: circulaire n° 2011–186 du 24 octobre 2011. http://www.education. gouv.fr/pid25535/bulletin_officiel.html?cid_bo=58234. Accessed 19 Mar 2014.

Ministère de l'éducation nationale. (2013a). *Textes de référence sur l'éducation au développement durable*. http://eduscol.education.fr/cid47919/textes-reference.html. Accessed 19 Mar 2014.

Ministère de l'éducation nationale. (2013b). *Des Écoles supérieures du professorat et de l'é ducation pour mieux former les enseignants*, 16–21. http://www.education.gouv.fr/cid73417/ annee-scolaire-2013-2014-refondation-ecole-fait-rentree.html. Accessed 19 Mar 2014.

Musset, M. (2010). L'éducation au développement durable. *Dossier d'actualité, 56*, IFé, ENS Lyon. http://ife.ens-lyon.fr/vst/LettreVST/56-septembre-2010.php. Accessed 19 Mar 2014.

Ricard, M. (2004a). Le développement durable à l'école. *Journal des instituteurs et des professeurs des écoles, 1581*, 59–60.

Ricard, M. (Ed.). (2004b). *Actes du colloque international sur l'éducation à l'environnement pour un développement durable*. Paris 14–15 avril 2004. ftp://trf.education.gouv.fr/pub/edutel/actu/ 2004/dvpt_durable/colloque.pdf. Accessed 19 Mar 2014.

Ricard, M. (Ed.). (2007). *Actes colloque international "Avancées et propositions en matière d'é ducation pour un développement durable"*. http://www.graine-idf.org/pmb_images/graine/lge/ pdf/Actes%20du%20colloque_1174467679.pdf. Accessed 19 Mar 2014.

Ricard, M. (Ed.). (2009). *Proceedings international symposium "working together on education for sustainable development"*. Bordeaux: M. Ricard Publ.

Ricard, M. (2012). The recognition of sustainable development in the French educational system. In U. Stoltenberg & V. Holz (Eds.), *Education for sustainable development: European approaches* (Higher education for sustainability, Vol. 7, pp. 109–121). Bad Homburg: VAS Verlag für Akademische Schriften.

STEF-ENS Cachan. (2013). *Programmes de recherche 2010–2013*. UMR Sciences Techniques Éducation Formation, STEF-ENS Cachan. http://www.stef.ens-cachan.fr/rech/axes.htm. Accessed 19 Mar 2014.

UVED (2012). *Digital university for environment and sustainable development*. http://www.uved.fr. Accessed 19 Mar 2014.

Chapter 18
Paving the Way to Education for Sustainable Development in Cyprus: Achievements, Findings and Challenges

Aravella Zachariou and Chrysanthi Kadji-Beltran

18.1 Observations Concerning the ESD Implementation Process in the Cypriot Educational System

Education for Sustainable Development (ESD) represents an ambitious, complex and wide ranging educational reform, presenting significant intellectual, pedagogical and strategic challenges for schools (Stevenson 2007). It is in essence a transition and upgrading from Environmental Education (EE), in a way that responds to current conditions and demands (Liarakou and Flogaitis 2007). While in most countries this transition was carried out gradually, Cyprus had no transition period. EE in Cyprus was at an embryonic stage at a time when in some other countries it had already been placed at the core of their educational systems and was moving towards ESD. The first steps towards EE in the educational system of Cyprus were taken in the early 1990s (Zachariou 2005a). During that decade the efforts for introducing EE consisted of initiatives that lacked coordination, continuity and planning and did not challenge any of the dominant orientation of the educational structures, regulations, curriculum and pedagogical practices of schooling, which focused on reproducing rather than transforming the social and economic structures of society (Zachariou and Valanides 2006; Kadji-Beltran et al. 2013a, b).

A. Zachariou (✉)
Environmental Education Unit, Cyprus Pedagogical Institute, Ministry of Education and Culture, Nicosia, Cyprus
e-mail: aravella@cytanet.com.cy

C. Kadji-Beltran
Department of Primary Education, School of Education, Frederick University, Frederick, Cyprus
e-mail: pre.kch@frederick.ac.cy

© Springer International Publishing Switzerland 2015
R. Jucker, R. Mathar (eds.), *Schooling for Sustainable Development in Europe*, Schooling for Sustainable Development 6, DOI 10.1007/978-3-319-09549-3_18

18.1.1   First Phase: Introductory Steps for ESD in Cyprus

The Cypriot educational system failed to follow the steps taken in some countries for integrating EE and moving towards ESD since environmental degradation on the island was not intense or obvious and therefore other national, socio-cultural issues were attributed a higher priority. Education in Cyprus, as elsewhere, is deeply ideological-political and it is structured according to the unique historical social-political conditions that formed the current state of the island (British rule, independence, Turkish invasion and occupation of part of the island) (Zachariou 2005a). Within this framework it is evident that any initiatives taken before the entry of Cyprus into the European Union in 2004, for restructuring the educational system in order to adjust and respond to the needs of the modern world, found no response. The EE initiatives were hindered by epistemological traditions of higher priority that dominated the Cypriot educational system including: the development of a national identity and raising students' national awareness; the Christian epistemology connected to the Greek Orthodox ideals; and the focus on encyclopaedic cognition, uniformity, theoretical and abstract knowledge, specialisation and rationalism (Persianis 1996). Additional obstacles were the centralised character of the educational system, along with the schools' obligation to implement the ministerial requirements and mandates that promoted uniformity and ignored the particularities and real needs of the schools and the communities (Eliam and Trop 2013). These conditions meant that when the country was finally officially introducing EE/ESD in its formal education system, around 2004, this was an innovation with piecemeal and superficial implementation, interchangeably using the two terms and with their content being shapeless and 'fuzzy' for the majority of the educational community (teachers, principals, inspectors, counsellors). The form that EE and ESD had taken at the time was still distant to ESD's description as "the most all-encompassing educational ideology" and "the most radical pedagogy shaping global society" (Spring 2004: 100).

Despite the obstacles, gaps and weaknesses, the efforts taken during the first phase (1996–2004) of the EE and ESD presence in Cyprus were particularly important since they constituted a starting point for: developing a critical mass of educators who later on became important agents for the effective promotion and implementation of the field in the schools; founding ESD throughout the educational system of Cyprus based on organised planning, approaches and structures, in a systematic way that ensured the continuity of its integration in all levels of education.

These efforts were manifested in three forms:

(a) *Mandates/Circulars*. The mandates/circulars—published by the Ministry of Education and addressed to schools—highlighted how various disciplines (such as language, environmental studies, geography, history, religion, arts) can be used as a pedagogical means for the study of environmental and sustainable development (SD) issues based on the principles of EE (MoEC 1993, 1996). These mandates/circulars provided teachers with suggestions and ideas about how to implement EE/ESD inside and outside the school grounds (e.g. field studies in the school, the neighbourhood or the local community, organising excursions and visits to places of environmental interest). They also enabled the development of communication and collaboration networks of the schools with other schools, organisations and institutions that could support the study of environmental issues at school as well as the implementation of EE programmes by an individual school (MoEC 1997). The mandates constituted a lateral and indirect way of supporting EE, given the absence of policy, and intended to rectify schools' and educators' inability to use the EE educational framework within the curriculum subjects. Research conducted by Zachariou (2005a), concerning EE in the different curriculum areas within primary education, reveals that the presence of EE in Cyprus "tended to follow a peripheral and marginalised approach, where, in spite of the innovative elements of EE, its implementation still nourished the norms and supported the existing social system by taking the form of a teacher-centred EE and promoting a mono-dimensional model of EE aligned with the technocratic nature of Cypriot education" (Zachariou 2005a: 295).

(b) *Projects.* EE programmes such as 'Eco-Schools', 'Learning about Forests' and programmes of bilateral cooperation such as 'Green Leaf' or 'SEED', enabled the development of EE projects in schools on all educational levels, on an optional basis (Papavlou et al. 2002; Kadji-Beltran 2002a; Zachariou 2005b). The introduction of these programmes was promoted by environmental organisations' initiatives. The programmes ran in parallel with the official ministerial instructions and were meant to compensate for the absence of an organised official educational policy for EE/ESD. They either had the form of whole school programmes or simple projects and their key investigation topics were connected to predefined themes such as the forests, waste management, water, energy and biodiversity.

Despite the programmes' dynamic educational content, the absence of an official, organised policy for EE/ESD resulted in confusion and vagueness in the field of EE. The lack of precedence and experience led to a uniform implementation of the programmes according to the instructions received, overlooking the particularities of schools, the local environment and the communities (Kadji-Beltran 2002b; Zachariou 2012).

The dominance of the natural environment aspect of SD within ESD was reflected in the findings of a nationwide research study that explored the school principals' perceptions and understanding of the concept of SD and sustainable schools (Zachariou and Kadji 2009). The results indicated that most school principals shared the view that sustainable schools are closely related to the content of EE and closely linked to the protection and conservation of the natural environment. Additionally, principals ignored any social and economic dimensions of SD. The same misconception was reflected in the topics on which the schools chose to focus. The majority of these topics were clearly connected to the topics of the EE programme (deforestation, drinking water shortage, sea pollution, shortage of natural resources, species loss), in other words the environmental aspect of SD. Few schools reported working with issues such as refugees, human rights, culture, and multicultural society. That was not because principals had an understanding of the holistic and systemic nature of SD, but because these topics constituted priority issues of the Cyprus Ministry of Education official policy for political and social reasons (MoEC 2006: 70).

(c) *Development of Environmental Education Centers.* The development on a pilot basis of the first public EE centre (EEC) in Lemithou was a means of connecting formal education to non-formal education and the transfer of the learning process and the study of environmental issues beyond the traditional classroom settings (Kadji-Beltran and Zachariou 2004). The first EEC opened the path for the development of an EEC network, geographically covering most of the island. The EEC network has worked towards reinforcing experiential and empirical learning, promoting actions that enable students and other public groups to interact with the local environment and facilitating social participation and connection of the students and the public with the local area, as a field for learning and acting.

18.1.2   Second Phase: The Implementation of ESD in the Education System of Cyprus

The initiatives that took place in Cypriot schools, combined with the global changes and movements that led to the UN Decade of ESD (DESD) 2005–2014, triggered the initiation of radical changes concerning the integration of ESD, its presence and its role in the educational system of Cyprus. The broader picture of ESD in Cyprus during this period is not very different from ESD in some other countries, where it was nothing more than an extension and development of EE (Bourn 2005: 15).

ESD integration at the heart of the educational system of Cyprus and its promotion at all levels of formal and non-formal education was triggered by the entry of Cyprus into the European Community in 2004. This required the alignment of the educational policy of Cyprus with international mandates. ESD in Cyprus evolved alongside the global processes that shaped ESD within the DESD 2005–2014, the vision of which can be achieved when the countries: increase the quality of teaching and learning in ESD; provide new opportunities to incorporate ESD into educational reform efforts; make progress and attain the UN millennium development goals through ESD efforts; and facilitate networking, exchange and interaction among stakeholders in ESD (Firth and Smith 2013: 171).

Simultaneously, the enactment of the ESD Strategy, which was founded on the vision, the context and the principles of the DESD (UNECE 2005a), supported the organizations', institutions' and the international stakeholders' efforts for establishing ESD as an educational priority for SD.

18.2   Decade of ESD and ESD Achievements in Cyprus Educational System

18.2.1   National Action Plan for EE Focusing on ESD

Along with the international processes of the evolution of ESD, Cyprus has adopted the UNECE's Strategy for ESD during the high level meeting of Education and Environment Ministries in Vilnius (UNECE 2005b), which became the basis for elaborating the *National Action Plan for Environmental Education with focus on Education for Sustainable Development* (NAPEESD) (CPI 2007) by the Ministry of Education and Culture. This was the starting point for establishing a framework for the implementation of ESD on all levels of formal and non-formal education. NAPEESD is considered the most important policy document on ESD in Cyprus. It was approved and enacted by the Ministerial Board (2007). It is included in the National Strategy for Sustainable Development (MoANRE 2010) and it is the result of stakeholders' (public sector, private sector and NGOs) pubic consultation. NAPEESD took account of the conditions of EE in Cyprus and acknowledged the need for re-constructing the educational policy for ESD through comprehensive planning. This was deemed necessary for providing the actors involved in the teaching and learning processes with the tools and skills needed for acquiring environmental literacy, taking action and participating in decision making for the improvement of the environment and achievement of SD (CPI 2007: 5).

NAPEESD intends to establish ESD on all levels of formal, informal and non-formal education in a unified, systemic and concise way, entailing a series of central actions evolving around the following:

- the organisational and institutional framework for supporting ESD;
- the national curriculum;
- the educational means and tools;
- research and evaluation;
- in service training of the educators;
- the use of technology,;
- national and international cooperation and participation;
- EE programmes;
- EECs; and
- sponsoring (ibid.: 16).

NAPEESD was the culmination of a long collective effort directed at dealing with ESD within the Cypriot educational system. It also triggered the changes in which ESD is perceived as a radical change of the educational system that "involves more than a line of unconnected activities" (Scott 2005: 4) and quickly set ESD as a focal objective and important challenge for the Cypriot educational system.

18.2.2   Integrating ESD in the Philosophy of the New National Curricula of Cyprus

The second phase of ESD evolution in Cyprus (2004–2013) is connected to the overall advances observed in parts of Europe and elsewhere as well as with the ongoing dialogue with respect to the need for radical changes and reform in the Cypriot educational system. The advances redefine the role of education as a prerequisite for promoting the behavioural changes needed to achieve SD.

The conclusive remarks of the UNESCO World Conference on ESD in Bonn (2009b) highlight that investment in ESD is an investment for the future and indeed, in some cases, a life saving measure. In the same vein, the strategic framework for European Cooperation in Education and Training "ET 2020" (CEU 2009) emphasises that education and training have a crucial role to play in meeting the many socio-economic, demographic, environmental and technological challenges for Europe and its citizens today and in the years ahead. Meanwhile the Council of the European Union for Education, Youth, Culture and Sport in its Council Conclusions for ESD (2010: 3) considers that education and training are indispensable for achieving a more sustainable Europe and World, emphasizing that sustainability can be used as a tool to enhance quality education on all levels of education and training.

The reform of the educational system in Cyprus focused on the transition from an antiquated, centralised, uniform and bureaucratic organisational framework to an educational system that would lead to the development of a democratic, humanist school that respects diversity and pluralism, inculcates skills, attitudes and political virtues and leads to citizens' education (MoEC 2004: 2–3). In essence the dialogue for the updating of the Cyprus educational system is founded on the principles of quality education, where its outcomes (learnt knowledge, skills, values and behaviours) can change the way students and others live to ensure a sustainable present and future (UNESCO 2006). In that framework education needs to be relevant to real life, address the social and other dimensions of learning and empower learners to become active participants in shaping the modern world. Therefore, quality education is closely linked to the objectives of ESD. ESD can provide a sound background and framework for achieving quality education (UNESCO 2005), since quality education as a term is often not clearly defined and difficult to put into practice (UNESCO 2009a).

The new National Curriculum in Cyprus was developed according to the principles of quality education. This philosophy is uniform for all educational levels and all subjects and seeks to transform the individuals into active citizens, motivated by democracy, boldness, courage, persistency, social justice and solidarity, people that respect the natural environment and promote SD, develop and experience gender equity, understand and interpret natural and social phenomena (MoEC 2008). The curriculum's objective is to create citizens that will be critical, open minded, capable of living together and working collectively, recognising and respecting cultural diversity and drawing from their own culture the self-esteem and identity needed for living in a contemporary multicultural society (CPI 2010).

The philosophy embedded in the rewritten National Curriculum (NC) promotes the purposes, policies and practices of a sustainable school, which requires a holistic reform of a school's operation, including its curriculum and teaching (Birney and Reed 2009: 13).

18.2.3   The National Curriculum (NC) for ESD—Sustainable School and ESD School Planning in Cypriot Education

The definition of a sustainable school becomes clearer following the decision of the National Curricula special committee to integrate ESD into all levels of formal education (CPI 2010). The development of new curricula evolving around the concept of SD and environmental issues was a central action required by the strategy. Environmental and SD issues are expected to be embedded in the school programme, to be systematically studied, in a holistic and systemic way, starting with the early years all the way through lifelong learning, provided by formal as well as non-formal education (CPI 2007).

The new NC aims at the development of sustainable schools, i.e. schools that constitute model organisations promoting and integrating SD within their operations. They can contribute systematically, on a long term basis, through all levels of formal education, to the formation of citizens who lead sustainable life styles, take responsibility for their choices, participate in decision making for the protection and conservation of the environment and ensure a high standard of quality of life based on the three SD dimensions of economy, environment and society (CPI 2012b: 4).

The new NC constitutes a hallmark for ESD in Cyprus since it highlights the transition from the marginalised and occasional study of environmental issues in schools to the holistic approach of these issues as a fundamental part of the educational vision and policy of each school in the country. Each school is flexible and free to develop its own policy so as to integrate local environmental and SD issues, sensitive to its own needs and objectives (MoEC 2009a: 5–6).

The implementation of the new NC integrates innovative elements. A key element is the promotion of substantial whole school changes at all three levels of the school's operation: the pedagogical (teaching and learning processes); social/organisational (culture, social environment, educational policy); the technical/economic (infrastructure, equipment, administrative practices) (Posch 1998; Sterling 2001; Gough 2006).

These elements include:

(a) the development of the schools' sustainable environmental education policy (SEEP) "ESD school plan" according to the school's particular circumstances and needs;
(b) the setting of targets concerning the students, the educators, the schools and the communities;

(c) the ESD curriculum's structure which is not organised according to subject matter but evolves around 12 thematic units (i.e. forests, energy, water, waste management, urban development, production and consumption, desertification, transport systems, poverty, culture and environment, biodiversity, tourism) of national, regional and global interest. The thematic units and the issues related to them interweave through specific expected learning outcomes explicitly defined in learning stages. Currently, it has only been defined for primary education, where the ESD Curriculum has been officially introduced. Since the ESD Curriculum follows a common and unified philosophy for all educational levels, its implementation at pre-primary level is imminent and at a later stage it will be integrated in secondary education.

(d) the role of the lessons that operate as tools for an interdisciplinary-holistic exploration of the thematic units. ESD is allocated time within the timetable of primary education (stages 1–4: two teaching periods per week [2 times 40 min] within the interdisciplinary area of "Life Education"; stages 5–6: one teaching period per week [1 times 40 min]). This time is to be used over and above the time used for activities within other curriculum lessons so as to facilitate additional actions (CPI 2012b);

(e) the preparation of suitable supportive educational material for the thematic areas of the ESD curriculum developed according to the principles of ESD, taking in account the needs of the Cypriot Educational System and approaching the SD issues holistically and in an interconnected way. The educational materials are not informational text books but multimodal tools which can explore different SD topics through educational-pedagogical proposals which are founded on the philosophy of the ESD teaching techniques and methods.

This framework enables educators to adjust the proposed material to their class's individual needs and objectives and use it in the context of formal, non-formal and informal education (Zachariou et al. 2011, 2012a).

The NC for ESD is therefore expected to play an important role in highlighting the principles, pedagogical processes, methodological approaches, organisational and social structures required for the promotion of a holistic school approach ("whole school approaches") (Henderson and Tilbury 2004) and the transformation of the school into a learning organisation: a dynamic system of learning, self-organising, interacting with the community, developing, transforming and evolving (Liarakou and Flogaiti 2007).

The promotion of holistic school approaches in the Cypriot educational system is achieved via planning and implementing the SEEP of the school since:

1. The SEEP is developed by the whole school;
2. The entire school works towards its implementation;
3. It responds to the needs and particularities of the school unit and the school's immediate environment;
4. The focal issues of the policy are agreed by the whole school and everyone in the school engages in their exploration;

5. These issues are studied through the curriculum's thematic units;
6. The SEEP requires the cooperation with the community and the formation of collaboration networks with organisations and institutions;
7. It enables changes in the school and the community through actions and interventions determined through its planning;
8. It integrates self-evaluative processes for its pedagogical, organisational and social level by means of indicators. Evaluation takes the form of schools' self-evaluation and the outcomes become the basis for its continuation;
9. The SEEP is characterised by coherence, consistency and continuity (Zachariou 2012).

The overall changes that have taken place in the area of ESD in Cyprus appear to be an example of good practice. They can contribute to reinforcing the on-going international dialogue on ESD and meeting the need for establishing ESD within school plans, "since this is considered necessary for ensuring the long-term implementation of ESD based on a progressive, coherent and adaptative approach (...) Without sustainable school plans/policies to establish the responsibility of and provide guidelines for schools to act upon, long-term and systemic ESD change might not be sustained" (UNECE 2013). The undergoing changes are simultaneous with changes in some other countries where the educational systems are working towards integrating ESD in a whole school approach that leads schools towards the development of sustainable schools.

The UK government, for example, reinforces the ESD School plans in England by stating that a plan on SD can build coherence among a range of initiatives and school practices (for details see Chap. 19 in this volume). It offers schools in England a wider framework in which they can connect their work to a range of policies and initiatives, such as Every Child Matters, school travel planning, healthy living, school food, extended services, citizenship and learning outside the classroom (DfES 2003, 2006a, b).

In Manitoba, Canada, all schools and communities seek to become sustainable and help their students become informed, responsible and active citizens of Canada and the world and contribute to the social, environmental and economic prosperity and fair life quality for everyone in the present and in the future. This effort is founded on whole school approaches and the promotion of ESD School plans. It leads to reinforcing what is learnt about SD within the classroom by transforming schools' operations into SD life skills workshops, minimising the schools ecological footprint and strengthening its connection to the community (Swayze et al. 2010).

The progress report for the second phase of the UNECE ESD Strategy implementation (2007–2010) testifies that out of the 36 countries that delivered a National Implementation Report, 63 % had adopted a "whole-institution approach" which tends to refer to the simultaneous infusion of sustainability in a school's curriculum, reduction of its institutional ecological footprint, strengthening students' participation, and improving school-community relationships (UNECE 2011: 10).

The integration of the ESD curriculum in Cyprus triggered a chain of changes in teacher education, non-formal education, research, technology and international collaborations in the field of ESD which are needed for its effective implementation in schools and constitute fundamental changes that strengthen and consolidate ESD at a growing number of schools and in civil society generally.

It is not possible to explore in depth here all the changes and progress that have occurred in Cyprus since 2004, therefore this chapter focuses on presenting teacher education in ESD in Cyprus and the links between formal and non-formal education as two key actions for the promotion of ESD connected to and in parallel development with the ESD curriculum.

18.2.4   ESD Teacher Education and Training

ESD teacher education and training in Cyprus is connected to the progressive integration of ESD in the Cypriot educational system. The first phase of the EE/ESD development in Cyprus (1996–2004) included a limited number of EE In-Service Training (INSET) courses and revealed the teachers' urgent need for training (Kadji-Beltran 2002b). The training offered on EE was not oriented towards meeting the specific needs of different groups of educators and was not based on peer learning and collective forms of professional development (Pashiardis 1997).

The INSET courses were offered by the Cyprus Pedagogical Institute (CPI), which is responsible for the official professional development of teachers and school principals in Cyprus. INSET in Cyprus is mainly informal and voluntary, and it is centrally determined, with limited input from the schools or the teachers (Karagiorgi and Symeou 2005). This picture was reflected in the EE INSET and professional development programmes which were limited to some short term seminars, took place mainly at the beginning of the school year and provided the information and basic training needed for teachers in order to implement specific EE programmes in schools (mainly the Eco-Schools programme which was widely implemented by the Cypriot schools), since these programmes constituted the main means for promoting EE (Kadji-Beltran 2002a) (not ESD) in schools.

Teacher education changed from 2004 onwards due to the official policy and curricula for ESD. The fact that the development of sustainable schools is a focal point of the new ESD Curriculum and that each school is expected to develop and implement its individual SEEP demands skilled and competent teachers. These new demands highlight the urgent need for effective and practical teacher education for ESD implementation, since both experienced as well as newly appointed teachers have limited experience in the new framework of ESD and need close and practical support (Kadji-Beltran et al. 2013b).

These facts lead to the development and promotion of various educational programmes and training courses, both obligatory and optional. With regard to compulsory education in ESD issues, there are two series of programmes of education and training for primary school teachers carried out throughout the year.

The first series of compulsory education and training courses is focused on teachers' training for the implementation of the NC of EE/ESD. It is implemented centrally on an annual basis and it is developed in three phases:

(a) The first phase informs teachers about the philosophy, principles, pedagogical framework of the NC, as well as the way of its effective implementation;
(b) The second phase explains the methodological framework for the implementation of the NC and the teaching techniques proposed, as well as the basic steps for planning the ESD School plan [Sustainable Environmental Education Policy (SEEP)] of the school;
(c) The third phase involves the implementation of quality standards for the assessment of the school unit in relation to the effective promotion of SEEP in their school.

The second series of compulsory teacher education and training is closely connected to the implementation of the NC for EE/ESD. This series is school-based and conducted through school networks in a practical way. This programme is also annual and it is addressed to primary school teachers. The schools in each city are divided into networks. Each network consists of ten schools which have common geographical, cultural and social characteristics. The second series is also delivered in three phases:

(a) In the first phase, teachers, one from each school, plan their school's ESD School plan (SEEP). They discuss and exchange opinions about difficulties that might arise, as well as examples of good practice concerning the organization, the issues of investigation, the objectives, interventions and changes promoted by each school.
(b) The second phase is about ESD classroom teaching. The lessons are connected to the ESD School plan. Colleagues can attend the lessons which are followed by discussion and feedback on content, teaching techniques, student participation, organization of the learning process, and so forth.
(c) The third phase is about self-assessment. During this phase, teachers assess the implementation of their ESD School plan (SEEP) under the guidance of the advisors of ESD (MoEC 2013; Zachariou 2013).

Optional ESD courses of various types, forms and duration are also offered. These address teachers of all formal education levels and take place outside school hours. Alternatively, they can take the form of school-based seminars and focus on the needs, interests and priorities of the schools. Most of the courses have a theoretical and practical structure and they are implemented in educational settings outside the CPI (EE Centers, environmental fields, museums) (CPI 2009a). The courses introduce participants to subjects such as sustainable schools, theory, methodology and pedagogical techniques of ESD, the use of outdoor settings as pedagogical means for promoting ESD through formal and non-formal education (2008a, 2009b), and designing ESD projects through parental involvement and local community collaborations (Zachariou and Symeou 2008).

It is also important to acknowledge the school principals' role in implementing ESD and sustainable school development and highlight the need of refocusing their professional development effectively towards ESD. For this purpose CPI has included elements of ESD in the principals' compulsory INSET. Nevertheless, these are limited to a 3-h course for newly appointed principals, and a 5-h training course for deputy head teachers. ESD elements are confined to introducing the main aspects of ESD, the concept of sustainable schools, providing them with guidelines for including establishing and implementing ESD School plans as well as aspects of ESD in their school's educational policy (CPI 2008b).

Despite the observed progress in the field of teacher education and training for ESD the Cypriot educational system still has a long way to go in order to move from the transmissive to a transformational model for INSET in ESD. This is supported by the findings of a research study that explored and identified principals' needs and the main approaches of Continuous Professional Development (CPD) that could enhance their competencies and empower them to promote ESD in schools effectively (Zachariou et al. 2012b). The findings indicate that ESD education and training of school principals appears to be inadequate in terms of content and form. Within their training, school principals believe that it is important to place an emphasis on holistic programmes based on compulsory training, not on educational leadership per se. They specifically argue for a focus on principals' education and training on ESD, pointing out that this education should be practical, experiential, liberating and emancipating. School principals indicated school-based seminars, experiential workshops and seminars in EE centres, as the most effective approaches to ESD education.

Similar conclusions were drawn in another nationwide research study, involving primary school teachers in order to explore the novice teachers' perceptions of ESD and their self-efficacy for teaching ESD issues (Zachariou and Kadji 2013). Research outcomes show that teachers have limited self-efficacy for working on ESD with their classes, particularly with planning ESD lessons, using teaching materials for ESD, using the ESD curriculum, establishing links with the local community and applying ESD educational approaches. More effective education and training in ESD is therefore imperative. It should be multilevel and offered through different forms and types of education, such as mentoring, discussing and reflecting upon model-lessons, attending experiential workshops, participating in INSET seminars on ESD teaching approaches, as well as school based INSET in ESD.

The gap between the agreed policy and teacher education in ESD is identified in several countries. Several research findings document teachers' inefficacy to deliver ESD or take action towards ESD due to limited and insufficient training (McKeown 2000; Powers 2004; Van Petegem et al. 2005). Even though ESD has become part of educational policies and teacher education programmes, there is still the need for exploring suitable and effective forms of preparing teachers for ESD. The report on the progress of the DESD 2005–2014 underlines this need and highlights that: "ESD's presence goes hand in hand with a rethinking of the kind of learning necessary to address sustainability issues" (UNESCO 2012: 29).

The need for teacher education and training programmes on ESD is augmented due to the gaps and weaknesses in terms of ESD in the teachers' university education in Cyprus (Zachariou and Valanides 2006). This is also indicated by Hopkins (2002) who stated that departments of education did not integrate ESD in their programmes of study in a dynamic and substantial way. Additionally, the second global evaluation report on ESD indicated that much emphasis was placed on CPD programmes in ESD and not during initial teacher training (UNECE 2011). Various researchers have demonstrated that ESD is not an important part of universities' teacher education programmes. Its presence in the programmes is isolated, limited and most of the time restricted to natural and environmental science modules, or it takes the form of short introductory programmes and projects (Haigh 2005; Heck 2005; Van Petegem et al. 2005).

18.2.5   ESD and Non-formal Education in the Cypriot Educational System

The promotion of non-formal education is important for the achievement of ESD. Non-formal education is related to the learning processes that take place in external settings beyond the school's boundaries. It seeks to motivate students to engage in the study of socio-environmental issues which are not attached to predefined concepts of the curriculum but emerge from their personal interests and need to acquire information, knowledge and skills, to question and reflect upon, participate in real life actions and raise their quality of life (Falk and Dierking 1998; Olusanya Kola 2005). The UNECE ESD strategy document underpins non-formal education as a key prerequisite for the achievement of ESD since "formal education is empowered by non-formal education which is more participative, learner-oriented and promotes lifelong learning" (UNECE 2005a: 7).

In the case of Cyprus, the National Action Plan for ESD explicitly states the need for implementing alternative ways of learning by employing pedagogical techniques connected to experiential practice, personal experience and interaction with the settings (CPI 2007). In line with this, the ESD curriculum places special emphasis on using the local community and environment as an educational tool for the design and delivery of educational activities. The use of settings, external to school, enables the engagement in active and experiential approaches that can reveal the interconnection between environmental, social, political and cultural parameters and promote the development of collaboration networks amongst the school, governmental and non-governmental organisations. These actions would lead to the development of sustainable schools on the one hand and promote SD within the local community on the other (MoEC 2009: 11–12).

The implementation of the aforementioned takes place through a number of actions such as:

(a) action community programmes based on collaboration networks between schools, parents and members of the local community (Zachariou et al. 2005);
(b) the promotion of educational excursions "that enable students to communicate and interact as a group, become familiar with the local natural, cultural and social environment, and offer voluntary services to the community" (MoEC 2008/2009: 1822);
(c) the introduction of the National Network of EE Centres, which will be discussed in the following section as this is probably the most extensive, organised and systematic action of non-formal education for EE/ESD promoted within the Educational System of Cyprus.

18.2.6 National Network of Environmental Education Centres (NEEC)

EECs were developed within the framework of the DESD principles as an alternative form of education that goes beyond traditional academic knowledge and the linear teaching models, in order to set ESD at the heart of the educational system in Cyprus as a transformative educational process.

Even though similar institutions had been developed in some other countries long before Cyprus (Zachariou and Kaila 2009), the formation of the NEEC in Cyprus started in 2004 as a tool for complementing and supporting the work conducted by schools on the study of environmental and SD issues beyond the school boundaries within various settings. It was intended that the EECs become places that can offer opportunities for multiple environmental actions and activities and actively engage all participants in the study of environmental and SD issues in the field. The EEC's programmes take advantage of the local settings and are intentionally connected to the EE/ESD NC (CPI 2013: 17).

Currently the NEEC incorporates four EECs developed in order to meet specific criteria: the island's unique environmental diversity; the place's potential for outdoor exploration of curriculum issues; the support of local communities; the communities' active engagement in the programmes offered, and their use as key educational and pedagogical tools (Zachariou and Kadis 2008). The EECs are not just an optional complementary EE/ESD activity for schools but, according to the EEC mandate, they "are connected to schools and are integrated in the official educational process. EECs facilitate the exploration of SD issues (e.g. water, biodiversity, local communities, local agriculture etc.) within suitable environmental settings and through different educational techniques (e.g. field study, scientific investigations, etc.)" (CPI 2012a: 1–2). The educational programmes offered by the NEEC address all educational levels and they follow a three stages structure:

introduction—field work—feedback. In addition, they propose educational activities for expanding the field activities within the school settings (CPI 2013).

EECs' educational programmes also address educators, university students, parents and other special groups. The activities can include the use of the facilities of the community in which the centres are located, (museums, local workshops, wineries, natural and cultural history monuments) and may also involve the local people in the EEC programmes as a valuable source for the propagation of the 'unwritten local wisdom' and past experience. EECs have, therefore, the potential to promote intergenerational interaction and to enable civic society to acquire first-hand experience and knowledge of the local environment, reflect, respect, become aware and finally become active for its protection whilst at the same time improving the quality of life.

The importance of EECs as providers of non-formal education for ESD is documented through the research of Amyrotou (2013) who found that, compared to formal education, non-formal educational interventions in the field have an important impact on the learners' cognition, awareness and their intention to adapt environmentally friendly behaviour concerning biodiversity issues. Similarly research conducted by Hadjiachilleos and Zachariou (2013), exploring the acquisition of scientific skills by pre-primary school students in non-formal educational environments (EECs), highlights the authenticity of the non-formal environment and the variety of stimuli that this provides during the data collection, observations, data processing and drawing conclusions. Furthermore, according to Petrou's research (2011), EECs appear to promote creativity, investigation skills, fun through learning, collaboration, respect of different opinions, the use of the senses and critical thinking. These results are consistent with international research, which found that outdoor learning in non-formal environments (e.g. EECs), facilitating experiential learning (Ballantyne and Packer 2009), is fundamental to nurturing the above competencies. Students' involvement in a variety of activities in the non-formal learning environment can transform their natural curiosity and enthusiasm to important learning experiences (Kisiel 2007; Scott and Mathiews 2011).

18.3 Challenges for ESD in the Educational System of Cyprus

As discussed in the previous sections, the development of ESD during the two initial phases of its evolution (1996–2003, 2004–2013) was rapid, impressive and innovative, especially if we consider that Cyprus does not share the same EE tradition and SD culture as some other European countries. Nevertheless, the instigation of the DESD clearly recognized the need for intensified efforts to achieve SD, and consequently, with regard to ESD, far reaching changes in the way education is practised (Firth and Smith 2013: 169).

At the end of the DESD it is important to reflect upon the degree to which these challenges have been met at a national level and look beyond 2014 to contemplate the way ahead. This threshold is highlighting the transition to the third phase of the ESD evolution in Cyprus (2014 and beyond), during which we have to acknowledge the forthcoming challenges ahead and lead to an ESD that can effectively contribute to the formation of a sustainable (or at least less unsustainable) society.

The challenges concern the educational system of Cyprus and its ability to provide substantial status and context to ESD in a way that would encourage schools to integrate the principles, values and practices of SD into all aspects of education and learning aiming to impact on the way people think. In order to meet this challenge, we should also consider how the policy documents (National Strategic Planning, SEEP of the schools, schools' self-evaluation) are perceived as dynamic sources of educational policy as well as means for transforming schools into communities of learning for ESD (Stevenson 2007) and sustainability examples for society.

It is therefore critical to understand clearly the concept, the context and the role of the sustainable school as a carrier of change and booster of social capital with key aspects of SD, concerning the interdependence of society, economy and environment, citizenship, participation and cooperation, social, cultural, biological diversity, quality of life, equity and justice, as well as the carrying capacity of Planet Earth for present and future generations (Scott 2013). This challenge becomes greater if we consider the key school players' (principals and teachers) perceptions of ESD and sustainable schools. Understanding the sustainable school is a key element for success, especially by those who are responsible for its achievement. Research has revealed that 'sustainable school' is a vague concept for the school principals and identified conceptual gaps about ESD since: this was perceived as the equilibrium between the environment, economy and the society; it was considered equivalent to EE; and it was connected to the protection and conservation of the environment (Zachariou & Kadji-Beltran 2013). School principals also presented low confidence in their skills to administer a school according to ESD principles (Kadji-Beltran et al. 2013a).

These challenges become more complex if we consider the liberal character and autonomy of sustainable schools. In the case of Cyprus, ESD was set as a priority of the educational system by the enacted policy. It is therefore a paradox that the schools' autonomy for ESD implementation is currently not promoted according to the particularities and special conditions and needs of the individual schools, or the unique elements of the local history, the political and social traditions and the overall local framework (UNESCO 2009a: 16). The reason for this conflict is the persistence of a centralized system in terms of the school's curriculum, timetable and funding (Pashiardis 2004), which limits the schools' capacity to operate outside the system's structural parameters. This has to be approached as a challenge since it is an impediment for sustainable schools in Cyprus and can restrict their development to stage zero and first stage where there might be initiatives in a school through the work of interested teachers with isolated curriculum inputs, and school leaders that do not acknowledge or do not support the idea of a sustainable school or understand its importance with regard to social, cultural and natural capital stocks (Scott 2013: 186).

Some important questions concerning sustainable schools and the third phase of ESD in Cyprus need to be answered:

- What kind of a sustainable school do we want to develop?
- Do we seek change for orienting schools towards learning for SD or to a change that combines learning "for" and "as" SD (Scott 2009)?
- What kind of school do we want to realise (e.g. radical in contributing to social reconstruction for a sustainable society or conservative in reproducing the current social and economic order) (Kadji et al. 2013)?
- What changes have to be made to the established norms of ESD, so that it will not preserve the status quo but lead towards the direction and implications of change (UNESCO 2006: 16)?

In the latter case, the policy discourse of ESD at national level should be (re) contextualised and transformed into practice and into pedagogical actions (Stevenson 2007). Such a transformation would enable active involvement and contextual, strategic and textual (Fien and Tilbury 2002) interpretations of the SD principles in a way that will support principals' and teachers' role for redirecting schools towards sustainability.

Another important challenge in ESD implementation is the INSET/professional development of teachers and school leaders. Despite the progress observed, teacher education still needs to be oriented towards a system-wide change and include multiple forms of education, which would provide the competencies needed for leading and implementing sustainable schools. Initial teacher education should also be empowered and connected to the educators' professional development in ESD. Most of the teacher education institutions in Cyprus fail to promote ESD within their programmes of study. This phenomenon is a global challenge for ESD: Ryan and Tilbury (2013) argue that higher education curricula remain an underdeveloped space in terms of the core impulse of ESD.

We have aimed to present briefly the ongoing processes of ESD evolution in Cyprus in a critical-analytical way, keeping in view the global parallel advances in the field. We acknowledge that within the ESD dynamic in Cyprus, we can still identify gaps and weaknesses. Yet, throughout the processes new precedents have been created about the role of ESD in the school and in the broader Cypriot society. It is important that we take advantage of everything that has been achieved so far, and use it as a tool to confront the intense economic crisis in the southern European countries and specifically in Cyprus. As politically oriented education, ESD can give answers and solutions through the learning processes: confront the dominant models of non-SD through sustainable local practices, enable students, educators and local populations to critically explore issues from their own local framework, change their views and approaches and seek to improve their life quality through meaningful experiences, actions and interventions that emerge from the triple "bottom line" of society, environment and economy.

18.4 Conclusion

It is imperative to view and respond to the above challenges, through the broader lens of changes that need to take place on a long term, systematic and unified basis. The end of the DESD (2005–2014) marks the beginning of a new cycle of ESD review, on national, regional, continental and global levels. In the case of Cyprus, this requires the establishment of an ESD coordinating body responsible for the implementation of the ESD strategy as well as for ensuring its long term viability by means of review and control mechanisms. Finally, the search for funding sources nationally and at the European level and the development of multi-level collaborations (national, regional, international levels) for the implementation of common policies and actions on areas of common interest, constitute focal axes which the state can use in order to meet the challenges that emerge from ESD and ensure the continuity, the empowerment, the evolution and the expansion of everything that has already been achieved in the field of ESD in Cyprus.

References

Amyrotou, G. (2013). Σύγκριση αποτελεσματικότητας εκπαιδευτικών παρεμβάσεων για την Περιβαλλοντική Εκπαίδευση σε τυπικά και μη τυπικά πλαίσια εκπαίδευσης με θέμα τη βιοποικιλότητα [Comparing the effectiveness of environmental education interventions for biodiversity in formal and non-formal settings]. Unpublished master thesis, Frederick University, Nicosia.

Ballantyne, R., & Packer, J. (2009). Introducing a fifth pedagogy: Experience-based strategies for facilitating learning in natural environments. *Environmental Education Research, 15*(2), 243–262.

Bourn, D. (2005). Education for sustainable development and global citizenship: The challenge of the UN-Decade. *Zeitschrift für Internationale Bildungsforschung und Entwicklungspadagogik, 28*(3), 15–19.

Council of European Union (CEU). (2009). *Council conclusions on strategic framework for European Cooperation in Education and Training* ("ET 2020") (OJC 119, 28.5.2009). http://eur-lex.europa.eu/LexUriServ/LexUriServ.do?uri=CELEX:52009XG0528(01):EN:NOT. Accessed 20 June 2010.

Council of European Union (CEU). (2010). *European Council conclusions for Europe 2020 strategy for jobs and growth* (EUCO 7/1/10 REV 1). http://www.consilium.europa.eu/ueDocs/cms_Data/docs/pressData/en/ec/113591.pdf. Accessed 20 July 2013.

Cyprus Pedagogical Institute (CPI). (2007). *National action plan for environmental education and education for sustainable development*. Nicosia: CPI/MoEC.

Cyprus Pedagogical Institute (CPI). (2008a). *Optional training seminars 2007–2008*. Nicosia: MoEC.

Cyprus Pedagogical Institute (CPI). (2008b). *Programmes for in-service training of school heads and deputy heads*. Nicosia: MoEC.

Cyprus Pedagogical Institute (CPI). (2009a). *Environmental education centers programmes guide*. Nicosia: MoEC.

Cyprus Pedagogical Institute (CPI). (2009b). *Optional training seminars 2008–2009*. Nicosia: MoEC.

Cyprus Pedagogical Institute (CPI). (2010). *Αναλυτικά Προγράμματα Προδημοτικής, Δημοτικής και Μέσης Εκπαίδευσης (τ.χ. Α′)* [Curriculum of pre-primary, primary and secondary education (Vol. A′)]. Nicosia: CPI.

Cyprus Pedagogical Institute (CPI). (2012a). *Έναρξη Προγραμμάτων του Δικτύου Κέντρων Περιβαλλοντικής Εκπαίδευσης* [Launching programs for the network of environmental education centers] (CPI 10.04.7, 25.9.2012). http://www.pi.ac.cy/pi/files/anakoinoseis/20120925_ kpe_egkyklios.pdf. Accessed 20 Sept 2013.

Cyprus Pedagogical Institute (CPI). (2012b). *Οδηγός Εφαρμογής Προγράμματος Σπουδών Περιβαλλοντικής Εκπαίδευσης/Εκπαίδευσης για την Αειφόρο Ανάπτυξη για τους Εκπαιδευτικούς της Δημοτικής Εκπαίδευσης* [Guide for primary teachers for implementing the curriculum for environmental education/education for sustainable development]. Nicosia: MoEC/CPI/CDU.

Cyprus Pedagogical Institute (CPI). (2013). *Οδηγός Προγραμμάτων Δικτύου Κέντρων Περιβαλλοντικής Εκπαίδευσης: ΚΠΕ Κοινότητας Πεδουλά* [Programmes guide of Network of Environmental Education Centers: EEC of Pedoulas Community]. Nicosia: CPI.

Department for Education and Skills (DfES). (2003). *Sustainable development action plan for education and skills*. London: DfES.

Department for Education and Skills (DfES). (2006a). *National framework for sustainable schools*. London: DfES.

Department for Education and Skills (DfES). (2006b). *Sustainable schools*. Nottingham: DfES.

Eliam, E., & Trop, T. (2013). Evaluating school-community participation in developing a local sustainability agenda. *International Journal of Environmental & Science Education, 8*(2), 359–380.

Falk, J. H., & Dierking, L. D. (1998). *Lessons without limit: How free-choice learning is transforming meaning*. Walnut Creek: Alta Mira Press.

Fien, J. & Tilbury, D. (2002). The global challenge of sustainability. In D. Tilbury, R. Stevenson, J. Fien & D. Schreuder (Eds.), *Education and sustainability: Responding to the global challenge* (pp. 1–12). Gland/Cambridge: IUCN. http://ibcperu.org/doc/isis/13028.pdf. Accessed 14 Apr 2014.

Firth, R., & Smith, M. (2013). As the UN Decade of education for sustainable development comes to an end: What has it achieved and what are the ways forward? *Curriculum Journal, 24*(2), 169–180. doi:10.1080/09585176.2013.802893.

Gough, A. (2006). Sustainable schools in the UN Decade of Education for Southern African. *Journal of Environmental Education, 23*, 48–63.

Haigh, M. (2005). Greening the university curriculum: Appraising an international movement. *Journal of Geography in Higher Education, 29*(1), 31–48.

Heck, D. (2005). Institutionalizing sustainability: The case of sustainability at Griffith University Australia. *Applied Environmental Education and Communication, 4*, 55–64.

Hadjiachilleos, S., & Zachariou, A. (2013). The impact of non-formal learning environment in developing scientific investigation skills in students of pre-primary education in Cyprus. In A. Dimitriou (Ed.), *Concepts for nature and environment in pre-primary education: Research data, methodological approaches and educational applications* (pp. 264–279). Thessaloniki: Epikentro.

Henderson, K., & Tilbury, D. (2004). *Whole-school approaches to sustainability: An international review of sustainable school programs*. Report prepared by the Australian Research Institute in Education for Sustainability (ARIES) for The Department of the Environment and Heritage, Australian Government.

Hopkins, C. (2002). The role of education in attaining a sustainable future. In *Conference on environmental management for sustainable universities* (pp. 1–3). Grahamstown: Rhodes University.

Kadji-Beltran, C. (2002a). *Evaluation of environmental education programs as a means for policy-making and implementation support. The case of Cyprus primary education*. Unpublished doctoral dissertation, University of Warwick, Warwick.

Kadji-Beltran, C. (2002b). Considering the teacher's profile for effective implementation of environmental education. In *Proceedings 2nd international conference of science education* (pp. 419–430). Nicosia: CPI/University of Cyprus.

Kadji-Beltran, C., & Zachariou, A. (2004). Experiential and active learning as a means of environmental cognition enhancement: The contribution of the environmental education centers in the development of environmentally responsible citizens. In W. L. Filho & M. Littledyke (Eds.), *International perspectives in environmental education* (pp. 167–182). Frankfurt: Peter Lang.

Kadji-Beltran, C., Zachariou, A., & Stevenson, R. B. (2013a). Leading sustainable schools: Exploring the role of primary school principals. *Environmental Education Research, 19*(3), 303–323. doi:10.1080/13504622.2012.692770.

Kadji-Beltran, C., Zachariou, A., Liarakou, G., & Flogaiti, E. (2013b). Empowering education for sustainable development in schools through mentoring. In *Professional development in education*. Abingdon: Taylor and Francis. doi:10.1080/19415257.2013.835276.

Karagiorgi, Y., & Symeou, L. (2005). Teachers' in-service training within the framework of lifelong professional development. *Newsletter of the Pedagogical Institute, 6*, 12–17.

Kisiel, J. (2007). Examining teacher choices for science museum worksheets. *Journal of Science Teacher Education, 18*(1), 29–43.

Liarakou, G., & Flogaiti, E. (2007). *Από την Περιβαλλοντική Εκπαίδευση στην Εκπαίδευση για την Αειφόρο Ανάπτυξη: Προβληματισμοί, Τάσεις και Προτάσεις* [From environmental education towards education for sustainable development: Speculations, trends and proposals]. Athens: Nisos.

McKeown, R. (2000). Environmental education in the United States: A survey of pre-service teacher education programs. *The Journal of Environmental Education, 32*(1), 4–11.

Ministerial Board (MB). (2007). *Απόσπασμα από τα πρακτικά της συνεδρίας του Υπουργικού Συμβουλίου "Στρατηγικό Σχέδιο για την Περιβαλλοντική Εκπαίδευση με επίκεντρο την Αειφόρο Ανάπτυξη"* [Extract from the proceedings of the Ministerial Board meeting "National Action Plan for Environmental Education focus on Sustainable Development"] (No. of decision 66/145/4/10/2007). Nicosia.

Ministry of Agriculture Natural Resources and Environment (MoANRE). (2010). *Αναθεωρημένη Εθνική Στρατηγική για την Αειφόρο Ανάπτυξη* [Revised national strategy for sustainable development]. http://www.moa.gov.cy/moa/environment/environment.nsf/69E3B0E74C4A5110 C225793C002CD199/$file/NSDS_revised.pdf. Accessed 10 Sept 2012.

Ministry of Education and Culture (MoEC) (1993). Ετοιμασία Εθνικού Προγράμματος Δράσης για το Περιβάλλον και την Ανάπτυξη με βάση τις δεσμεύσεις της Διάσκεψης του Ριο [Preparation of national action plan for the environment and the development according to the Rio declaration] (Circular 127/69–17/3/1993). Nicosia.

Ministry of Education and Culture (MoEC). (1996). Συμπεράσματα της συνεδρίας του Συμβουλίου Περιβάλλοντος [Conclusions of environmental council meeting] (Circular 127/69/17–24/5/1996). Nicosia.

Ministry of Education and Culture (MoEC). (1997). Ενδυνάμωση της Περιβαλλοντικής Οικολογικής συνείδησης των παιδιών [Empowering the environmental awareness of students] (Circular 127/69/19–23/9/1997). Nicosia.

Ministry of Education and Culture (MoEC). (2004). *Democratic and humanistic education in the eurocypriot society: Perspectives of reorganisation and modernisation*. Nicosia: MoEC.

Ministry of Education and Culture (MoEC). (2006). *Annual report*. Nicosia: MoEC.

Ministry of Education and Culture (MoEC). (2008). *Αναλυτικό Πρόγραμμα για τα Δημόσια Σχολεία της Κυπριακής Δημοκρατίας* [Curriculum for public schools of Cyprus Republic]. Nicosia: MoEC.

Ministry of Education and Culture (MoEC). (2008/2009). Οι Περί Λειτουργίας των Δημοσίων Σχολείων Δημοτικής Εκπαίδευσης Κανονισμοί του 2008 και 2009 [Legislations for the operation of primary public schools] (MoEC C.R. 225/2008, 276/2009). http://www.moec. gov.cy/dde/kanonismoi/kanonismoi_e1_ekdromes_episkepsis.pdf. Accessed 21 June 2011.

Ministry of Education and Culture (MoEC). (2009). *Curriculum for environmental education/ education for sustainable development*. Nicosia: MoEC.

Ministry of Education and Culture (MoEC). (2013). Cyprus informal report on progress in the implementation of the UNECE strategy for education for sustainable development. http://www.unece.org/fileadmin/DAM/env/esd/8thMeetSC/Cyprus.pdf. Accessed 15 Sept 2013.

Olusanya Kola, A. (2005). Free-choice environmental education: Understanding where children learn outside of school. *Environmental Education Research, 11*(3), 297–307.

Papavlou, Th., Psallidas, V., Kalaitzidis, D., Perikleous, E., Kourouzidis, Th., Stathopoulos, P., & Skoullos, M. (2002). *Εγχειρίδιο παιδαγωγικών δραστηριοτήτων "το σποράκι πηγή ζωής"* [Tool for educational activities "SEED a source of life"]. Athens: Elliniki Etaireia-Society for the Environment and Cultural Heritage.

Pashiardis, P. (1997). Towards effectiveness: What do secondary school leaders need in Cyprus? *British Journal of In-Service Education, 23*(2), 267–282.

Pashiardis, P. (2004). Democracy and leadership in the educational system of Cyprus. *Journal of Educational Administration, 42*(6), 656–668.

Persianis, P. (1996). Η *εκπαίδευση της Κύπρου μπροστά στην πρόκληση της Ευρώπης* [Cyprus Education in front of European challenge]. Nicosia: Author.

Petrou, M. (2011). Τα Κέντρα Περιβαλλοντικής Εκπαίδευσης ως χώροι μη-τυπικής Εκπαίδευσης στην Κύπρο: Ο ρόλος τους στη διαμόρφωση των περιβαλλοντικών στάσεων και δράσεων των μαθητών/τριων του δημοτικού σχολείου [Environmental education centers in Cyprus as places of non-formal education: Their role in promoting primary school students environmental attitudes and actions]. Unpublished master thesis, Frederick University, Nicosia.

Posch, P. (1998). The ecologisation of schools and its implications for educational policy. In J. Elliot (Ed.), *Environmental education: On the way to a sustainable future. Report on Linz international conference* (pp. 50–54). Vienna: ENSI.

Powers, A. (2004). Teacher preparation for environmental education: Faculty perspectives on the infusion of environmental education into pre-service methods programs. *The Journal of Environmental Education, 35*(3), 3–11.

Ryan, A., & Tilbury, D. (2013). Unchartered waters: Voyages for ESD in the higher education curriculum. *Curriculum Journal, 24*(2), 2–13. doi:10.1080/09585176.2013.779287.

Scott, W. A. H. (2005). *ESD: What sort of decade? What sort of learning?* http://www.ubu10.dk/downloadfiles/What_sort_of_Education.pdf. Accessed 25 Feb 2007.

Scott, W. (2009). Judging the effectiveness of a sustainable school: A brief exploration of issues. *Journal of Education for Sustainable Development, 3*(1), 33–39.

Scott, W. (2013). Developing the sustainable school: Thinking the issues through. *Curriculum Journal, 24*(2), 181–205. doi:10.1080/09585176.2013.781375.

Scott, C., & Mathiews, C. (2011). The "science" behind a successful trip to the zoo. *Science Activities Classroom Projects and Curriculum Ideas, 48*(1), 29–38.

Spring, J. (2004). *How educational ideologies are shaping global society: Intergovernmental organizations, NGO's, and the decline of the nation-state*. New York: Routledge.

Sterling, S. (2001). *Sustainable education: Re-visioning learning and change*. Bristol: Green Books.

Stevenson, R. B. (2007). Schooling and environmental/sustainability education: From discourses of policy and practices to discourses of professional learning. *Environmental Education Research, 13*(2), 265–285.

Swayze, N., Buckler, C., & MacDiarmid, A. (2010). *Guide for sustainable schools in Manitoba*. Winnipeg: IISD. http://www.iisd.org/publications/pub.aspx?id=1381. Accessed 20 Aug 2013.

United Nations Economic Commission for Europe (UNECE). (2005a). *UNECE strategy for education for sustainable development* (CEP/AC. 13/2005/3/Rev. 1). http://www.unece.org/fileadmin/DAM/env/documents/2005/cep/ac.13/cep.ac.13.2005.3.rev.1.e.pdf. Accessed 14 Apr 2014.

United Nations Economic Commission of Europe (UNECE). (2005b). *Vilnius framework for the implementation of the UNECE strategy for education for sustainable development* (CEP/AC.13/2005/4/Rev.1). http://www.unece.org/fileadmin/DAM/env/documents/2005/cep/ac.13/cep.ac.13.2005.4.rev.1.e.pdf. Accessed 8 Sept 2013.

United Nations Economic Commission for Europe (UNECE). (2011). *Learning from each other: Achievements, challenges and ways forward—Second evaluation report of the United Nations economic commission for Europe strategy for education for sustainable development* (Information Paper No. 8). http://www.unece.org/fileadmin/DAM/env/esd/6thMeetSC/Informal%20Documents/PhaseIIProgressReport_IP.8.pdf. Accessed 5 July 2012.

United Nations Economic Commission for Europe (UNECE). (2013). *Concept paper on priority action areas* (Information Paper No. 5). http://www.unece.org/fileadmin/DAM/env/esd/8thMeetSC/Information_Paper_5_Final.pdf. Accessed 23 July 2013.

United Nations Educational, Scientific and Cultural Organisation (UNESCO). (2005). *Draft consolidated international implementation scheme.* http://unescodoc.unesco.org/images/0014/001403/140372e.pdf. Accessed 20 Oct 2006.

United Nations Educational, Scientific and Cultural Organisation (UNESCO). (2006). *Framework for the UNDESD international implementation scheme.* http://unesdoc.unesco.org/images/0014/001486/148650E.pdf. Accessed 1 June 2013.

United Nations Educational, Scientific and Cultural Organisation (UNESCO). (2009a). *United Nations Decade of Education for Sustainable Development (DESD, 2005–2014): Review of contexts and structures for Education for Sustainable Development.* Paris: UNESCO.

United Nations Educational, Scientific and Cultural Organisation (UNESCO). (2009b). *Bonn Declaration for education for sustainable development.* http://www.esd-esd-world-conference2009.org/fileadmin/download/BonnDeclarationFinalFR.pdf. Accessed 20 Apr 2009.

United Nations Educational, Scientific and Cultural Organisation (UNESCO). (2012). *DESD monitoring and evaluation: Shaping the education of tomorrow. 2012 report on the UN Decade of Education for Sustainable Development. Abridged.* Paris: UNESCO.

Van Petegem, P., Blieck, A., Imbrecht, I., & Van Hout, T. (2005). Implementing environmental education in pre-service teacher training. *Environmental Education Research, 11*, 161–171.

Zachariou, A. (2005a). *Περιβαλλοντική Εκπαίδευση και Αναλυτικό Πρόγραμμα: Θεωρητικό πλαίσιο και αρχές στη Δημοτική Εκπαίδευση της Κύπρου* [Environmental education and national curriculum: Theoretical framework and principles in national curriculum of Cyprus primary education]. Athens: Hellin Publications.

Zachariou, A. (2005b). *Δάσος: παιδαγωγικές δραστηριότητες διαθεματικής εφαρμογής* [Forest: Educational activities of interdisciplinary application]. Nicosia: MoEC/CPI.

Zachariou, A. (2012). *Basic steps for implementing the curriculum of environmental education/ education for sustainable development.* Paper presented in inspectors training courses for New National Curriculum, Nicosia.

Zachariou, A. (2013). *Teacher education for sustainable development: Cyprus example.* http://www.unece.org/fileadmin/DAM/env/esd/8thMeetSC/Presentations/Cyprus_teacher_education.pdf. Accessed 28 Aug 2013.

Zachariou, A., & Kadis, K. (2008, December). Environmental education centers and education for sustainability: The network of Environmental Education Centers in Cyprus. In *Proceedings of the 4th panhellenic conference of PEEKPE*, Nafplio, Greece.

Zachariou, A., & Kadji, C. (2009). Cypriot primary school principals' understanding of education for sustainable development key terms and their opinions about factors affecting its implementation. *Environmental Education Research, 15*(3), 315–342.

Zachariou, A., & Kadji-Beltran, C. (2013). *Emerging challenges for ESD teaching and learning for novice teachers.* Paper presented at 7th World Environmental Education Congress, Marrakesh, Morocco.

Zachariou, A., & Kaila, M. (2009). Η συμβολή της μη τυπικής εκπαίδευσης στην Εκπαίδευση για την Αειφόρο Ανάπτυξη: Το παράδειγμα των Κέντρων Περιβαλλοντικής Εκπαίδευσης [The contribution of non-formal education in Education for Sustainable Development: The example of Environmental Education Centers]. In M. Kaila, A. Katsikis, P. Fokiali, & A. Zachariou (Eds.), *Education for sustainable development: New data and orientations* (pp. 311–343). Athens: Atrapos.

Zachariou, A., & Symeou, L. (2008). The local community as a means for promoting education for sustainable development. *Applied Environmental Education & Communication, 7*(4), 129–143.

Zachariou, A., & Valanides, N. (2006). Education for sustainable development: The impact of an out-door program on student teachers. *Science Education International, 17*(3), 187–203.

Zachariou, A., Symeou, L., & Katsikis, A. (2005, September). *Action community programs: An alternative proposal for promoting the social-critical orientation of environmental education in school process.* Paper presented at the 1st conference of school environmental education programs, Isthmos Korinthou, Greece.

Zachariou, A., Kadis, K., & Theodosiou, S. (2011). *Εκπαιδευτικό Υλικό για την Ορθολογιστική Διαχείριση των Απορριμμάτων* [Educational material for sustainable management of waste]. Nicosia.

Zachariou, A., Kadis, K. & Nikolaou, A. (2012a). *Θέματα Αειφόρου Ανάπτυξης στην Εκπαίδευση* [Sustainable development issues in education]. Nicosia: Bank of Cyprus Cultural Foundation.

Zachariou, A., Kadji-Belrtan, C., & Manoli, C. (2012b). School principals' professional development in the framework of sustainable schools in Cyprus: A matter of refocusing. *Professional Development in Education, 39*(5), 712–731. doi:10.1080/19415257.2012.736085.

Chapter 19
Reflections on ESD in UK Schools

Stephen Martin, James Dillon, Peter Higgins,
Glenn Strachan, and Paul Vare

19.1 Introduction

The UK Government, with responsibility for education only in England, believes that "sustainable development is a key responsibility for all of us and everyone has to play their part in making it a reality" (Office for Standards in Education (Ofsted) 2009: 7), and that schools, as places of teaching and learning, have a particularly important role to play in helping pupils understand the impact they have on the planet. They argue that, as models of good practice, schools can be places where sustainable living and working are demonstrated to young people and the local community. All of which is embraced by the idea of education for sustainable development (ESD). ESD can be thought of as a process of learning how to make decisions that take into account the long-term future of the economy,

S. Martin (✉)
University of the West of England, Bristol, UK
e-mail: esmartin@talktalk.net

J. Dillon
Department of Education, University of Ulster, Derry, UK
e-mail: j.dillon@ulster.ac.uk

P. Higgins
University of Edinburgh, Edinburgh, UK
e-mail: Pete.Higgins@ed.ac.uk

G. Strachan
London South Bank University, London, UK
e-mail: Glenn@glennstrachan.co.uk

P. Vare
University of Gloucestershire, Gloucestershire, UK
e-mail: Learning4L@aol.com

© Springer International Publishing Switzerland 2015   335
R. Jucker, R. Mathar (eds.), *Schooling for Sustainable Development in Europe*,
Schooling for Sustainable Development 6, DOI 10.1007/978-3-319-09549-3_19

ecology and equity of all communities. As UNESCO (United Nations Education, Scientific and Cultural Organization), puts it: "Building the capacity for such futures-oriented thinking is a key task of education" (2003: 4).

Future proofing our children's education is not an easy task. Any child starting primary school in September 2014 will complete his or her secondary education in around 2028. No one can predict with any accuracy how the world will change over this period, but it is likely to change in many significant ways. An expanding population, increasing globalisation and advances in technology will bring colossal societal and ecological changes, particularly if our unsustainable practices and lifestyles prevail (Pretty 2013). Without significant policy interventions, more people will be consuming more resources; climate change will cause global temperatures to increase; demand for food will double globally; more than four million people in the UK will have diabetes and there will be an ageing population. This is just a taste of what children's future might look like.

This chapter provides an account of the current status of ESD in schools across the UK's four national and administrative jurisdictions (England, Wales, Scotland and Northern Ireland) and sets out some of the characteristics of best practice in each of these. It provides an analysis of current barriers to progress, and outlines potential opportunities for enhancing the core role of education and learning in the pursuit of a more sustainable future.

The most recent surveys of progress on the implementation of ESD in the UK were undertaken in 2008, 2010 and 2013 by the UK National Commission for UNESCO (UNESCO UK 2008, 2010, 2013). This chapter builds on that work and sets out a succinct account of the current status of ESD across the UK. It draws on evidence from various sources from England, Wales, Northern Ireland and Scotland, some of which is set out in the 2013 Policy Brief published by the UK National Commission of UNESCO (UNESCO UK 2013). The Brief's main purpose was to inform the UK government of progress on the integration of ESD across all of the learning contexts in which issues relating to sustainability can be taught and learned. It also assessed how far the UK had realised the objectives of the UN Decade of Education for Sustainable Development (DESD).

The DESD had four key objectives:

- facilitating networking and collaboration among stakeholders in ESD;
- fostering greater quality of teaching and learning of sustainability topics;
- supporting countries in achieving their millennium development goals through ESD efforts; and
- providing countries with new opportunities and tools to reform education.

The UK government signed up to the DESD in 2005, sharing the belief that education has a key role in the development of the values, behaviour and lifestyles required for a sustainable future.

In addition, this chapter sets out to provide an analysis of progress in schools in support of the UK government's objective for sharing best practice in ESD in all learning contexts. This breadth of view has now assumed a much more important policy priority given the UK coalition government's current focus on stimulating economic growth by creating a substantial green economy linked to climate change adaptation in the UK (DBIS 2010; DECC 2010; Luna et al. 2012). The enhanced focus on quality and standards in the devolved national governments in all forms of educational provision is also highly relevant since contemporary evidence indicates that good practice in ESD leads to better learner outcomes (Barratt Hacking et al. 2010; Martin et al. 2009).

An increasing number of schools are now positioning 'sustainability' as a central guiding principle for all of their activities. Reports by Ofsted (2009), the Department for Children, Schools and Families (DCSF 2010), the Co-operative Group (2011), the Scottish Government (2010), all strongly suggest a link between the adoption of sustainability as a guiding principle and the improvement of schools as a whole. In these Schools there is good empirical evidence of improvements in standards, behaviour, teacher and student motivations, attendance, examination results, community linking and environmental performance. A recent review of a small sample of the 17,000 schools registered for the Eco-Schools programme in the UK found evidence of positive impacts on wellbeing, behaviour, motivation and cognitive skills that benefited the whole school (Keep Britain Tidy (KBT) 2013).

19.2 The UK Political Context

The UK has a partially devolved political constitution which comprises the London based UK government itself, and the devolved administrations and directly-elected parliaments and governments in Northern Ireland, Scotland and Wales. The extent of their devolved policy responsibilities varies significantly, but each has responsibility for primary, secondary and tertiary education provision and their funding, and each can add to the UK's policies on sustainable development (SD) with specific arrangements. These devolved responsibilities preceded the start of the DESD by 5 years. Consequently, different policy emphases on education and different progress on implementing policy on ESD exist across four countries and political jurisdictions.

England has no separate devolved administration and is governed through the UK parliament and civil service. The current policy remit for schools in England is held by the Department for Education (DfE) which states that it is "committed to sustainable development" (The Prime Minister's Office n.d.) and believes it important to prepare young people for the future. But education's role in supporting the UK's SD policy has nearly always been a 'below the radar' issue for the UK government. DfE's approach is based on the belief that schools perform better when they take responsibility for their own improvement and want schools to make their own judgments on how SD should be reflected in their ethos, day-to-day operations and through ESD.

19.3   The UK ESD Context

The analysis of the current status of ESD in schools across the UK's four devolved administrations, which follows, sets out key differences along with recommendations to enhance the key role of education in furthering the UK's objectives for a more prosperous and sustainable future.

19.3.1   Reflections on Education for Sustainable Development and Global Citizenship (ESDGC) in Schools in Wales

Policy Context

> The UNESCO Decade of ESD (2005–2014) offers a forum for Wales to share the progress to date, the developments and the uniqueness of ESDGC in Wales and to learn from others. (DCELLS 2008a: 11)

SD was enshrined as the central organising principle of the Welsh Assembly Government (WAG) when the devolution of powers, which included responsibility for education, took place in 1999. This commitment to SD has been re-affirmed at regular intervals in policy documents, including *One Wales: One Planet* (WAG 2009). In 2013 the renamed Welsh Government brought forward legislation to further reinforce its commitment to SD in the "Future Generations Bill". Prior to the start of the DESD in 2005 the WAG was being challenged by those with an interest in ESD, including Oxfam Cymru and the RSPB Cymru, to elucidate on what its commitment to SD meant for education in Wales. In 2002 the Qualifications, Curriculum and Assessment Authority for Wales (known by the acronym for its Welsh language title, ACCAC) published a document on ESDGC which listed the following nine key concepts that it believed underpinned ESDGC:

*   Interdependence
*   Citizenship and stewardship
*   Needs and rights
*   Diversity
*   Sustainable change
*   Quality of life
*   Uncertainty and precaution
*   Values and concepts
*   Conflict resolution.

In 2004 there was already an ESD panel in Wales, but one which had a strong environmental education (EE) bias, and there was also a separate Global Citizenship (GC) panel. Both panels were conscious of the common aspects in their work and agreed to form a joint ESDGC Panel made up of representatives from WAG, from the formal sectors of education and from NGOs. The title ESDGC, while it overtly recognised the importance of GC, indicated the coming together of two adjectival education traditions rather than making a step change that would lead teachers to seeing this approach to education as a fully integrated whole.

The ESDGC Panel, supported by funds from WAG, produced *Education for Sustainable Development and Global Citizenship: A Strategy for Action* in 2006 (DCELLS 2006). It covered five sectors of education: schools, youth, further education and work-based learning, higher education, and adult and community education. To ensure a whole-institutional approach the Strategy identified actions across the following five "Common Areas", applicable to all the formal sectors of education:

* Commitment to Leadership
* Teaching and Learning
* Institutional Management
* Partnerships
* Research and Monitoring.

Interplay Between Policy and Practice

In 2006 Estyn (the education inspectorate in Wales) commissioned research from which the outcomes highlighted a lack of understanding about ESDGC at classroom level (Estyn 2006a). In response to this research Estyn published *Update in Inspecting ESDGC* in September 2006, resulting in all school inspections being required to report on ESDGC (Estyn 2006b). In January 2007 the WAG appointed a "Champion" for ESDGC on a 3 year contract, accountable to the Panel, with a brief to drive the implementation of ESDGC across all the education sectors.

The initial focus on schools following the publication of the ESDGC Strategy was to engage with teachers and achieve a better understanding of what ESDGC meant for primary and secondary schools across Wales. The ESDGC Panel commissioned the research and development of *Education for Sustainable Development and Global Citizenship: A Common Understanding for Schools* (DCELLS 2008b), published in July 2008 and sent to all schools in Wales—similar documents were subsequently produced for the other sectors of education in Wales. While the nine concepts identified by ACCAC (2002) were retained in the *Common Understanding* it was felt that a set of themes, which could cover the broad scope of ESDGC, would provide accessible entry points for teachers to integrate ESDGC into the curriculum and school life generally as well as linking to existing WAG documents such as the *Skills Framework for 3–19 Year Olds in Wales* (DCELLS 2008c). The following set of seven themes presented in the *Common Understanding* document were used to map the content, skills and values associated with ESDGC.

* Wealth and Poverty
* The Natural Environment
* Identity and Culture
* Health

- Climate Change
- Choices and Decisions
- Consumption and Waste. (DCELLS 2008b: 14)

By 2008 there were a number of supporting policies, guidance, and related drivers to encourage teachers and schools in Wales to engage with and implement ESDGC, including the following examples:

- The legitimacy offered by the WAG's commitment to SD;
- Estyn inspecting and reporting on ESDGC;
- The ESDGC Panel, including the Minister for Children, Education, Lifelong Learning and Skills, setting strategy and administering the ESDGC Action Plan and awarding small grants for ESDGC projects;
- An ESDGC Champion delivering the Action Plan and providing a central point of contact for all aspects of ESDGC in Wales;
- A common document in all schools outlining the content and approach of ESDGC (DCELLS 2008b);
- The Directors of Education in the 22 local authorities in Wales nominating a representative to be a conduit for disseminating ESDGC information.

While there was central support and drivers to implement ESDGC, it was left to individual local authorities and schools to decide how to respond to the ESDGC agenda, with some local authorities giving it a higher priority than others. There were other initiatives that supported various aspects of ESDGC, such as EcoSchools, Forest Schools and international school linking. What ESDGC added to these initiatives was the fact that it was broader than any one individual initiative and it was an on-going approach to education that was not completed when an award was achieved.

The "Enabling Effective Support" project in Wales, funded by the UK Department for International Development, was a good example of cooperative working on the ESDGC agenda. It ran ten regional forums across Wales for educators which linked with the ESDGC Champion and obviated the need for a separate ESDGC network. During this period the WAG continued to produce guidance and support materials that contributed to ESDGC in schools as well as other sectors of education. These included *Out of Classroom Learning* (DCELLS 2007) and *ESDGC: Information for Teacher Trainees and New Teachers in Wales* (DCELLS 2008d).

The ESDGC Panel submitted a response to the consultation on the review of the National Curriculum in Wales conducted in 2007. The new Curriculum integrated ESDGC into Science and Geography (DCELLS 2008e, f) and featured it prominently in Personal and Social Education. ESDGC also figured in the Learning Pathways 14–19 and the Welsh Baccalaureate. The ESDGC in the curriculum was a significant contribution to Teaching and Learning, but that was only one of the five Common Areas in the 2006 Strategy and the overall aim for schools was to embed ESDGC across all the five Common Areas. The broader scope of ESDGC was reflected in Estyn's inspection of schools and evidence from the inspection reports illustrates that this has been achieved, at least in some cases, as shown by the following extracts referring to individual schools:

> The provision for education for sustainable development and global citizenship (ESDGC) is outstanding and fully embedded in the life and work of the school. The school makes every effort to act in a sustainable way and pupils regularly monitor energy and water consumption and are involved in re-cycling, composting and waste minimisation schemes. The school's commitment to the Fair Trade ethos is excellent and pupils have a clear understanding that the actions of people in one country can have a direct, beneficial impact on the lives of those in other countries. (. . .) Global citizenship is further promoted through initiatives, such as the links programme with Bangladesh, and it has gained the Foundation Level of the International School Award. (. . .) The school is part of the eco-schools award scheme and is justly proud of achieving the European Green Flag in recognition of its commitment to conservation and the environment. Pupils are very proud of their school grounds and local community and genuinely feel they can make a real difference, both locally and globally, through active citizenship and care for the environment. (. . .) The school is fully committed to the national priorities for lifelong learning and community regeneration. The school is at the heart of the regeneration of the local community and the working relationships forged with a range of agencies and personnel, including Communities First, are exemplary and of great benefit to the children and their families. (Estyn 2010: 18–19)
>
> The school has firmly embedded and strategically planned ESDGC thoroughly across the whole school; it is an outstanding feature of the school. (Estyn 2009: 6)
>
> The ESDGC co-ordinator actively promotes and champions the ESDGC agenda with support from members of the school council, SLT (Senior Leadership Team) and staff. The school has been involved in a large number of initiatives aimed at raising pupils' awareness of sustainability and their role as citizens of the world. (Estyn 2009: 23)

The end of the ESDGC Champion's contract in December 2010 coincided with the appointment of a new Minister for Children, Education and Lifelong Learning and there was a decline in the priority and resource given to ESDGC by the WAG, a decline which has continued through the latter half of the DESD. The results from 2010 PISA (the OECD Programme for International School Assessment), which focused on Literacy, Numeracy and Science, placed Wales well down the rankings. The response of the new Minister was to boost the policies and resources in these subject areas, in part at the expense of other initiatives such as ESDGC. The 'push' by the WAG between 2006 and 2010 to embed ESDGC in schools was only partially successful as borne out by Estyn inspections. According to Her Majesty's Chief Inspector of Education and Training in Wales, between 2010 and 2013 judgements on the standards of ESDGC in schools show 76 % of primary and 66 % of secondary schools are either "excellent" or "good", while the rest are described as "adequate" or "unsatisfactory".[1]

The continued profile of ESDGC in schools up to the present day is a result of the extent to which ESDGC has been genuinely embedded, the enthusiasm for this approach to education by some teachers, headteachers and local authority leaders, and the continued inclusion of ESDGC in Estyn inspections.

---

[1] Figures presented by Her Majesty's Chief Inspector of Education and Training in Wales at an ESDGC Forum in Cardiff on 17.06.2013.

A practitioner based ESDGC Schools Network was established in 2011, which has facilitated conferences and forums, but without core funding the future of the Network is constantly in doubt.

The Welsh Government has recently consulted on introducing a legal requirement for larger public bodies in Wales to address specific issues relating to SD. The "Future Generations Bill", if it is adopted, may add impetus to certain aspects of ESDGC at school level through its legal impact on local authorities and on higher education institutions in Wales.

A key debate that would have significance for the future of ESDGC in Wales is around the differences in the approach to learning inherent in ESDGC compared to the approaches being promoted by the Welsh Government to drive up standards in Literacy and Numeracy, and whether ESDGC offers longer term benefits to the quality and standards of learning in Wales. These issues require a much wider debate at all levels, from the Welsh Government to school classrooms, than they are currently receiving.

There is pride in Wales with regard to achievements associated with ESDGC and SD generally. The commitment to SD as a central organising principle of the Government in Cardiff means that the majority of the developments in ESDGC in Wales would have happened regardless of 2005–2014 being identified by the UN as the DESD. One of the UNESCO initiatives for the Decade, the Regional Centres of Expertise (RCE) in ESD, was slow to be established in Wales, primarily because the initiative taken by the WAG and the appointment of the ESDGC Champion meant that the functions of an RCE were largely being covered.

Some of the findings of the original Estyn research in 2006 (Estyn 2006a), such as the variation in the level of understanding about ESDGC at classroom level, still persist and development across the schools in Wales has been uneven. During the academic year 2013–2014 Estyn has been tasked by the Welsh Government to conduct a review of the impact of ESDGC in schools, using the 2006 research as a baseline. The outcome of this research will be published in June 2014 and it is likely to have a significant effect on the future of Welsh Government policy with regard to ESDGC. However, while the extent to which policies and initiatives over the last decade have embedded ESDGC is still debatable, there are sufficient committed practitioners in Welsh schools to ensure ESDGC persists at some level into the future.

19.3.2   Critical Reflections on ESD in Schools in Scotland

Policy Context

In Scotland, interest amongst practitioners in what later became 'ESD' began in the 1960s and 1970s, and in 1974 Her Majesty's Inspectorate of Schools (HMIE) in Scotland published a significant and progressive report on Environmental Education (EE) which was well ahead of its time (HMIE 1974). Further reports, such as the highly significant *Learning for Life* which fully located EE as a process involving formal, informal and non-formal education (Scottish Office Environment Department 1993) followed. Analysis of this and other reports in the period to 2007 can be found in Lavery and Smyth (2003) and McNaughton (2007). Throughout this period there was no lack of interest in ESD and the educational community made efforts to raise the profile to encourage policy commitment. However, this support was not forthcoming for a variety of reasons, such as inflexible educational structures and at least, prior to the establishment of the Scottish Parliament in 1999, a lack of autonomy in educational policy formation. Higgins and Lavery (2013) have detailed the research and policy advice, the stages in policy development, the slow pace of change and the reasons for this.

However, paradoxically, from the 1990s, Scotland already had a significant influence on international conceptualisation and practice in the field, initially through the involvement of John Smyth in the 1992 Rio de Janeiro "Earth Summit" who was a key architect of the strong focus on education in the report. Similarly, Scotland was one of the small group of countries participating, from the outset in 1998, in the UNESCO group on "Reorienting Teacher Education to Address Sustainable Development".

One feature of the early conceptualisation of Sustainable Development Education (SDE), until recently the favoured term in Scotland, was the centrality of outdoor learning experiences to the concept, certainly through the 1990s (Higgins and Lavery 2013). This has again become a key feature of the current philosophical and policy discussion. However, throughout the past two decades, the curricular location of ESD has been uncertain, and this has been the result of two primary factors; the range of academic and 'personal attitudes' fields that sustainability can be justifiably linked with, and its fundamentally interdisciplinary nature. Furthermore, as it frequently challenges the political *status quo*, a historical lack of political support for a core curricular place for ESD seems unsurprising.

Whilst the curriculum in Scotland has always been distinct and separate from the other jurisdictions, the establishment of the Scottish Parliament in 1999 fostered greater confidence in developing a new approach to educational autonomy, and a new 3–18 core curriculum—*Curriculum for Excellence* (CfE) (Scottish Government 2009a)—was launched across Scotland in 2009. This has placed great emphasis on the "capacities" of learners, and in the past decade, changes in both the social and political acceptance of the need to address sustainability have paved the way for its inclusion in CfE.

Climate change has been a significant driver in the perceived relevance of ESD, and the Scottish Government has made this issue a priority through the Climate Change (Scotland) Act 2009 (Scottish Government 2009b) and supporting strategies for reducing waste and increasing renewable energy production. The Act is "regarded as the most ambitious legislation of its kind in the world" (Education Scotland 2011: 2). The government's current economic strategy now includes the "Transition to a Low Carbon Economy" as one of six priorities, and highlights the ambition that all Scotland's demand for electricity should be met by renewables by 2020, and that a greener economy could support 130,000 jobs by

2020 (Scottish Government 2011). Preparing the general public and a workforce for this transition gives specific focus and urgency to the role of ESD in all education sectors.

The Interplay Between Policy on ESD and Its Impact on Practice

As well as the reports noted earlier, the period from 2000 onwards was characterised by a number of advisory and working groups which had variable degrees of influence and duration. Notably, however, the announcement of the DESD and the changes to the curriculum spurred Scotland's learning and teaching advisory agency (then called Learning and Teaching Scotland) to convene a Sustainable Development Education Liaison Group (SDELG) to advise on ESD for schools and government (Higgins and Kirk 2009), and this was a key influence on policy until 2012. It contributed to two key Scottish Government policy responses to DESD—*Learning for our Future* (Scottish Government 2008) and *Learning for Change* (Scottish Government 2010), both of which outlined plans and targets for all education sectors and communities. Perhaps the most significant legacy of the Scottish response to the DESD was that global citizenship education and ESD were embedded in the Curriculum for Excellence (Scottish Government 2009a).

For schools, curricular inclusion has meant that aspects of ESD are located in specific "experiences and outcomes" of the curriculum 'subjects' Social Studies, Science and Technologies and in the senior phase of the national qualifications and examinations (Higgins and Lavery 2013). Before and since these inclusions a strong external contribution has been made by Eco-Schools Scotland, with almost all schools being registered, and about half holding its "Green Flag" (Eco-Schools Scotland 2014). Its popularity is due in part to Scottish Government support and funding to the Eco-Schools programme and to the enthusiasm of education professionals and school staff. Other awards and support for ESD and related areas such as "Rights Respecting Schools" exist but as with Eco-Schools these are not mainstream curricular arrangements and therefore not available to all schools and students. This situation may lead to the hazardous assumption that ESD is appropriately covered in schools, whereas charitable-sector and optional provision and partial curricular coverage are no substitute for a *central* focus in the curriculum.

The period since the last Scottish Parliamentary elections in 2011 has seen some distinctive developments and as we approach the end of the DESD there is a sense of real progress being made. The incoming government made a manifesto commitment to exploring the concept of "One Planet Schools" and this led to the establishment of a Ministerial Advisory Group. It was given the remit to explore "One Planet Schools" in the context of *Learning for Change* (Scottish Government 2010) and CfE (Scottish Government 2009a). The ensuing report to Scottish Ministers on *Learning for Sustainability* (LfS) (Scottish Government 2012) made 31 recommendations relating to the whole student 3–18 experience within CfE, including curricular, community and campus elements. It established the concept of

LfS as having an equal focus on ESD, global citizenship and outdoor learning. The Government accepted the report in full (Scottish Government 2013) and has established an implementation group with a 2-year remit to conclude its work in 2015. The focus on the whole school environment and pupil experience is distinctive as is the intent to address sustainability through the integration of three equally important facets—ESD, Global Citizenship Education and Outdoor Learning. The inclusion of outdoor educational experiences as a core feature acknowledges the importance of both intellectual understanding of planetary ecological and geophysical systems, and the significance of affective and sensory experiences in developing a values orientation towards sustainability (Christie and Higgins 2012).

For the "Learning for Sustainability Implementation Group"—established in February 2014 by the Scottish Government and tasked with driving forward the above mentioned 31 recommendations of the *Learning for Sustainability* Report— to be effective it will need close links with education professionals in formal education and the private sector, and the recently established UN University Regional Centre of Expertise in ESD for Scotland (http://rcescotland.org, accessed 15 April 2014) that has been identified by the Government as a key means of providing this support and two-way communication.

In 2013 the General Teaching Council for Scotland (GTCS) published its revised "Professional Standards", part of a national framework for teachers' professional development. These standards relate to "registration", "career-long professional learning" and "leadership and management". "Learning for sustainability" is embedded in the professional values and personal commitments sections of the standards, and all registered education professionals such as teachers and lecturers will be expected to demonstrate this in their practice. Whilst this is a highly significant and distinctive feature of the new structure of Scottish education, establishing the standards, and in particular a positive commitment to learning for sustainability throughout the teaching profession is a major undertaking that will require a focus in pre-service training and extensive in-service provision, and this will need commitment by the GTCS, Education Scotland[2], Education Authorities and Teacher Education Institutions (TEIs).

The Practice Context: The Formal and Informal Curriculum, Impacts on Learner Outcomes, Quality Standards and Inspection

Whilst the curricular foundations are in place to support ESD becoming a central feature of all schooling, inspection and evaluation processes are necessary to monitor progress and establish the most effective ways of achieving positive change. *Learning for Change* committed Her Majesty's Inspectors of Education

---

[2] Education Scotland is the government educational advisory body that replaced Learning and Teaching Scotland following its merger with Her Majesty's Inspectors of Education in 2011.

(HMIE) to supporting "the development of sustainable development education within Curriculum for Excellence through self-evaluation and the school inspection process" (Scottish Government 2010: 16). Whilst in 2012 a new inspection process was introduced, HMIE are expected to continue to include ESD in evaluations as an aspect of learning, and a new self-evaluation framework for ESD and outdoor learning in educational institutions is under development. Whilst welcome, this is still a long way from ensuring that ESD is a central feature of the inspection process.

It is a proposition clearly worth consideration that addressing complex interdisciplinary issues, such as SD and social responsibility and using a whole school approach to do so, might be correlated with broader improvements in pupil development. Scott has proposed a set of holistic characteristics of school leavers that would indicate that the school has properly prepared them for life in a complex modern society, and has suggested that educational programmes in such schools would "have a wide range of (...) the characteristics of effective ESD, that are congruent with the Decade aims" (Scott 2013: 17–18). Rather more instrumentally, Scotland is one of 11 nations involved in a study sponsored by UNESCO (Hopkins 2013) investigating links between ESD and educational attainment and achievement. However, without some standardised reporting by teachers, headmasters and other stakeholders, any potential relationship will be difficult to discern.

A central emphasis on ESD in inspections would not guarantee that schools committed enthusiastically to ESD. But it would emphasise the significance of ESD both in schools and their 32 regional authorities. In Scotland education budgets are raised through taxation at a local level and these funds are diminishing in real terms so it seems unlikely that spending on ESD will be a priority. Other approaches are necessary. For example Education Scotland is helping "teachers who have succeeded in enhancing SDE [=ESD] in their schools by encouraging them to act as mentors for staff in other schools, building peer-support and collaboration" (Higgins and Lavery 2013: 342). However, as Higgins and Lavery point out "such approaches are far from being a national policy or even expectation", and unless appropriate structural provision is made in terms of policy expectations, training and inspection future progress is far from assured.

Initial Teacher Training and Continuing Professional Development

Whilst across the higher education sector there is considerable growth in demand for SD and social responsibility oriented undergraduate degrees with many more courses and healthy uptake, this is not the case in teacher education institutions (TEIs). Despite the GTCS having agreed the revised professional standards to include 'learning for sustainability' Scottish TEIs are not *required* to include detailed coverage of SD in their programmes, and any inclusions are dependent on local institutional and staff interest. This is a clear impediment to deeper embedding of ESD in Scottish schools at a time when an encouraging message to potential applicants for courses, current students and the profession should be clear—that 'learning for sustainability' as a core element of the professional standards is the responsibility of all education professionals.

As will be evident, the historical context, current policy framework, political will and professional structures are all in place for a transformative phase in ESD or 'learning for sustainability' in Scotland. However, such change needs to be signalled as an imperative to all those currently involved in all aspects of 3–18 education, and all those considering education as a career. Overall, at the level of political and administrative rhetoric, some progress has been made in official documents, but much now rests on the collective impact of the various initiatives outlined above.

19.3.3 Reflections on ESD in Schools in England

In providing an account of ESD in England's schools it is tempting to identify the UK General Election of 2010 as a pivotal moment. For many supporters of ESD this was when "the clock was turned back" and Government support for sustainable schools declined. At the school level however, change is rarely so drastic.

Despite recent policy reversals, the situation in relation to ESD in English schools remains little changed; indeed it has evolved over decades, growing from rural studies and agricultural education to Environmental Education (EE), supplemented with development education and thence to various permutations of education with/about/for sustainability (ULSF 2001). A brief account of the policy background will inform the snapshot that follows.

Policy Context

According to Disinger, interest in EE had grown by 1969 to "occasion the development of definitional statements" (1985: 61–62) such as those formalised by IUCN (1970) and at Tbilisi in 1977 (UNESCO-UNEP 1978). At that time central government exercised little control over what was taught in England's schools beyond broad guidance and largely supportive visits by Her Majesty's Inspectors (HMIs). EE itself was supported through the distribution of Tbilisi Conference papers to local education authorities with some advice and guidance being offered by HMIs (Hansard 1978).

The 1988 Education Reform Act ushered in England's (and Wales's) first National Curriculum; henceforth advocates of EE would have to fight their own corner within an increasingly politicised system. EE was omitted from the original government-defined National Curriculum included in the Act but political lobbying helped it to become one of five cross-curricular themes to be covered by non-statutory guidance (NCC 1990). EE remained popular at classroom level throughout to such an extent that "the document [National Curriculum] itself was perceived as being redundant by many schools" (Palmer 1998: 25).

In 1997, under the New Labour Government, the Departments of Education and Environment established the Sustainable Development Education Panel (SDEP) whose first report linked education outcomes to seven SD principles. This suggested coherence between ESD and SD but its failure to define ESD in terms of *educational* principles or structures obstructed integration into mainstream education (Vare and Scott 2013). Meanwhile a report on "citizenship education" (QCA 1998) referred to EE and SD as "important contexts and content to support the aim and purpose of citizenship education in schools" (ibid.: 41).

When the term 'sustainable development' first appeared in *The National Curriculum: Handbook for Primary Teachers in England Key Stages 1 and 2* (QCA 1999), a contemporary analysis (Chiatzifitou 2002) observed a lack of clarity in defining terms such as 'environment' and 'sustainable development' while noting how values-based components only received attention in the non-statutory sections.

In 2003 the Department for Education and Skills (DfES) published its Sustainable Development Action Plan with the first objective being ESD. Shortly afterwards a new national SD strategy *Securing the Future* (HM Government 2005) appeared with a chapter on education featuring a highly supportive statement from the Prime Minister. The schools inspection service, the Office for Standards in Education (Ofsted), captured the *zeitgeist* with a survey of ESD practice in schools that led to the publication of its report *Taking the first step forward* (Ofsted 2003). This later on fed into more comprehensive guidance for schools inspectors (Ofsted 2010), highlighting ways in which English schools might address SD through positive behaviour, community links and cross-curricular working. The judgements of Ofsted inspectors have huge consequences for schools, so this level of interest in ESD was potentially a highly significant development.

By comparison, the DESD (2005–2014) (UNESCO 2004), had little impact at the school level, unsurprising given the lack of enthusiasm demonstrated by the then Secretary of State for Education, Charles Clarke, in an interview with the UK Parliament's Environmental Audit Committee (EAC):

> (Mr Clarke) . . . I believe strongly. . . that statements and declaratory remarks do not take us very far, including in terms of the UN Decade. I think it is a question of what we actually do. I think we are absolutely full up to here with declaratory statements . . . if I was to say that I gave major priority to our location within the UN Decade, the truth is I do not. I give major priority to try to sort out our school transport policy, to try to get a curriculum which moves forward. . . (EAC 2003, § 208).

Beyond revealing the marginal status of international documents, this extract demonstrates the extent to which political vision appears to have given way to a focus on individual components within the system. As David notes, despite the rhetoric of ESD and citizenship education, educational debate since the 1990s has been "preoccupied with 'what works' with respect to 'raising standards'" (David 2007: 431).

The National Framework for Sustainable Schools

A pragmatic approach is evident in England's *National Framework for Sustainable Schools* (DCSF 2008). Launched in 2006 this non-statutory guidance comprises three interlocking parts: (a) a commitment to care; (b) an integrated approach linking campus, curriculum and community; (c) eight 'doorways' or thematic entry points. While the second part promotes integrative thinking, the doorways have been seen as problematic:

> ... there are risks inherent in a doorways approach; for example, presenting sustainability as a series of fragmented and unrelated ideas in what is a rather conservative and limited approach to the issues we face. (CREE 2009: 10)

Furthermore, the doorways omit biodiversity, a crucial ESD component in terms of ecological understanding and first-hand experiences of nature. Despite these concerns, the Framework helps participating schools to build upon their existing efforts and importantly provides a system of monitoring through the Sustainable Schools Self-evaluation tool (S3) (DfE 2011).

Interplay Between Policy and Practice

Evidence of the positive impact of tackling sustainability in schools suggests improved student engagement and attendance (Gayford 2009) and enhanced well-being across a number of indicators (KBT 2013). Further evidence of impact is available from a wide-ranging review conducted by Centre for Research in Education and the Environment at the University of Bath (DCSF 2010).

The UK Coalition Government elected in 2010 has demonstrated an antipathy towards ESD by shelving proposed ESD inspection guidelines (Ofsted 2010) and dropping the previous Government target of all schools becoming 'sustainable schools' by 2020. The National Framework remains available on a voluntary basis although much of the support material has been removed from the Department for Education webpages (Martin et al. 2013).

In 2013, initial proposals for a revised National Curriculum made no reference to sustainability and even omitted climate change; the latter issue was restored after a public outcry and intervention by the Government's own Energy Secretary (Wintour 2013).

Other areas of Government offer some support; funding for "global learning programs" is available to local authorities (DfID n.d.) and the *Natural Environment White Paper* (Defra 2011) emphasises the importance of connecting people with nature and endorses the work of NGOs, including the Sustainable Schools Alliance (SSA 2012), although no additional resources are available to this national network or to schools.

Local and national NGOs continue to offer a variety of programmes on specific themes. A comprehensive ESD-focused service to schools has been provided by the NGO "Sustainability and Environmental Education" (SEEd), offering conferences, webinars and a policy forum. The Ellen MacArthur Foundation (EMF 2012) offers a refreshingly critical approach with its focus on the 'circular economy'. Numerically most significant is the Eco-Schools programme with over 70 % of schools in England (17,000) registered (KBT 2013). Not all registered schools participate while those that do so may focus on practical activity with teams of enthusiastic pupils rather than on embedding the learning across the curriculum (Vare forthcoming). Keep Britain Tidy itself, the organisation running the programme, has enjoyed long-standing Government support for promoting Eco-Schools although this is currently in question.

Forest Schools (Learning Outside the Classroom, LOtC 2014) is increasingly popular at primary level but, like Eco-Schools, this usefully tangible framework carries the danger of limiting practice as schools focus on the programme rather than the wider ethos of learning outside the classroom or sustainability.

Indeed, Sayer (2000) suggests that a clear framework of what *should be* can lead to empty moralising by those who 'follow the script' or lead to the false assumption that what is proposed will somehow come into being. Thus supportive Government directives could lead many to think that the job has been done when nothing could be further from the case. We might take heart therefore, that a 'clear message' from Government need not be the most critical factor in embedding ESD. South Africa's curriculum, for example, has been underpinned by social and environmental concerns for 20 years yet transformation is slow (Lotz-Sisikta and Schudel 2007).

Schools in England as elsewhere are inherently conservative institutions responding incrementally to curriculum reform but also adapting any innovations to their own ethos, their locality, media preoccupations and occasionally, Government-sponsored non-statutory guidance. If governments are frustrated by the uneven way in which their reforms are adopted (BBC 2012), proponents of international strategies, such as the DESD, should be even more wary of expecting rapid change, estranged as they often are from classroom practice.

For better or worse, England's schools *do* respond to the Ofsted inspection framework; this reflects Government priorities, which in turn respond to popular concerns. Thus, as SD issues increasingly arrest popular consciousness, we can expect these to be absorbed into the fabric of the English education system, regardless of the political preferences of the Government of the day.

19.3.4   Reflections on ESD in Schools in Northern Ireland

Policy Context

Responsibility for SD policy in Northern Ireland currently resides within the Office of the First and Deputy First Minister (OFMDFM), which has a convening role for policy formation and strategic direction. In 2010 OFMDFM published the latest Northern Ireland SD strategy, *Everyone's Involved* (OFMDFM 2010), and its related Implementation Plan, into which all government departments in Northern Ireland made contributions and commitments. Since 2007 ESD has been a statutory requirement within the school curriculum and falls under the aegis of the Department of Education.

At primary level (students aged 4–10 years) ESD is found in the World Around Us area of learning, whilst for 11–14 year olds, it is included in the statutory areas of Learning for Life and Work and Environment and Society. Related key aspects of the Northern Ireland curriculum (CCEA 2014) are Mutual Understanding, and Local and Global Citizenship. Schools and teachers are provided with resources designed to help explain and encourage SD principles within a pupil's overall learning experience. A good practice guide, developed by the Interboard Education for Sustainable Development Group (2005), was produced partly as a response to the beginning of the DESD. However, this useful guide has not been updated since 2005. The continued delay in setting up a single Education and Skills Authority in Northern Ireland (DENI 2012) may be hindering further progress on this issue. Undoubtedly, such a single authority could play an important coordinating role in increasing the ESD resources available to schools in Northern Ireland.

In January 2009, the Department of Education published *Schools for the Future: A Policy for Sustainable Schools* (DENI 2009). Whilst this document does reference SD principles and the Northern Ireland SD Strategy, it is clearly describing SD terms of infrastructure and capacity, not in terms of directly engaging and encouraging ESD. As the authors of the document itself state: "The focus of this document is on the longer term viability of schools provision" (DENI 2009: 3).

It was hoped by many, including the SD Commission in Northern Ireland (Northern Ireland SD Commission 2010), that a mandatory commitment to ESD would be included in the 2010 Education (school development plans) Regulations. However, despite the inclusion of an ESD clause in the consultation paper, this did not happen. Rather, an appendix to the final regulations allows for a more ad hoc, school-by-school, approach to develop (DENI 2010).

A commitment to highlight and encourage ESD principles was contained within the 2008–2011 Northern Ireland Programme for Government (NIE 2008) through a SD and Environment awards scheme. The Office of the First and Deputy First Minister (OFMDFM 2012), in partnership with the Department of the Environment has hosted IMPACT Awards in 2010 and 2012 recognising and rewarding young people from schools and community groups across Northern Ireland. The Eco-Schools programme has over 930 registered schools in Northern Ireland, accounting for 75 % of all schools (Eco-Schools Northern Ireland 2014).

Interplay Between Policy and Practice

Such success as there has been in advancing ESD in Northern Ireland schools has been marshalled by the NGO sector. The Royal Society for the Protection of Birds (RSPB) and the Red Cross have been especially active and a broad coalition of organisations, under the banner of the ESD Forum (ESD Forum n.d.) which embraces the universities, several local authorities and government agencies, as well as a range of NGOs. The Forum provides valuable networking opportunities, training events and regular communications for its wide membership base. Their influence and resources enable ESD committed teachers to deliver some excellent work in schools.

Schools in Northern Ireland are encouraged to adopt an approach to ESD that takes on more than a curriculum focus. The school buildings and building regulations, management of school resources, waste management, and sustainable transport for example, should all be actively encompassed within a framework for SD, alongside the development of strong links with the local community, other local schools and businesses, and the community and voluntary sector. A 2010 Northern Ireland Education and Training Inspectorate report in relation to ESD across a sample of schools in Northern Ireland highlighted the central role of strong and inspirational leadership in establishing ESD as part of a whole school ethos, ensuring that ESD is effectively integrated into school development plans, giving a clearly defined role to an environmental coordinator in schools, and involving all staff (teaching and non-teaching) in taking the lead in developing ESD. The Report highlighted a number of case studies from schools across Northern Ireland which the authors felt represented effective practice in ESD. These are some of the key outcomes of this report:

> Environmental education is now very much embedded in the whole culture and ethos of the school with learner pledges, classroom charters, whole school assemblies, events and competitions all dedicated to promoting environmental awareness and action. There is a clear overview of where ESD issues are explored through the curriculum. (Northern Ireland Education and Training Inspectorate 2010: 6)
>
> The school has recognised that awards can be an important vehicle for the development of environmental initiatives within the school but that they cannot just be an add-on and need to be explicitly embedded within the curriculum. To facilitate this, the role of Eco-coordinator has now been distributed to the senior management team (SMT) and the World Around Us co-ordinators. (ibid.: 8)
>
> ...learners in Key Stage 2 started a school based campaign to raise awareness about energy use and its impact on climate change. In ICT class, the learners designed posters and stickers urging staff and learners to "switch off and turn down". Eco monitors go around turning off lights and screens in classes, and shutting doors and windows; points are given to the most energy saving classroom. (ibid.: 9)
>
> The ESD work in the school is managed and driven by the learners. The school's Eco-Committee, which consists of elected representatives from each class, meets weekly to co-ordinate the ESD work of the school. (ibid.: 15)

Crucially much good ESD practice in Northern Ireland takes place in partnership with learners and the wider school community, such as NGOs and community and voluntary organisations. Schools that encourage learners to contribute and lead on ESD projects report that this sense of ownership leads to more engagement with the principles of ESD, more understanding of the long term behaviour changes required, and more willingness to carry on ESD related activities outside of the confines of the school and the curriculum. This active participation and ownership has increased positive interaction between learners, staff and the wider community, building powerful coalitions of interest and action.

A particular aspect of ESD in Northern Ireland which is of particular interest is its positioning within the curriculum. As a cross-curricular issue ESD is taught across a range of subjects, for example, geography and science in secondary schools (World Around Us in primary schools), as are the themes of Mutual Understanding and Local and Global Citizenship. These themes have particular relevance and importance in Northern Ireland as they address some of the issues learners and communities face as members of a divided society emerging from conflict. A combination of early evaluations and research into Education for Mutual Understanding provided a critique that underlined the need for a stronger focus on human rights, civic responsibility, justice and democracy, particularly in the context of the signing of the Good Friday Agreement in 1998 (see http://en.wikipedia.org/wiki/ Good_Friday_Agreement, accessed 5 May 2014). Smith (2003) argued for the inclusion of an inquiry-based approach to citizenship education that is defined in terms of citizens' rights and responsibilities rather than their national identities.

The 2010 Inspection Report makes a number of recommendations which would help to further cement ESD as a fundamental part of the Northern Ireland Curriculum, including, increased training and support for teachers, sharing of good practice, and improved access to information and resources. Above all, the report recommends that all schools understand the importance of ESD and that the curriculum also reflects this.

It is interesting to note that a 2009 Evaluation of the introduction of Local and Global Citizenship into the Northern Ireland Curriculum (UNESCO Centre, University of Ulster 2009) reached many of the same conclusions as the 2010 Report on ESD in Northern Ireland, calling for senior management engagement and a 'whole school' approach to the issue, along with an enhanced pupil voice, greater connection to other school policies, and the need for increased status and sustained professional development for teachers. It seems that an approach which emphasises the importance and centrality of ESD for teachers, management and the school community (as with Local and Global Citizenship) is key to successful delivery.

The Five Nations Network (n.d.), a forum sharing practice in education for citizenship and values in England, Ireland, Northern Ireland, Scotland and Wales, formed in 2000, has enabled dialogue between teachers, educators, policy makers, curriculum planners, members of the inspectorate, representatives of NGOs and young people from across the UK and Ireland. Its work is overseen by a Strategy Group with country representatives from each of the five nations. Perhaps an equivalent body for ESD across Britain and Ireland could provide the coordinated impetus that is required. The SEEd Charity (SEEd 2012) also provides a good example of ongoing collaborations.

Some schools in Northern Ireland are using ESD projects to raise awareness of cultural diversity and to promote good relations and mutual understanding between people of different races, religions and political opinions. The 2010 Northern Ireland Education and Training Inspectorate Report on Effective ESD practice in Northern Ireland comments: "Environmental Initiatives help schools to deliver the Northern Ireland Curriculum requirement to foster attitudes and dispositions such as community spirit, concern for others, inclusion and respect" (2010: 13). The primary school outlined in the following case study delivered part of its ESD programme through the lens of Personal Development and Mutual Understanding and The World Around Us:

> The learners have researched ethnic groups in the area and considered the differences in dress, diet and beliefs that exist. In religious education classes, they considered the values and ethics taught in Christianity and in Hinduism. Subsequently, they discussed how they treat others and how they would like to be treated by others. The school has placed a

particular emphasis on the value of cultural diversity and stresses the importance of helping learners to become informed about the local and global dimensions of the world in which they live. The programme encourages the learners to broaden their horizons and to adopt the attitudes of respect and acceptance which will help them become responsible citizens, better able to contribute positively to their own communities and to society as a whole. (Northern Ireland Education and Training Inspectorate 2010: 13)

Whilst this joined up approach is commendable in terms of its approach to community understanding in Northern Ireland, it is important to ensure that a balance is maintained and that the crucial aspects of ESD as a driver for a lifelong understanding and commitment to SD principles are maintained. The 2011–2015 Northern Ireland Programme for Government (NIE 2011) commitment to ensure that all learners in Northern Ireland have the opportunity to take part in 'shared' education programmes by 2015 provides an opportunity to ensure that sharing in education can also be part of a wider strategy to deliver ESD purposefully.

There are some excellent examples of ESD in schools in Northern Ireland resulting from the drive, commitment and understanding by school communities, aided by clear curriculum instructions and informed support from NGOs and other outside bodies. However, in the absence of a single Education and Skills Authority and the vacuum that this may have created, there is a need for a more coordinated approach to maximising the potential of the subject in Northern Ireland. There is also a need to ensure that ESD is understood and taught on its own merits, fully supported by educational authorities.

19.4 Conclusions

In 2010 the education landscape of England changed as a consequence of the formation of a coalition government for the UK. For England, education and SD policies were significantly influenced by a political ideology and advocacy which supported smaller and less directive government in some fields and substantial central, rather than local, controls in others. This led to the withdrawal of central policy support for a range of issues in England of which SD and ESD were important examples. The most visible expressions of such changes can be found on the DfE website which illustrates some of the benefits of a sustainable school, and contains relevant resources, but the policy emphasis is now on individual schools deciding whether ESD is important to them. In the devolved administrations of Scotland, Wales and Northern Ireland arrangements for school organisation and curriculum are different. Much of the development of policy and practice on SD and ESD has continued albeit at a different pace and with clearly differentiated national activities and priorities. Likewise, the coalition government withdrew central funding from the Sustainable Development Commission (SDC) which had led much of the SD agenda across the UK administrations under the previous UK-Labour government for the best part of 9 years (from 2000 to 2009). One of the singularly most successful outcomes of the work of the SDC was its role in countering the lack of buy-in from different parts of government in part by supporting the development of SD Action Plans across whole departments in Belfast, Cardiff, Edinburgh and London. Since 2010 much of this progress and central policy impetus has disappeared which has diminished the wider UK development and adoption of ESD in all forms of education and learning provision.

Whilst this chapter is not an exhaustive account of current UK ESD development and activity it illuminates the good practice in ESD that exists in many schools and at all levels across the UK, which is characterized by good teaching and enhanced learner outcomes. It is linked to the professional standards and qualifications of teachers who are part of innovative communities and networks of ESD practice. The overriding conclusion from many sources suggests that: "schools that embrace ESD are also schools which succeed and do well" (KBT 2013: 2).

In England and Northern Ireland there is currently less policy emphasis on SD and this has inhibited the wider adoption of good practice in ESD. In Wales the prominence given to ESDGC in national policy has diminished in spite of the significant emphasis placed on SD by the Welsh (Assembly) Government. In Scotland there is a greater focus on a more integrated and coherent approach to SD and ESD, with education being recognised by policy makers and practitioners as a key enabler in the transition to a sustainable society. However, good practice is not widespread in UK schools largely because there is no national strategic framework which puts it at the core of the education policy agenda in all the UK's administrative jurisdictions. Without it the much needed coherence, direction and impetus of existing activity is limited and patchy. The absence of an overarching UK Strategy for SD which sets out a clear vision about the contribution learning can make to its SD goals is a major barrier in scaling up existing good practice, as well as helping to prevent unnecessary duplication of effort and resources.

References

ACCAC. (2002). *Education for global citizenship and sustainable development.* Cardiff: Qualifications, Curriculum and Assessment Authority for Wales.

Barratt Hacking, E. C., Scott, W., & Lee, E. (2010). *Evidence of impact of sustainable schools.* London: Department for Children, Schools and Families. https://www.education.gov.uk/publications/standard/publicationDetail/Page1/DCSF-00344-2010. Accessed 15 Apr 2014.

BBC. (2012). *Gove tells head teachers school reforms need to be accelerated.* Report by Judith Burns, 24th March. http://www.bbc.co.uk/news/education-17481888. Accessed 15 Apr 2014.

CCEA (Council for the Curriculum, Examinations and Assessment Northern Ireland). (2014). *Northern Ireland curriculum.* http://www.nicurriculum.org.uk/. Accessed 15 Apr 2014.

Chatzifotiou, A. (2002). An imperfect match? The structure of the national curriculum and education for sustainable development. *Curriculum Journal, 13*(3), 289–301.

Christie, E., & Higgins, P. (2012). *The impact of outdoor learning on attitudes to sustainability.* Commissioned report for the Field Studies Council, UK.

CREE (Centre for Research in Education and the Environment). (2009). *Supporting DCSF's sustainable schools agenda: Networking sustainable schools in the South West—2008/09; External evaluation report to the Government Office for the South West.* Bath: University of Bath.

David, M. E. (2007). Changing the educational climate: Children, citizenship and learning contexts? *Environmental Education Research, 13*(4), 425–436.

DBIS (Department for Business, Innovation and Skills). (2010). *Improving the quality of further education and skills training*. London: DBIS. https://www.gov.uk/government/policies/improving-the-quality-of-further-education-and-skills-training. Accessed 15 Apr 2014.

DCELLS. (2006). *Education for sustainable development and global citizenship: A strategy for action* (Information Document No. 017/2006). Cardiff: Welsh Assembly Government.

DCELLS. (2007). *Out of classroom learning: Making the most of first hand experiences in the natural environment (Information Document No. 022/2007)*. Cardiff: Welsh Assembly Government.

DCELLS. (2008a). *Education for sustainable development and global citizenship: A strategy for action—Updates January 2008* (Information Document No. 055/2008). Cardiff: Welsh Assembly Government.

DCELLS. (2008b). *Education for sustainable development and global citizenship: A common understanding for schools* (Information Document No. 065/2008). Cardiff: Welsh Assembly Government.

DCELLS. (2008c). *Skills framework for 3 to 19-year-olds in Wales*. Cardiff: Welsh Assembly Government.

DCELLS. (2008d). *Education for sustainable development and global citizenship: Information for teacher trainees and New teachers in wales* (Information Document No. 066/2008). Cardiff: Welsh Assembly Government.

DCELLS. (2008e). *Science in the national curriculum for Wales: Key stages 2–4*. Cardiff: Welsh Assembly Government.

DCELLS. (2008f). *Geography in the national curriculum for Wales: Key stages 2–4*. Cardiff: Welsh Assembly Government.

DCSF. (2010). *Evidence of impact of sustainable schools*. London: Department for Children, Schools and Families.

DCSF (Department for Children, Schools and Families). (2008). *Planning a sustainable school: Driving school improvement through sustainable development*. London: Department for Children, Schools and Families.

DECC (Department for Energy and Climate Change). (2010). *Meeting the low carbon skills challenge—A government response*. London: DECC. https://www.gov.uk/government/uploads/system/uploads/attachment_data/file/48807/1116-meeting-low-carbon-skills-consresponse.pdf. Accessed 15 Apr 2014

Defra. (2011). *The natural choice: securing the value of nature*. London: Department for Environment, Food and Rural Affairs. http://www.defra.gov.uk/environment/natural/whitepaper. Accessed 15 Apr 2014.

DENI (Department of Education Northern Ireland). (2009). *Schools for the future: A policy for sustainable schools*. http://dera.ioe.ac.uk/8856/. Accessed 19 Sept 2013.

DENI. (2010). *Education (school development plans) regulations (Northern Ireland)*. http://www.legislation.gov.uk/nisr/2010/395/pdfs/nisr_20100395_en.pdf. Accessed 17 Sept 2013.

DENI. (2012). Education and Skills Authority. http://www.deni.gov.uk/index/about-the-department/8-admin-of-education-pg/education-and-skills-authority.htm. Accessed 24 Sept 2013.

DfE. (2011). *s3: Sustainable school self-evaluation—Version final, February 2011*. London: Department for Education.

DfES. (2003). *Sustainable development action plan for education and skills*. London: Department for Education and Skills.

DfID (Department for International Development). (n.d.). *Global school partnerships*. http://www.coe.int/t/dg4/nscentre/ge/WAEA/WAEALaureate-DGSP-Summary.pdf. Accessed 29 Apr 2014.

Disinger, J. F. (1985). What research says: Environmental education's definitional problem. *School Science and Mathematics, 85*(1), 59–68.

EAC (Environmental Audit Committee. (2003). *Learning the sustainability lesson, tenth report of session 2002–03* (Vol. II: Interviews and written evidence). London: House of Commons.

Eco-Schools Northern Ireland. (2014). http://www.eco-schoolsni.org/about.aspx?dataid=399762. Accessed 15 Apr 2014.

Eco-Schools Scotland. (2014). http://www.keepscotlandbeautiful.org/sustainable-development-education/eco-schools. Accessed 5 May 2014.

Education Scotland. (2011). *Exploring climate change*. Edinburgh: Scottish Government. http://www.educationscotland.gov.uk/exploringclimatechange/response/scotland.asp. Accessed 25 Apr 2014.

EMF (Ellen MacArthur Foundation). (2012). http://www.ellenmacarthurfoundation.org/. Accessed 15 Apr 2014.

ESD Forum. (n.d.). http://www.eefni.org.uk/aboutus/. Accessed 15 Apr 2014.

Estyn. (2006a). *Establishing a position statement on education for sustainable development and global citizenship in Wales*. Cardiff: Estyn.

Estyn. (2006b). *Update on inspecting education for sustainable development and global citizenship*. Cardiff: Estyn.

Estyn. (2009). *A report on the quality of education at Bryn Celynnog comprehensive school*. Cardiff: Estyn.

Estyn. (2010). *A report into the quality of education at Craigfelan primary school*. Cardiff: Estyn.

Gayford, C. (2009). *Learning for sustainability: From the pupils' perspective*. Godalming: WWF.

Hansard. (1978). *Written question by Janet Fookes MP*. http://hansard.millbanksystems.com/written_answers/1978/dec/12/environmental-studies#S5CV0960P0_19781212_CWA_122. Accessed 15 Apr 2014.

Higgins, P., & Kirk, G. (2009). Sustainability education in Scotland: The impact of national and international initiatives on teacher education and outdoor education. In B. Chalkley, M. Haigh, & D. Higgitt (Eds.), *Education for sustainable development. Papers in honour of the united nations decade of education for sustainable development (2005–2014)* (pp. 161–174). London: Routledge.

Higgins, P., & Lavery, A. (2013). Sustainable development education. In T. Bryce, W. Humes, D. Gillies, & A. Kennedy (Eds.), *Scottish education* (4th ed., pp. 337–342). Edinburgh: Edinburgh University Press.

HM Government. (2005). *Securing the future: The UK government sustainable development strategy*. London: TSO.

HMIE (Her Majesty's Inspectorate of Schools). (1974). *Environmental education: A report by H.M. Inspectors of schools*. Edinburgh: HMSO.

Hopkins, C. (2013). Educating for sustainability: An emerging purpose of education. *Kappa Delta Pi Record, 49*(3), 122–125. doi:10.1080/00228958.2013.819193.

Interboard Education for Sustainable Development Group. (2005). *Good practice guide for primary, secondary and special schools*. Belfast: Northern Ireland Executive. http://www.belb.org.uk/downloads/esd_good_practice_guide.pdf. Accessed 17 Sept 2013.

IUCN. (1970). *International working meeting on EE in the school curriculum* (Final report). Gland: IUCN/UNEP/WWF.

KBT (Keep Britain Tidy). (2013). *Eco-schools England: Exploring success to inform a new horizon*. Wigan: Keep Britain Tidy.

Lavery, A., & Smyth, J. (2003). Developing environmental education, a review of a Scottish project: International and political influences. *Environmental Education Research, 9*(3), 361–383.

LOtC (Council for Learning Outside the Classroom). (2014). *Forest schools network*. http://www.lotc.org.uk/fen/. Accessed 15 Apr 2014.

Lotz-Sisikta, H., & Schudel, I. (2007). Exploring the practical adequacy of the normative framework guiding south Africa's national curriculum statement. *Environmental Education Research, 13*(2), 245–263.

Luna, H., Martin, S., Scott, W., Kemp, S., & Robertson, A. (2012). *Universities and the green economy: Graduates for the future. Policy think tank report*. York: The Higher Education Academy. http://www.heacademy.ac.uk/assets/documents/esd/Graduates_For_The_Future_Print_130812_1322.pdf. Accessed 15 Apr 2014.

Martin, S., Jucker, R., & Martin, M. (2009). Quality and education for sustainable development: Current context and future opportunities. In L. E. Kattington (Ed.), *Handbook of curriculum development* (pp. 443–453). New York: Nova Science Publishers, Inc.

Martin, S., Dillon, J., Higgins, P., Peters, C., & Scott, W. (2013). Divergent evolution in education for sustainable development policy in the United Kingdom: Current status, best practice, and opportunities for the future. *Sustainability, 5*(4), 1522–1544. doi:10.3390/su5041522. http://www.mdpi.com/2071-1050/5/4/1522. Accessed 15 Apr 2014.

McNaughton, M. J. (2007). Sustainable development education in Scottish schools: The sleeping beauty syndrome. *Environmental Education Research, 13*(5), 621–635.

NCC (National Curriculum Council). (1990). *Curriculum guidance seven: Environmental education*. York: NCC.

Northern Ireland Education and Training Inspectorate. (2010). *Effective practice in education for sustainable development in a sample of primary, post primary and special schools in Northern Ireland*. Bangor: Northern Ireland Education and Training Inspectorate. http://dera.ioe.ac.uk/938/1/effective-practice-in-education-for-sustainable-development-in-a-sample-of-primary-post-primary-and-special-schools-in-northern-ireland.pdf. Accessed 17 Sept 2013.

Northern Ireland Executive (NIE). (2008). *Programme for government 2008–2011: Driving investment and sustainable development*. http://www.northernireland.gov.uk/pfgfinal.pdf. Accessed 19 Sept 2013.

Northern Ireland Executive (NIE). (2011). *Programme for government 2011–2015: Building a better future*. Belfast: OFMDFM. http://www.northernireland.gov.uk/pfg-2011-2015-final-report.pdf. Accessed 23 Sept 2013.

Northern Ireland Sustainable Development Commission. (2010). *SDC welcomes new school development plans for Northern Ireland*. http://www.sd-commission.org.uk/news.php/340/sdc-welcomes-new-school-development-plans-for-northern-ireland. Accessed 15 Apr 2014.

OFMDFM (Office of the First and Deputy First Minister). (2010). *Everyone's involved: Sustainable development strategy*. Northern Ireland Executive. http://www.ofmdfmni.gov.uk/sustainable-development-strategy-lowres__2_.pdf. Accessed 17 Sept 2013.

OFMDFM. (2012). *Young people's achievements celebrated at the IMPACT 2012 awards*. http://www.northernireland.gov.uk/index/media-centre/news-departments/news-ofmdfm/news-ofmdfm-march-2012/news-ofmdfm-210312-young-peoples-achievement_s-.htm. Accessed 19 Sept 2013.

Ofsted (Office for Standards in Education). (2003). *Taking the first step forward… towards an education for sustainable development: Good practice in primary and secondary schools*. London: Office for Standards in Education.

Ofsted. (2009). *Education for sustainable development: Improving schools—improving lives*. http://www.ofsted.gov.uk/resources/education-for-sustainable-development-improving-schools-improving-lives. Accessed 15 Apr 2014.

Ofsted. (2010). *Sustainable development: Briefing for section 5 inspectors*. London: Office for Standards in Education.

Palmer, J. (1998). *Environmental education in the 21st century: Theory, practice, progress and promise*. London: Routledge.

Pretty, J. (2013). The consumption of a finite planet: Well-being, convergence, divergence and the nascent green economy. *Environmental & Resource Economics, 55*(4), 475–499. doi:10.1007/s10640-013-9680-9. http://www.julespretty.com/wp-content/uploads/2013/09/5.-Env-Res-Econ-2013-Pretty.pdf. Accessed 15 Apr 2014.

QCA. (1998). *Education for citizenship and the teaching of democracy in schools: Final report of the advisory group on citizenship (the crick report)*. London: Qualifications and Curriculum Authority.

QCA. (1999). *The National Curriculum: Handbook for primary teachers in England Key Stages 1 and 2*. London: Department for Education and Employment (DfEE) and the Qualifications and Curriculum Authority (QCA).

Sayer, A. (2000). *Realism and social science*. London: Sage.

Scott, W. (2013). Developing the sustainable school: Thinking the issues through. *Curriculum Journal, 24*(2), 181–205. doi:10.1080/09585176.2013.781375.

Scottish Government. (2008). *Learning for our future: Scotland's first action plan for the decade of education for sustainable development.* Edinburgh: Scottish Government. http://www.scotland.gov.uk/Resource/Doc/137705/0034170.pdf. Accessed 29 Apr 2014.

Scottish Government. (2009a). *Curriculum for excellence building the curriculum 4: Skills for learning, skills for life and skills for work.* Edinburgh: Scottish Government. http://www.educationscotland.gov.uk/Images/BtC4_Skills_tcm4-569141.pdf. Accessed 25 Apr 2014.

Scottish Government. (2009b). *Climate change (Scotland) Act 2009.* Edinburgh: Scottish Government. http://www.legislation.gov.uk/asp/2009/12/contents. Accessed 25 Apr 2014.

Scottish Government. (2010). *Learning for change: Scotland's action plan for the second half of the UN decade of education for sustainable development.* Edinburgh: Scottish Government. http://www.scotland.gov.uk/Publications/2010/05/20152453/5. Accessed 25 Apr 2014.

Scottish Government. (2011). *The government economic strategy.* Edinburgh: Scottish Government. http://www.scotland.gov.uk/Resource/Doc/357756/0120893.pdf. Accessed 25 Apr 2014.

Scottish Government. (2012). *Learning for sustainability. The report of the One Planet Schools Working Group.* Edinburgh: Scottish Government. http://www.scotland.gov.uk/Topics/Education/Schools/curriculum/ACE/OnePlanetSchools/LearningforSustainabilitreport. Accessed 25 Apr 2014.

Scottish Government. (2013). *Learning for sustainability. The Scottish Government's response to the report of the One Planet Schools Working Group.* Edinburgh: Scottish Government. http://www.scotland.gov.uk/Topics/Education/Schools/curriculum/ACE/OnePlanetSchools/. Accessed 25 Apr 2014.

Scottish Office Environment Department. (1993). *Learning for life: A national strategy for environmental education in Scotland.* Edinburgh: HMSO.

SEEd (Sustainability and Environmental Education). (2012). http://se-ed.co.uk/edu/. Accessed 15 Apr 2014.

Smith, A. (2003). Citizenship education in Northern Ireland: Beyond national identity? *Cambridge Journal of Education, 33*(1), 15–31.

SSA (Sustainable School Alliance). (2012). http://se-ed.co.uk/edu/sustainable-schools/sustainable-schools-alliance/. Accessed 5 May 2014.

The Co-operative Group. (2011). *Your guide to becoming a more sustainable school.* http://www.se-ed.org.uk/Sustainability_Guide%20.pdf. Accessed 15 Apr 2014.

The Five Nations Network. (n.d.). http://www.fivenations.net/. Accessed 24 Sept 2013.

The Prime Minister's Office. (n.d.). *Sustainable development.* http://transparency.number10.gov.uk/content/cross-government-priority/sustainable-development. Accessed 29 Apr 2014.

ULSF (University Leaders for a Sustainable Future). (2001). *History and definitions of higher education for sustainable development.* http://www.ulsf.org/dernbach/history.htm. Accessed 5 May 2014.

UNESCO. (2004). *United Nations Decade of Education for Sustainable Development: Draft international implementation scheme.* Paris: UNESCO.

UNESCO UK. (2008). *ESD in the UK: A survey of action.* London: UK National Commission for UNESCO. http://www.unesco.org.uk/uploads/ESD%20in%20the%20UK%20in%202008%20-%20May%202008.pdf. Accessed 15 Apr 2014.

UNESCO UK. (2013). *Education for Sustainable Development (ESD) in the UK: Current status, best practice and opportunities for the future.* London: UK National Commission for UNESCO. http://www.unesco.org.uk/uploads/Brief%209%20ESD%20March%202013.pdf. Accessed 15 Apr 2014.

UNESCO Centre University of Ulster. (2009). *Evaluation of the pilot introduction of education for local and global citizenship into the revised Northern Ireland curriculum.* http://cass.welbni.org/downloads/28/267_25_UU%20Evaluation%20Summary%20Report.pdf. Accessed 24 Sept 2013.

UNESCO-UNEP. (1978). *Inter-governmental conference on environmental education, 14–26 October 1977, Tbilisi.* Paris: UNESCO-UNEP.

Vare, P. (Forthcoming). Are there inherent contradictions in attempting to implement education for sustainable development in schools? (working title). To be submitted to *Environmental Education Research*.

Vare, P., & Scott, W. (2013). From environmental education to ESD: Evolving policy and practice. In C. Russell, J. Dillon, & M. Breunig (Eds.), *Environmental education reader*. New York: Peter Lang (in press).

WAG (Welsh Assembly Government). (2009). *One wales: One planet, The sustainable development scheme of the Welsh Assembly Government*. Cardiff: Welsh Assembly Government.

Wintour, P. (2013, May 2). Energy secretary urges Michael Gove to reinstate climate change on curriculum. *The Guardian*. http://www.theguardian.com/politics/2013/may/02/michael-gove-climate-change-curriculum. Accessed 15 Apr 2014.

UNESCO. (2003). *Framework for a draft international implementation scheme*. Paris: United Nations Education, Scientific and Cultural Organisation (UNESCO). http://www.google.ch/url?sa=t&rct=j&q=&esrc=s&source=web&cd=2&ved=0CC8QFjAB&url=http%3A%2F%2Fportal.unesco.org%2Feducation%2Fen%2Ffile_download.php%2F9a1f87e671e925e0df28d8d5bc71b85fJF%2BDESD%2BFramework3.doc&ei=ECpNU8CTGYGg0QXt74FQ&usg=AFQjCNGBs2U9G6m16fjJtsB-Tl4Tr1ESYw&bvm=bv.64764171,d.d2k&cad=rja. Accessed 15 Apr 2014.

Chapter 20
Education for Sustainable Development in the Netherlands

Martin de Wolf and André de Hamer

20.1 Characteristics of the Dutch Educational System

20.1.1 Public and Private Schools

The Dutch educational system can be characterised as a system with much freedom for staff in primary and secondary schools to design their own curriculum. To illustrate this: since 1917, every Dutch citizen has the right to start his or her own school, based on his or her own ideologies. These private schools are treated equally by the state as public schools. As a consequence, public and private schools receive the same financial governmental support, as long as they meet the qualitative standards, demanded by the Ministry of Education, Culture and Science (UNESCO 2012). This is a recognition that parents had to have the possibility to send their children to a school with the values and beliefs that fit with their religion. So, freedom of education, as described in Article 23 of the Dutch constitution, is mainly based on the principle that the government does not provide schools based on religions or specific ideologies themselves, but it creates the possibility for others to found and run these schools. But in order to take full responsibility for education in the Netherlands, the government does run and support schools, called public schools (Luijkx and De Heus 2008).

In 2010/2011, about one-third of the primary schools in the Netherlands were public schools. At secondary level, only a quarter of Dutch students went to public

M. de Wolf (✉)
Fontys University of Applied Sciences, Eindhoven, The Netherlands
e-mail: m.dewolf@fontys.nl

A. de Hamer
Duurzame PABO, Utrecht, The Netherlands
e-mail: andre.dehamer@gmail.com

© Springer International Publishing Switzerland 2015
R. Jucker, R. Mathar (eds.), *Schooling for Sustainable Development in Europe*,
Schooling for Sustainable Development 6, DOI 10.1007/978-3-319-09549-3_20

schools. The majority of Dutch students get their primary and secondary education in private schools. From 70 % (secondary education) up to 90 % (primary education) of these private schools have a Christian background (CBS 2013).

20.1.2 Pedagogy

Besides the freedom to run private schools, the Dutch educational system also provides the freedom for schools to choose their own pedagogy. It is up to the board of a school to decide with which pedagogy they are willing to work. Public schools can equally choose their own pedagogy. Contemporary pedagogies are based on the ideas of such educationists as Dalton, Montessori, Jena and Steiner (Rijksoverheid n.d.a).

20.1.3 Curriculum Development

In 2004, the Dutch government required secondary schools to position themselves on curriculum development. Doing 'the same as usual' was not an option, since the prevailing view was that curricula were too fragmented. Schools had to choose between four different scenarios or a variation on one of these scenarios. In short, Scenario 1 is a conventional school, Scenario 4 is an innovative school. This latter type of school does not have a traditional timetable and gives students the opportunity to choose their own learning arrangements (Taakgroep Vernieuwing Basisvorming 2004).

Most schools chose Scenario 2 (more focused on project-based learning). Only about 15 % of the secondary schools had the ambition to be a Scenario 4 school (Busman et al. 2006). This illustrates the fact that change in the Dutch educational system is, for one reason or another, slow. The implications of this in the context of Education for Sustainable Development (ESD) will be described in the second part of this chapter.

20.2 Design of Dutch School Curricula

20.2.1 A Model for Curriculum Development

As we have seen, schools in the Netherlands have much freedom: private schools are equal to public schools, they can choose their own pedagogy and they are free to decide how they introduce innovations to their programmes. But if you take a closer

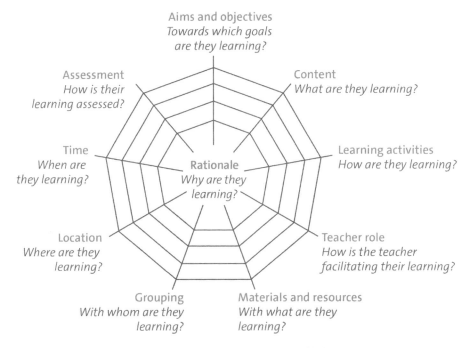

Fig. 20.1   The curricular spiderweb (Thijs and Van den Akker 2009)

look at the way the Dutch school curricula are designed, it becomes clear that this freedom is limited.

Van den Akker (Thijs and Van den Akker 2009) developed a model for curriculum development (Fig. 20.1), which can be used to describe how free schools are to design their curricula. Initially, this model was designed to show how interdependent different aspects of a curriculum are: if you change something, it will have consequences for other aspects of the curriculum.

From Objectives to Testing

The model (Fig. 20.1) starts with aims and objectives. In the Netherlands these are described by the Dutch Ministry of Education, Culture and Science (ECS). The so-called core objectives limit what schools can do, but are formulated in a very general way. This allows schools to make their own decisions for other elements in the model: specifically content, learning activities, the role of the teacher, classroom materials and grouping (classes can differ in size and composition of age-groups).

For the final 2 or 3 years of secondary schooling (depending on the level of education), the government has formulated aims and objectives that are more specific, called final objectives. These final objectives are assessed in two different ways: partly by tests designed by the school and partly by a national examination, carried out under the responsibility of the government. This national examination counts for 50 % of the total score for the examination of the students. Consequently, there is a very strong focus during the final years of secondary schooling on ensuring that students are well prepared to pass this national punctuation only a low percentage of students in a specific school pass this examination, that school will face problems as 'a bad school'.

At primary schools, there is a comparable national examination, called the Citotest, named after the organisation that is responsible it. Primary Schools have not been obliged to offer this examination to their students, but parents consider the test an important one (Luijkx and De Heus 2008). If primary schools do not offer the Citotest or when students have a very low average score on this test, the school might be out of favour with parents. The consequences of this for ESD will be discussed below.

Time and Location

There are two additional aspects of the model (Fig. 20.1, Thijs and Van den Akker 2009) that must be mentioned: time and location. For both aspects of Dutch school curricula, there are some governmental regulations. Students must receive 7,520 h of lessons during the 8 years of primary schooling. At secondary school, the standard is 1,040 h of lessons per year (Rijksoverheid n.d.b). There have been discussions in the Netherlands about these standards: should there be such a strict quantitative standard or should standards be more qualitative?

There is also limited freedom to choose the location for educational activities. If someone wants to construct a school building (for instance a 100 % recyclable building), the government needs to give permission, based on spatial planning regulations.

20.3   ESD in Dutch School Curricula

20.3.1   Characterising ESD in the Netherlands

Since the beginning of this century, the Dutch government has asked society, including schools, to pay attention to sustainable development (SD). Because of the constitutional right to freedom of education (despite its limitations outlined above), schools can choose whether and how they pay attention to SD.

Three institutions that have made a significant impact on Dutch primary and secondary education, have contributed, to a certain extent, to the exploration and implementation of SD in Dutch school programmes. These institutions are SLO

(Stichting Leerplan Ontwikkeling: Curriculum Development Foundation), Cito (assessment center) and Kennisnet (knowledge sharing and dissemination through IT). The following gives a short overview over how ESD is understood in the Netherlands, including the influence of these three institutions.

The SLO published two documents on ESD on behalf of the Dutch Ministry of Education, Science and Culture (ESC). In these documents, ESD is defined as education that enables children: to make their own choices within complex situations where different possibilities and opinions exist; to take a position; to take responsibility for their own behaviour (Remmers 2007; Bron et al. 2009).

Cito has published four basic concepts of ESD in relation to existing subjects. These concepts do not introduce new knowledge, but existing knowledge is integrated in a new and socially relevant way. Out of these basic concepts a few pilot lessons have been developed (Wagenaar 2007).

Kennisnet developed an online portal for ESD in primary and secondary education: "An educational starting point for teachers and students who are interested in nature, sustainability, food, energy and the environment" (Kennisnet n.d.).

Apart from these three institutions, many other organisations have contributed to the implementation of ESD in the Netherlands. One example is the publishing of a book titled *Duurzame ontwikkeling op de basisschool* (*Sustainable development in primary education*, de Hamer et al. 2008), which has been used by a large number of teachers and curriculum developers.

To summarise, ESD in the Netherlands can be described as a way of learning which makes students more aware of their involvement in complex issues in society and which enables them to judge which solution fits best with the analysed situation. In other words, ESD in the Netherlands can be characterised as learner-oriented, future-oriented, action-based, problem-solving and inclusive education (de Wolf et al. 2011). This also involves skills related to systems thinking and ethics (Sleurs 2008).

ESD and Transformative Social Learning

Dutch schools have the freedom to choose their own pedagogy and scenario for educational innovation. Most secondary schools have chosen Scenario 2 to qualify how they want to integrate learning processes. Only a minority of schools have chosen Scenario 4, where students can choose their own learning arrangements. So the majority of schools merely have adapted their existing pedagogy to allow for more innovative forms of learning. According to Wals, ESD means the creation of time and space for social and transformative learning. Social learning can be considered as "a learning system in which people learn from each other and collectively become more capable of dealing with setbacks, stress, insecurity, complexity and risks" (Wageningen UR n.d.). Transformative social learning includes space for alternative paths of development, new ways of thinking, pluralism, consensus and respectful disagreement, autonomous thinking, self-determination and contextual differences (Wals 2006). If this way of learning can be considered a condition for ESD-learning, the Dutch educational system needs a paradigm-shift.

ESD in Aims and Objectives

Although the government asked the Curriculum Development Foundation (SLO) to publish ideas for an integrated curriculum for ESD, the Dutch Ministry for ECS does not want to force schools to implement ESD in their curricula (Roorda 2010). The Dutch programme for ESD (see below) is also not the Education Ministry's responsibility; it has been initiated by the Dutch Ministry of Economic Affairs. Implementation of ESD in Dutch school curricula could be more successful if the Ministry of ECS would support it more specifically.

On the other hand, SD is indirectly part of the core objectives for primary and secondary education. In the core objectives for primary education, SD is explicitly mentioned in the preamble.

There are also elements in the description of core and final objectives for primary and secondary schools that are strongly linked to (E)SD. Ten out of the 58 core objectives for primary education contain aspects of (E)SD. At secondary education, 20 out of 58 core objectives give the opportunity for schools to design an ESD-curriculum (de Wolf et al. 2011). In the preamble of the syllabi for the final objectives for the last 2 years of pre-vocational secondary school, the following general objective is formulated: "the candidate learns (...) to get insight into relations between personal and societal environment. This contains specific attention to the relationship between human beings, nature and the concept of sustainable development" (Rijksoverheid 2008: 2). Teachers of all subjects at these schools are required to pay attention to this, in one or another way. More specific elements in the final objectives can be found in syllabi of the subjects science, chemistry, biology, nature, life and technology, history, geography, economy, civics and philosophy. For example, final objective 11.7 of the geography examination programme for pre-vocational schools reads: "the candidate is able to describe and explain the resource, occurrence, quality and usage of water in the Netherlands and to describe measures for a more sustainable usage of water and its effects" (CVE 2013: 18).

Much ESD?

Do these objectives automatically lead to an ESD-oriented programme? Not really since the elements in the core and final objectives are all fragmented aspects of ESD. This fosters a sustainability-approach related to one subject instead of the interrelationship between subjects. Or the objectives are formulated in a way which requires students to describe and explain a certain situation, while ESD should focus on problem-solving. Beside this fragmentation, the way Dutch schools assess pupils' achievements, is experienced as strongly focused on the reproduction of knowledge and application of rules. Characteristics of ESD, as mentioned earlier, are not automatically part of the design of a traditional school curriculum.

On the other hand, freedom of education offers staff in schools the possibility of implementing ESD as a core concept in their curriculum. As a consequence, schools have presented themselves as ECO-schools or Energetic schools (Van der Waal 2011). But these initiatives often depend on the individual efforts of school leaders or teachers and they are not embedded in the organisation of the school. An overview of such efforts is also limited and difficult. In 2011 a group of organisations took the initiative to list all schools claiming to be working explicitly on ESD: www.duurzaamonderwijs.nl (accessed 19 April 2014). This list is not complete but currently includes almost 200 primary schools.

It needs to be mentioned, though, that a large number of vocational schools is very active in the field of ESD, cooperating in a national network, called "Duurzaam Middelbaar Beroepsonderwijs" (Sustainable Vocational Education, DMBO). However, the aim of this chapter is to get an overview of ESD in primary and secondary education and teacher training institutions.

20.4   Government Programmes Giving ESD a Boost

We have seen that ESD can be found in regular curricula, but not universally so. That is why the government started national programmes to initiate, stimulate and support ESD in the Netherlands. In what follows, these national programmes are briefly described and evaluated.

20.4.1   LvDO

In 2004 the Dutch government, including almost all Ministries, provinces and municipalities, initiated a national ESD-programme called "Leren voor Duurzame Ontwikkeling" (Learning for Sustainable Development, LvDO), with the subtitle "from margin to mainstream." The programme was intended to embed learning processes into mainstream decision-making processes (Rijksoverheid 2004).

In 2008 the programme was continued with the title "LvDO: from agenda to action". This programme continued to support the aspirations, wishes and needs of the government and society. The programme connected social activities and policy processes in this area. The programme was based on three pillars. Pillar 1 "learning individuals" aimed at the following: "all students are able to actively contribute to sustainable development at the end of their school career" (Programmabureau Leren voor Duurzame Ontwikkeling 2008: 3). Pillar 2 "learning organisations" aimed to encourage and support a shift to SD within government organisations. Pillar 3, "learning society" was all about involving more citizens, professionals and organisations as participants in sustainability projects. The main target here was that better (more sustainable) decisions would be made in interactive (policy) processes (Programmabureau Leren voor Duurzame Ontwikkeling 2008).

As a result of the LvDO programme, the Dutch government spent five million euros a year (2008–2011) on ESD. With this budget, many pilot projects were financed. Unfortunately, they were financed independently in each province, with the risk of overlap between provinces. Where knowledge transfer between projects and provinces has added value, this took place mostly on an informal and irregular basis (Van der Waal 2011).

20.4.2   NME

Before the Dutch government introduced these ESD-programmes, a national EE-programme (Environmental Education) already existed, called "Natuur en Milieu Educatie" (NME). The first EE-programme was introduced in 1988, with the aim "to strengthen nature and the environment in society and ensuring the quality of nature and the environment by stimulating awareness, value, knowledge and skills within the field of nature and the environment" (Sollart and Vreke 2008: 11).

The more recent EE-programme (2008–2011) aimed to stimulate collaboration between school directors, EE-organisations, users and new partners for nature and the environment, within a so-called "arrangement-model" (i.e. cooperations between policymakers, developers and users of educational projects and innovators). Collaborative knowledge, quality and professionalism were supposed to make EE future-proof. The central themes were: green issues, water and energy (Smit et al. 2006).

20.4.3   Evaluation of LvDO and NME

The programmes LvDO and NME were evaluated in 2012 (Remmerswaal et al. 2012). Some of the conclusions were:

- At the state level the agenda-setting influence of the LvDO programme has been limited by the political context in recent years. After a strong start with the government-wide approach to SD, the programme was more or less reduced to creating the conditions for SD within central government, including human resources, in the second half. At provincial level, the LvDO programme managed to support educational pilot projects.
- Reducing the number of themes and projects has created focus. Part of this focus was to make the LvDO-networks part of existing organisations. The question will be for the upcoming years if these networks are sufficiently anchored.

- Within the EE-programme the "arrangement model" has taken a central place. This has reoriented the EE-sector towards demand-oriented work. Professionals within the EE-sector were introduced to a new way of working, with new forms of cooperation between organisations and EE. As a result, the EE-sector became smaller, but more different organisations got involved, like local governments and also commercial stakeholders.
- The evaluation of the EE-programme showed that clearly articulating demand and embedding this new way of working will remain challenges for the future.

Duurzaam Door

In 2013 the Dutch government decided to continue with ESD support, but by merging LvDO and NME into one programme, called "Duurzaam Door" (Continuing Sustainability). "Duurzaam Door" is an inter-governmental and interdepartmental programme with the subtitle "social innovation for a green economy". The purpose of this programme is to build up a green economy by developing knowledge, strengthening skills and by creating synergies between stakeholders. The programme aims therefore to equip actors (children, adults, schools, businesses, organisations) with the necessary awareness, knowledge, attitudes and perspectives in order to make sustainability a well-informed choice (Rijksoverheid 2013).

20.5 ESD in Primary Education

As a result of the national governmental programmes LvDO and NME some ESD-activities have taken place in the Dutch educational system. We already mentioned the publishing of ideas for an integrated curriculum for ESD by SLO (2007), which was initiated by the government via the programme LvDO.

A number of activities in primary education have been initiated or stimulated by "Duurzame PABO", the Dutch ESD network for teacher training institutions for primary education. This network was founded in 2005 and, among others, funded by the LvDO-programme. Some of its activities are:

- *Reading about Sustainability Day.* This reading event is organised by Duurzame PABO, along with Urgenda, Missing Chapter Foundation, AgentschapNL and other partners. Every year a book is specially written or selected for this day, in which sustainability is always the central theme. On Sustainability Day (October 10), readers (varying from mayors to school cleaners, from CEOs to student teachers) read to pupils of approximately 1,000 primary schools. Afterwards, the readers and students discuss and work further on sustainability. The schools and readers receive tips on how to move towards action. In many cases this leads to more attention being given to sustainability in these schools (www. dagvandeduurzaamheid.nl, accessed 17 April 2014).

- *Expeditie Geluk.* Duurzame PABO uses as a definition of SD: "take good care of yourself, others and your environment. Here and now, but also later and elsewhere". Expedition Geluk (happiness) focuses on taking good care of yourself and others, especially those close by. When children learn to take good care of themselves this forms the best basis for taking care of one another and the environment. Therefore Duurzame PABO is one of the initiators of Expedition Geluk, an expedition along the borders of the Netherlands when all schools and teacher training institutions are called upon to work for the happiness of children. Also, many companies, NGOs and individuals have participated in this expedition (www.expeditiegeluk.nl, accessed 17 April 2014).
- *School portraits.* Duurzame PABO and *Veldwerk Nederland* (Fieldwork Netherlands) have described more examples of ESD at individual schools in a number of school portraits. The portraits show how schools emphasise different aspects of sustainability. De Sokkerwei in Castricum for example was, in the late 1990s, the first primary school in the Netherlands with a sustainable building. Meanwhile, there are schools that implement all aspects of sustainability in all areas of the school (organisation and education). A good example is De Kariboe in Heemskerk (Veldwerk Nederland 2009).
- Seven more school portraits were compiled of so-called traditional renovation schools (e.g. Dalton, EGO and OGO). These schools ensure that the students are trained in such a way that they develop and build their lives in a changing world and contribute to a sustainable society (Meijer et al. 2010).
- *Competences.* On the initiative of Duurzame PABO and LvDO, the UNECE competences on ESD for teachers have been translated and edited for Dutch teachers and student teachers. These competences are being used at a growing number of teacher training institutions and primary schools (de Hamer and Leussink 2012).
- *PRISE.* How do you measure ESD at your school? How do you weigh vision, policy, environmental management, implementation and evaluation in education? How do you keep focus to take coherent steps towards SD in your school? PRISE (an auditing instrument for Primary Sustainable Education) is a derivative of AISHE (see below), a measuring instrument for sustainability in higher education (Roorda 2001). Duurzame PABO adapted and edited this instrument for primary schools (de Vries 2013).

There are also numerous projects and activities that are organised without support of the network Duurzame PABO. Some examples:

- *Eco-Schools.* A growing number of schools have joined Eco-Schools. Eco-Schools provide clear criteria to measure how sustainable a school is. The schools are guided through seven steps that lead them to the Green Flag. The Green Flag is the UN international certification for sustainable schools. Currently there are about 30 participating schools in the Netherlands and this number is rising steadily (www.sme.nl, accessed 17 April 2014).
- *De Vreedzame school* (*The Peaceful school*). This is a complete programme for primary schools for social competence and democratic citizenship. It considers

the class and the school as a community in which children feel heard and are seen, in which children have a voice, and where children learn to make decisions together and resolve conflicts. Children feel responsible for each other and for the community, and are open to the differences between people (www. devreedzameschool.net, accessed 17 April 2014).

20.6   ESD in Secondary Education

The national ESD-programmes LvDO and NME have had less effect on secondary than on primary education in the Netherlands. For example, the website "Groen Gelinkt" (where teachers can find fully prepared, instantly usable lessons about environmental issues, a cooperative outcome of the programmes NME and LvDO) counts almost 2,000 documents with learning activities for primary education and 'only' 66 documents for secondary education (www.groengelinkt.nl, accessed 17 April 2014). Nevertheless, there are some examples of outcomes of the national governmental programmes for secondary education:

- *Schools for sustainability.* In many provinces, the budget for the programme LvDO was spent to develop new ideas. One of these initiatives was Schools for Sustainability, where the provincial consultancy-departments of the Instituut voor Natuureducatie (Institute for Nature Education, IVN) offer this educational programme to secondary schools. Its aim is to ask pupils for their advice on the development of biodiversity and environmental issues in their region. The students' advice is then handed over to local or provincial governments, which makes the asking for advice authentic. The programme is supported by a website and ICT-rich learning materials (www.scholenvoorduurzaamheid.nl, accessed 17 April 2014). In different provinces, the programme Schools for Sustainability has faced difficulties in finding schools to conduct the programme. One of the reasons is that the programme offers learning activities that need to be adapted to the needs and the programmes of particular schools. Apparently, there have been just a few schools that see the possibilities to implement Schools for Sustainability within their own programme.
- *Pearls of biodiversity.* Following the principles of the arrangement-model of the NME-programme, the provincial arrangement Pearls of Biodiversity had the task of interlinking initiatives on biodiversity with each other and strengthening these initiatives within organisational or societal contexts. Beside these aims, the arrangement aimed to stimulate more balance between demand and supply within biodiversity-activities (www.biodiversiteitbrabant.nl, accessed 17 April 2014). The arrangement was organised around six sectors, of which secondary education was one. The main idea of the sub-arrangement for secondary education was to understand the needs of secondary schools with regard to education for biodiversity. Directors and teachers of different schools at all educational levels in secondary education were invited to make their needs explicit.

The result was a mission-statement of the involved schools which made clear what they wanted to achieve with education for biodiversity: "to make students aware of the way they are connected to biodiversity from their daily life, that they are part of the ecological system, can make well-argued decisions and realize that they have influence" (Biodiversiteit in Brabant 2011: 2). After an inquiry to find the educational supply for this demand, the conclusion was that it did not yet exist. A budget was available to create a new programme and an educational publisher was asked to deliver it (www.educatiefeneetbaargroen.nl, accessed 17 April 2014). This programme seems to face the same problem as that mentioned with Schools for Sustainability: schools find it hard to implement in their existing curriculum. The consequence is that the number of schools working with this programme is limited.

- *The Sustainable Chain Gang:* The most recent focus of LvDO is to bridge knowledge and experiences through networks. One of the initiatives of LvDO (and the present programme Duurzaam Door) is the start of an ESD-network for secondary schools in the Netherlands. In 2013, the network had not started its activities, but a growing number of schools have registered with this network and will exchange knowledge and ideas to put ESD in practice. The question is in what way this network will be able to strengthen ESD in Dutch secondary schools (NME Podium 2013).

In secondary education, a number of ESD-projects happen without the support of governmental programmes. In the following part of this paragraph, some examples of ESD-projects are described, initiated by a school, a non-governmental organization and a company.

- *Koning Willem II College.* Many secondary schools in the Netherlands have introduced SD by using project work in their yearly curriculum. These projects are often, however, isolated from the rest of the curriculum. The Koning Willem II College in Tilburg (a public secondary school) is an exception to this. They combine incidental project-activities with lessons in their curriculum, based on an integrated world citizenship programme.

  - In the first year of schooling, students learn about the Millennium Development Goals (MDG) and focus on the importance of good education and access to drinking water. In the second year, students are involved in an exchange programme. Students learn about the culture and society of one of the involved countries, but also about the realisation of the MDG's. This second year programme is linked to activities of the school to support projects in these countries. The third year is about refugees. Students learn about conflicts, the causes and consequences of being a refugee, and the students also meet a refugee and a war veteran.
  - The aim of this programme is to make students more aware of their interconnectedness with global issues. The lessons are part of the so-called mentor lessons, where students are supported to focus on their own identity, their future study or on work and society (Koning Willem II College n.d.).

- *Switch, sustainable food.* Switch has the mission to stimulate behaviour that contributes to a sustainable and fair world. To realise that, Switch advises and supports local governments, organisations and companies with their activities.

  - One of the educational activities of Switch is a programme about sustainable food. The lessons that are offered by Switch are originally for students at vocational schools (who are trained to become cooks or work in middle-management for hotels and restaurants). The lessons can also be used at pre-vocational schools, where the theme wellbeing and care is offered. Besides lessons about sustainable food, students learn to make a 'worldmeal' (made with a list of ingredients, based on an equal division of food throughout the whole world) and organise so-called qualify-dinners, where students stimulate cooks to serve sustainable meals in their restaurants.
  - In this case, Switch developed the lessons in cooperation with one of the involved vocational schools, to meet the needs of the school in a targeted way (Switch n.d.).

- *Fujifilm's Most Sustainable Class Award:* Fujifilm Manufacturing Europe B.V. (located in the Netherlands) has had great involvement in education, as part of their CSR (Corporate Social Responsibility) programme. Fujifilm is a Japanese company by origin and they consider 'care for your surrounding' as an important value. That is why Fujifilm sponsors different local activities in Brabant, one of the provinces in the southern part of the Netherlands. Fujifilm's Most Sustainable Class Award is an example of these activities.

  - Fujifilm offers lessons about waste, energy and climate change for secondary schools. Students learn about these subjects and are challenged to come up with solutions for one of the related problems that can be linked to their own life at school or at home. The winner of the Most Sustainable Class Award wins 7,500 euro to invest in renewable energy (Fujifilm 2013).
  - For Fujifilm, CSR is not the only reason to initiate and sponsor this educational activity. They consider sustainability and innovation as crucial elements in their mission and want to share this with young people. Apart from that, Fujifilm might face problems with enough, well-trained technical labour in the near future and hopes that this educational programme will stimulate young people to start studying in related fields.

20.7 The Status of ESD in Dutch Teacher Training Institutions

In the Netherlands there are several ways of obtaining a degree to be a teacher. Students who take a masters' degree (MA or MSc) course at university can get their Masters of Education (MEd) within 1 year. Most students, however, start with a teacher training programme immediately after attending secondary or vocational schools, at a so-called university of applied sciences. These teacher training institutions offer a Bachelor of Education (BEd) degree in a fulltime programme of 4 years. Teacher training institutions for primary education are called PABOs (Pedagogical Academy for Basis Onderwijs = Primary Education) in the Netherlands. Teacher training institutions for secondary education are simply called teacher training institutions.

PABOs and teacher training institutions need to work with a set of competencies for teachers, based on the Law on Professions in Education (Wet BIO). Furthermore, there is a set of aims and objectives in a body of knowledge, which is described for general knowledge (pedagogical) and knowledge per subject (e.g. biology, geography or Spanish). Currently, PABOs and teacher training institutions give new students a national examination to assess if they have achieved an acceptable standard in all the areas listed in the body of knowledge (Ministerie van OCW n.d.). This measure is intended by the national government to ensure certain comparable quality standards between PABOs or teacher training institutions. However, it might have a similar effect as the national examinations at primary and secondary schools, namely an emphasis on these examinations, taking away attention from less well-defined subjects, as could be the case with ESD.

The question here is in what ways students are trained to put ESD in practice as soon as they start working as teachers in primary and secondary schools?

PABOs (Teacher Training Institutions for Primary Education)

A student at PABO is prepared and trained to become a primary school teacher. Teachers, clearly, play a key role in determining the quality of education. The teacher establishes contact with students, provides a safe environment, organises meaningful learning activities and disseminates knowledge, helps those who find learning difficult, and gives everyone a feeling of personal worth. This requires complex skills, strongly associated with the individual person (Stevens 2010).

The teacher training institutions for primary education are free to develop their own curricula, but the state requires that all PABOs must implement a core curriculum for all primary school subjects by 2014–2015. The accreditation is undertaken by NVAO (Dutch Flemish Accreditation Organisation). PABOs are free to implement ESD (or not to do so).

Duurzame PABO (see above) is the network of teacher training colleges and schools actively engaged in ESD and also the organisation of teacher training institutions and primary schools. The mission of Duurzame PABO is that in 2020, each primary school and PABO in the Netherlands has worked visibly and noticeably on "care of yourself, others and your environment" (www.duurzamepabo.nl, accessed 17 April 2014).

One of the initiatives Duurzame PABO has organised is a Community of Practice (CoP) for teachers and managers of teacher training colleges. In the academic year 2012–2013 five teacher training institutions participated in this CoP. All five were able to develop towards SD, based on the Auditing Instrument for Sustainability in Higher Education (AISHE), which in turn is based on the EFQM (European Foundation for Quality Management)-model for organisational development (Roorda 2001). AISHE gives good insights into the extent to which sustainability is implemented. It helps institutions to understand the extent to which vision, policy, implementation, testing, evaluation and assurance of SD are secured in the organisation, contained in Plan-Do-Check-Act strategies. In the Netherlands, the Marnix Acedemie in Utrecht currently has the highest ranking of all teacher training institutions, based on AISHE:

- their goals on ESD are student oriented;
- they have an ESD-policy related to (middle and) long term goals;
- ESD-goals are formulated explicitly, measured and evaluated. There is feedback from the results.

At the same time, a number of PABOs are not so intensively involved in the network of Duurzame PABO and pay less attention to ESD in their programme.

Teacher Training Institutions for Secondary Education

To be a teacher at secondary school, students choose a particular school subject, such as French, history or biology. There is a comparable situation to that in secondary schools and PABOs: teacher training institutions can design their own curricula, but they need to ensure that students learn what is described in the national body of knowledge (http://10voordeleraar.nl, accessed 17 April 2014).

About half of the programmes offered at teacher training institutions contain aspects of ESD in their curriculum. That is the case with biology, health and lifestyle, civics, history, science, chemistry, geography, economics and technology. To what degree they pay attention to ESD depends mainly on the enthusiasm of teacher trainers and directors. Teacher training programmes for other subjects (languages and mathematics) might offer aspects of ESD too, but that is not directly based on a national body of knowledge.

Furthermore, the description of SD in the national body of knowledge for a subject does not guarantee that all characteristics of ESD are integrated. For instance, the body of knowledge for geography teachers reads: "a geography teacher knows and shows insight into sustainable development as a concept and is capable to apply this concept to different situations in society. Sustainability contains all forms of the search for a balance between the natural environment and the human usage of this environment" (Vereniging Hogescholen 2009: 16). It sounds like a neat description, but it does not go any further than replication of knowledge and application of related terms and rules (i.e. 'low order thinking'), instead of analysing relevant issues or evaluating related solutions (i.e. 'systemic thinking'), as one would expect in the context of ESD.

One of the teacher training institutions with experience of ESD in their curriculum, is Fontys University of Applied Sciences. Students of different subjects have been offered an interdisciplinary course on SD, followed by a project-based course. For the project-based course, students have to choose between projects in secondary schools, non-governmental organisations or local governments. They are asked to design and develop educational products designed to meet the needs of these organisations. By doing so, students have to apply characteristics of ESD to a new educational product, together with knowledge of the specific subject related to the project. In 2012, a group of 20 students developed educational arrangements for 10 secondary schools on behalf of a Dutch educational publisher, the result being called Codename Future. Another group of students developed lessons about food waste, in cooperation with a comedian. The idea was to integrate ESD and humour. A third example of this project work was the design of a GPS-route for a secondary school on behalf of an environmental organisation.

De Wolf (2011) studied the effects of this project-based approach and showed that this approach had more learning effects on general competences of students (communication, cooperation, reflection) than on specific ESD-competences. The study also showed that students learned, for instance, to relate subjects to each other, to come up with possible solutions for complex issues and to judge adequately these solutions.

The example of this teacher training institute might seem to signify good practice, but in fact it is rather a coincidence that a group of enthusiastic teachers, supported by their director, started this ESD-initiative as part of the regular curriculum. Big steps need to be taken before ESD is really implemented in the core curricula of Dutch teacher training institutions.

20.8 Possibilities to Improve ESD in the Netherlands in the Near Future

- First, we described the amount of freedom within the Dutch educational system. It is relatively easy to start a private school and both private and public schools can choose with which pedagogies they like to work. Also specific content, learning activities, the role of the teacher, materials and grouping (see above) can be chosen by schools themselves. But core and final objectives are formulated by the Dutch Ministry of ESC. For these objectives, national examinations exist. This limits the freedom of schools. Therefore, and due to the fact that ESD is not fully implemented in the core and final objectives of the Ministry of ESC, the number of schools that embrace ESD depends largely on the enthusiasm of individual teachers and directors. We would like to recommend that policy-makers in the Netherlands to give more attention to ESD in the formulation of national objectives for primary and secondary schools. There are many ideas on how to do so, for instance those described by the SLO in their publication about ideas for a core curriculum for ESD (Remmers 2007).

- Secondly, we found that many schools have chosen Scenario 2 for educational innovation. A minority of schools has the ambition to be as innovative as is described in Scenario 4. ESD, however, needs an innovative pedagogical approach like transformative social learning (Wals 2006). One idea to establish a pedagogical approach that fits the characteristics of ESD is the concept of five different minds for learning (Gardner 2007). Education in the Netherlands is too focused on the disciplined mind (reproduction of knowledge and application of rules). More attention needs to be paid to the synthesising mind, the creating mind, the respectful mind and the ethical mind. In a recent publication, the following was stated:

> If we want to achieve our goals, then we really do need to have the courage to set high standards. At the moment only one of the minds is addressed. Children can do better, especially if the learning process is modelled in such way that what children do also really makes a difference. Meaningful learning also fits nicely at this time. The 21st century has its own problems for the longer term, as the global population growth and the use of resources that is no longer in balance with the possibilities of our earth. The mission of education in the 21st century suits the need to be sustainable. This again fits nicely with the task of education to let children learn best: by meaningful learning. (Geisen 2013: 5)

- A third conclusion is related to the Dutch national ESD-programmes. Although it is good that these programmes exist and that they have supported many ESD initiatives, their impact has been limited. The national programmes have not reached secondary schools as much as primary schools, creating a big gap between the impact of the ESD-programmes on primary schools and secondary schools. To remedy this a number of changes are needed: these programmes need to work from the sideline of Dutch curriculum development (core and final objectives from the government are not formulated in cooperation with these programmes); initiatives for ESD often come from individual teachers and directors; we would recommend policy-owners of the present national ESD-programme to focus even more on secondary schools and individual teachers, where the influence of a governmental ESD-programme can be maximised.
- A fourth aspect is that many educational projects are not demand-oriented. Schools find it hard to get these projects implemented in their curriculum, despite the efforts that have been put into the design and development of these, often very inspiring, projects. It is important for those who are involved in the design and development of ESD-projects to work closely in the design stage with the users of these projects, i.e. the primary and secondary schools.
- A final conclusion concerns teacher training institutions, where a changing mindset to education is necessary. There is little emphasis on learning for the future or ESD at teacher training institutions. Future teachers need to be trained

with more attention to ESD and learning for the future. Some important relevant didactic principles are:

- Focus on a vision and perspective, focusing on a desired social development (not on a disaster scenario);
- Systems thinking, thus making connections between local and global, past, present and future involvement, multi-perspective thinking;
- Focus on participation (students take democratic decisions, for which they are jointly responsible);
- Meta-cognitive learning, so a focus on action, evaluation and reflection (think first, then act, then think again);
- Complex content made accessible for students by the teacher;
- Social, personal and methodical learning objectives are connected; they will make learning meaningful (de Vries and de Hamer 2010).

References

Biodiversiteit in Brabant. (2011). *Ambitiedocument Biodiversiteit in het VO*. http://biodiversiteitbrabant.nl/images/bestanden/Ambitiedocument%20Biodiversiteit%20in%20het%20voortgezet%20onderwijs.pdf. Accessed 20 Sept 2013.

Bron, J., Haandrikman, M., & Langberg, M. (2009). *Leren voor Duurzame Ontwikkeling; een praktische leidraad*. Enschede: Stichting Leerplan Ontwikkeling.

Busman, L., Klein, T., & Oomen, C. (2006). *Beweging in beeld. Feiten en cijfers over innovatie in het voortgezet Onderwijs*. Utrecht: Schoolmanagers VO.

CBS. (2013). *Schoolgrootte; onderwijssoort en levensbeschouwelijke grondslag*. http://statline.cbs.nl/StatWeb/publication/?DM=SLNL&PA=03753&D1=a&D2=1,8&D3=1-5&D4=0&D5=0,10,20&HDR=T,G3,G4&STB=G2,G1&VW=T. Accessed 16 Sept 2013.

CVE (College voor Examens). (2013). *Aardrijkskunde VMBO. Syllabus BB, KB en GT centraal examen 2015*. Utrecht: CVE.

Fujifilm. (2013). *Duurzaamste klas van Brabant bekend*. http://www.fujifilm.eu/nl/over-ons/bedrijfsprofiel/fujifilm-company-sites/fujifilm-manufacturing-europe-bv/nieuws/article/news/duurzaamste-klas-van-brabant-bekend/. Accessed 20 Sept 2013.

Gardner, H. (2007). *Five minds for the future*. Boston: MIT.

Geisen, G. (2013). *Autopoiesis. Perspectief op duurzaam, betekenisvol onderwijs. Samenvatting*. Utrecht: Agentschap NL.

Hamer, A., de Jansen P., Louman, E., Roorda, N., & de Vries, G. (2008). *Duurzame ontwikkeling op de basisschool. Praktische en didactische handreikingen*. Utrecht: Duurzame PABO en Veldwerk Nederland. http://www.google.ch/url?sa=t&rct=j&q=&esrc=s&source=web&cd=2&ved=0CDIQFjAB&url=http%3A%2F%2Fwww.leraar24.nl%2Fleraar24-portlets%2Fservlet%2Fdocument%3Fid%3D41&ei=EONPU_iWK9D07Ab8oYHwCg&usg=AFQjCNGBF1W7dEVyG5p9iKP_UYxk1ifGJQ&bvm=bv.64764171,d.ZGU&cad=rja. Accessed 17 Apr 2014.

Hamer, A. de & Leussink, E. (2012). *Leerkrachtcompetenties duurzaamheid*. Utrecht: Duurzame PABO.

International Bureau of Education of UNESCO. (2012). *World data on education*. The Netherlands. Geneva: UNESCO http://www.ibe.unesco.org/fileadmin/user_upload/Publications/WDE/2010/pdf-versions/Netherlands.pdf. Accessed 16 Sept 2013.

Kennisnet. (n.d.). *Duurzame school: Leren over natuur, milieu, duurzaamheid en voeding*. http://www.kennisnet.nl/themas/duurzame-school/. Accessed 16 Sept 2013.

Koning Willem II College. (n.d.). *Wereldburgerschap*. http://www.willem2.nl/LeftMenu/Algemeen/Wereldburgerschap. Accessed 20 Sept 2013.

Luijkx, R., & de Heus, M. (2008). *The educational system of the Netherlands*. http://www.mzes. uni-mannheim.de/publications/misc/isced_97/luij08_the_educational_system_of_the_nether lands.pdf. Accessed 16 Sept 2013.

Meijer, J., Veneman, H., Guerin, L., Oldersma, F., Steegstra, F. Penders, T., Schreurs, E., & ten Thije, A. (2010). *Eén aarde is genoeg voor de hele wereld. Duurzame ontwikkeling op vernieuwingsscholen*. Den Haag: SOVO. http://www.google.ch/url?sa=t&rct=j&q=& esrc=s&source=web&cd=1&ved=0CCoQFjAA&url=http%3A%2F%2Fwww.duurzamepabo .nl%2Fcomponent%2Frokdownloads%2Fdownloads%2Fpublicaties%2F30-een-aarde-is-genoeg -voor-de-hele-wereld%2Fdownload.html&ei=juVPU8aIIKry7Ab4-IC4Dw&usg=AFQjCNH8 T3_ b3JzGFYD_FjuLZJAu80YSpQ&bvm=bv.64764171,d.ZGU&cad=rja. Accessed 17 Apr 2014.

Ministerie van OCW. (n.d.). *Werken in het onderwijs*. http://www.werkeninhetonderwijs.nl/index. php. Accessed 20 Sept 2013.

NME Podium. (2013). *Doe je mee met Duurzaam Voortgezet Onderwijs?* http://www.nmepodium. nl/Agenda/Doe-je-mee-met-Duurzaam-Voortgezet-Onderwijs%3F. Accessed 20 Sept 2013.

Programmabureau Leren voor duurzame ontwikkeling. (2008). *Van agenderen naar doen!* Utrecht: LvDO.

Remmers, T. (2007). *Duurzame ontwikkeling is leren vooruitzien*. Enschede: Stichting Leerplan Ontwikkeling.

Remmerswaal, A. H., Willems, M. P. J., Vader, J., Wals, A. E. J., Schouten, A. D., & Weterings, R. (2012). *Duurzaam doen! Leren in vitale coalities; Monitoring en evaluatie van de programma's Leren voor Duurzame Ontwikkeling en Natuur-en Milieueducatie*. Wageningen: Stichting Dienst Landbouwkundig Onderzoek.

Rijksoverheid. (2004). *2ᵉ kamer nota LvDO. Van marge naar Mainstream*. http://www. lerenvoorduurzameontwikkeling.nl/sites/default/files/downloads/nota_lvdo_marge_mainstream _tcm24-117455.pdf. Accessed 17 Sept 2013.

Rijksoverheid. (2008). *Examenprogramma's VO. Aanvulling beroepsgerichte vakken*. http://www. examenblad.nl/examenstof/intersectoraal-vmbo/2014/f=/examenprogramma_beroepsgerichte _vakken_intersectoraal.pdf. Accessed 16 Sept 2013.

Rijksoverheid. (2013). *Kennisprogramma Duurzaam door*. http://www.rijksoverheid.nl/ documenten-en-publicaties/brochures/2013/06/26/kennisprogramma-duurzaam-door.html. Accessed 18 Sept 2013.

Rijksoverheid. (n.d.a). *Openbaar en bijzonder onderwijs*. http://www.rijksoverheid.nl/onderwerpen/ vrijheid-van-onderwijs/openbaar-en-bijzonder-onderwijs. Accessed 16 Sept 2013.

Rijksoverheid. (n.d.b). *Schooltijden en onderwijstijd*. http://www.rijksoverheid.nl/onderwerpen/ schooltijden-en-onderwijstijd/vraag-en-antwoord/hoe-regelen-basisscholen-de-schooltijden-en- lesuren.html. Accessed 16 Sept 2013.

Roorda, N. (2001). *Auditing instrument on sustainability in higher education*. Amsterdam: Commissie Duurzaam Hoger Onderwijs.

Roorda, N. (2010). *Onderwijs, kwaliteit en duurzame ontwikkeling*. http://www.plado.nl/content/ onderwijs-kwaliteit-en-duurzame-ontwikkeling. Accessed 16 Sept 2013.

Sleurs, W (Ed.). (2008). *Competencies for ESD teachers*. http://www.unece.org/fileadmin/DAM/env/ esd/inf.meeting.docs/EGonInd/8mtg/CSCT%20Handbook_Extract.pdf. Accessed 16 Sept 2013.

Smit, W., Jansen, P., Koppen, C. S. A., van Bulten, M., Damen, M. L. C., & Custers, C. (2006). *Hoe duurzaam is NME? Een explorerend kwantitatief onderzoek naar langetermijneffecten van Natuur—en Milieueducatie op basisscholen*. Apeldoorn: Stichting Veldwerk.

Sollart, K. M., & Vreke, J. (2008). *Het faciliteren van natuur—en milieueducatie in het basisonderwijs*. Wageningen: Wageningen UR.

Stevens, L. (2010). *Zin in onderwijs*. Antwerpen/Apeldoorn: Garant-Uitgevers.

Switch. (n.d.). *Voedsel wereldwijd*. http://www.maakdeswitch.nl/programmas/voedsel-wereldwijd. Accessed 20 Sept 2013.

Taakgroep Vernieuwing Basisvorming. (2004). *Beweging in de onderbouw*. http://www.vecon.nl/ onderwijs/onderbouw/eindrapport_10jun2004.pdf. Accessed 16 Sept 2013.

Thijs, A., & van den Akker, J. (2009). *Curriculum in development*. Enschede: Stichting Leerplan Ontwikkeling.

Veldwerk Nederland. (2009). *Schoolportretten*. http://www.duurzamepabo.nl/publicaties/18-schoolportretten.html. Accessed 18 Sept 2013.

Vereniging Hogescholen. (2009). *Kennisbasis Aardrijkskunde, Geschiedenis, Economie, Gezondheidszorg & Welzijn, Godsdienst & Levensbeschouwing, Maatschappijleer, Omgangskunde. Lerarenopleiding voortgezet onderwijs*. http://www.vereniginghogescholen.nl/standpunten/doc_download/1065-kennisbasis-lerarenopleiding-voortgezet-onderwijs-gamma-studies. Accessed 2 May 2014.

Vries, G. de. (2013). *Primary sustainable education (PRISE)*. Utrecht: Duurzame PABO.

Vries, G. de & de Hamer, A. (2010). *Op weg naar een didactiek voor duurzame ontwikkeling*. 's-Hertogenbosch: Praxis bulletin the manummer Op=op.

Waal, M. van der. (2011). The Netherlands. In I. Mulà & D., Tilbury (Eds.), *National journeys towards Education for Sustainable Development* (pp. 77–102). Paris: UNESCO. http://unesdoc.unesco.org/images/0019/001921/192183e.pdf. Accessed 17 Apr 2014.

Wagenaar, H. (2007). *Duurzame ontwikkeling voor de basisschool. Domeinbeschrijving en voorbeeldlessen*. Arnhem: Cito.

Wageningen UR (University & Research Centre). (n.d.). *Education and Learning for* Sustainable Development. http://www.wageningenur.nl/en/Expertise-Services/Chair-groups/Social-Sciences/Education-and-Competence-Studies-Group/Education/Teaching-learning-and-competence-development/Education-and-Learning-for-Sustainable-Development.htm. Accessed 2 May 2014.

Wals, A. E. J. (2006, February 15). The end of ESD... the beginning of transformative learning. Emphasizing the 'E' in ESD. In M. Cantell (Ed.), *Proceedings of the seminar on Education for Sustainable Development* (pp. 42–59), held in Helsinki. Helsinki: Finnish UNESCO Commission.

Wolf, M. de (2011). *Ons werk—werkt het? Projectwerk duurzaamheid bij FLOT*. Tilburg: Fontys Hogescholen (unpublished).

Wolf, M. de (Ed.), van Otterdijk, R., Pennartz, P., Hurkxkens, P., & Toebes, T. (2011). *Lesgeven over duurzame ontwikkeling*. Antwerpen/Apeldoorn: Garant-Uitgevers.

Printed by Printforce, the Netherlands